The Palgrave Macmillan Animal Ethics Series

Series Editors

Andrew Linzey, Oxford Centre for Animal Ethics, Oxford, UK
Clair Linzey, Oxford Centre for Animal Ethics, Oxford, UK

In recent years, there has been a growing interest in the ethics of our treatment of animals. Philosophers have led the way, and now a range of other scholars have followed from historians to social scientists. From being a marginal issue, animals have become an emerging issue in ethics and in multidisciplinary inquiry. This series will explore the challenges that Animal Ethics poses, both conceptually and practically, to traditional understandings of human-animal relations. Specifically, the Series will:

- provide a range of key introductory and advanced texts that map out ethical positions on animals;
- publish pioneering work written by new, as well as accomplished, scholars;
- produce texts from a variety of disciplines that are multidisciplinary in character or have multidisciplinary relevance.

For further information or to submit a proposal for consideration, please contact Amy Invernizzi, amy.invernizzi@palgrave-usa.com.

Clair Linzey · Andrew Linzey
Editors

The Palgrave Companion to Humane Education

palgrave
macmillan

Editors
Clair Linzey
University of Oxford
Oxford, UK

Andrew Linzey
Oxford Centre for Animal Ethics
Oxford, UK

ISSN 2634-6672 ISSN 2634-6680 (electronic)
The Palgrave Macmillan Animal Ethics Series
ISBN 978-3-032-19794-8 ISBN 978-3-032-19795-5 (eBook)
https://doi.org/10.1007/978-3-032-19795-5

Cover illustration: Jo-Anne McArthur/We Animals

This Palgrave Macmillan imprint is published by the registered company Springer Nature Switzerland AG
The registered company address is: Gewerbestrasse 11, 6330 Cham, Switzerland

For Louise van der Merwe
with admiration for her courageous, pioneering work

Series Editors' Preface

This is a new book series for a new field of inquiry: Animal Ethics.

In recent years, there has been a growing interest in the ethics of our treatment of animals. Philosophers have led the way, and now a range of other scholars have followed from historians to social scientists. From being a marginal issue, animals have become an emerging issue in ethics and in multidisciplinary inquiry.

In addition, a rethink of the status of animals has been fuelled by a range of scientific investigations which have revealed the complexity of animal sentiency, cognition and awareness. The ethical implications of this new knowledge have yet to be properly evaluated, but it is becoming clear that the old view that animals are mere things, tools, machines or commodities cannot be sustained ethically.

But it is not only philosophy and science that are putting animals on the agenda. Increasingly, in Europe and the United States, animals are becoming a political issue as political parties vie for the "green" and "animal" vote. In turn, political scientists are beginning to look again at the history of political thought in relation to animals, and historians are beginning to revisit the political history of animal protection.

As animals grow as an issue of importance, so there have been more collaborative academic ventures leading to conference volumes, special journal issues, indeed new academic animal journals as well. Moreover, we have witnessed the growth of academic courses, as well as university posts, in Animal Ethics, Animal Welfare, Animal Rights, Animal Law, Animals and Philosophy, Human-Animal Studies, Critical Animal Studies, Animals and Society, Animals in Literature, Animals and Religion—tangible signs that a new academic discipline is emerging.

"Animal Ethics" is the new term for the academic exploration of the moral status of the non-human—an exploration that explicitly involves a focus on what we owe animals morally, and which also helps us to understand the influences—social, legal, cultural, religious and political—that legitimate animal abuse. This series explores the challenges that Animal Ethics poses, both conceptually and practically, to traditional understandings of human-animal relations.

The series is needed for three reasons: (i) to provide the texts that will service the new university courses on animals; (ii) to support the increasing number of students studying and academics researching in animal related fields, and (iii) because there is currently no book series that is a focus for multidisciplinary research in the field.

Specifically, the series will

- provide a range of key introductory and advanced texts that map out ethical positions on animals;
- publish pioneering work written by new, as well as accomplished, scholars, and
- produce texts from a variety of disciplines that are multidisciplinary in character or have multidisciplinary relevance.

The new Palgrave Macmillan Series on Animal Ethics is the result of a unique partnership between Palgrave Macmillan and the Ferrater Mora Oxford Centre for Animal Ethics. The series is an integral part of the mission of the Centre to put animals on the intellectual agenda by facilitating academic research and publication. The series is also a natural complement to one of the Centre's other major projects, the *Journal of Animal Ethics*. The Centre is an independent "think tank" for the advancement of progressive thought about animals, and is the first Centre of its kind in the world. It aims to demonstrate rigorous intellectual enquiry and the highest standards of scholarship. It strives to be a world-class centre of academic excellence in its field.

We invite academics to visit the Centre's website www.oxfordanimalethics.com and to contact us with new book proposals for the series.

Oxford, UK Andrew Linzey
Oxford, UK Clair Linzey
General Editors

By the Same Authors

Animal Ethics for Veterinarians
The Ethical Case against Animal Experiments
The Palgrave Handbook of Practical Animal Ethics
The Routledge Handbook of Religion and Animal Ethics
Ethical Vegetarianism and Veganism
An Ethical Critique of Fur Factory Farming
Animal Ethics and Animal Law
Animal Theologians
The Ethics of Fur: Religion, Culture, and Law

Contents

than thirty books, including *Animal Theology* (SCM Press/University of Illinois Press, 1994); *Why Animal Suffering Matters* (Oxford University Press, 2009); *The Global Guide to Animal Protection* (University of Illinois Press, 2013); and *The Palgrave Handbook of Practical Animal Ethics* (Palgrave Macmillan, 2018). In 2001, he was awarded a Doctor of Divinity degree by the Archbishop of Canterbury in recognition of his "unique and massive pioneering work at a scholarly level in the area of the theology of creation with particular reference to the rights and welfare of God's sentient creatures."

Clair Linzey (editor) is a research fellow in animal ethics at Wycliffe Hall in the University of Oxford and deputy director of the Oxford Centre for Animal Ethics. She is also the Frances Power Cobbe Professor of Animal Theology at the Graduate Theological Foundation. In addition, she is director of the Annual Oxford Animal Ethics Summer School. She serves as coeditor of the *Journal of Animal Ethics* and coeditor of the Palgrave Macmillan Animal Ethics book series, which has published more than forty books on animal-related issues. She is the author of *Developing Animal Theology* (Routledge, 2021) and coauthor of *An Ethical Critique of Fur Factory Farming* (Palgrave Macmillan, 2022). Her coedited volumes include *Animal Ethics for Veterinarians* (University of Illinois Press, 2017), *The Palgrave Handbook of Practical Animal Ethics* (Palgrave Macmillan, 2018), *The Routledge Handbook of Religion and Animal Ethics* (Routledge, 2018), and *Animal Ethics and Animal Law* (Lexington, 2023).

Melissa Logan is the director of education at the Alberta SPCA. She has worked in the education department for over eleven years, developing curriculum-relevant programs to inspire compassion for all living beings and enhance animal welfare across Alberta. She is particularly interested in the intersection of humane education and character education to improve the well-being of both animals and people. She holds a B.Sc., B.Ed. and is pursuing an M.Sc. in international animal welfare, ethics, and law at the University of Edinburgh.

Louisa McCune is the executive director of the Kirkpatrick Foundation. She currently sits on the Animal Grantmakers board of directors, having served as its 2020 president, and on the boards of the Oklahoma Contemporary Arts Center and the Kirkpatrick Family Fund as an advisory trustee. In 2020, she cofounded the Oklahoma Killer Whale Project and served as executive producer of the Odyssey Project (a livestream complete reading of Emily Wilson's *Odyssey* translation) with Oklahoma Contemporary. McCune is editor in chief and cofounder of *ArtDesk* and a member and cofounder of Common Bonds, which aims to end the needless euthanasia of cats and dogs in municipal shelters. She also serves on the board of the Kirkpatrick Policy Group, as its treasurer. McCune is coeditor of *Love Can Be: A Literary Collection about Our Animals* (University of Oklahoma Press, 2018) and editor of *The Oklahoma Animal Study* and was editor in chief of *Oklahoma Today* magazine from 1997 to 2011. She has a bachelor's degree from San Francisco State University.

Amanda McKibbon is senior director of affiliate and community outreach for the Ontario SPCA and Humane Society. Throughout her career, Amanda has focused on animal welfare, education, and child mental health. Amanda started with the Ontario SPCA in 2012 and in her current role leads and supports community and affiliate relations and humane education programming for children, the community, and the sector. Amanda is passionate about providing a voice for those who cannot always speak for themselves and believes in the power of education to shape and transform people and communities.

Rebecca McLean completed her Ph.D. in 2019 and is currently an unaffiliated independent researcher. Her interests include Scottish and Victorian literature, literary psycho-geography, and the Gothic.

Christina Nellist is an Eastern Orthodox theologian specializing in animal suffering and human soteriology. She is a fellow of the Oxford Centre for Animal Ethics; former visiting research fellow at the University of Winchester; guest lecturer on veterinary ethics at Chile's Iberia America University Veterinary School; and advisor to the Chilean government, receiving an award for her work on stray dog control via public health education. Her book *Eastern Orthodox Christianity and Animal Suffering: Ancient Voices in Modern Theology* was published in 2018 by Cambridge Scholars Publishing. She organized the first-ever session on animal suffering at the International Orthodox Theological Association conference in 2019 and was invited by His All Holiness Bartholomew to speak on behalf of animals at the 2019 Halki III Summit. She is founder and editor of the charity Pan-Orthodox Concern for Animals, which partners with the Christian Ethics of Farmed Animal Welfare

access to education. Wiessner attended the University of Denver Graduate School of Social Work, where she focused in the areas of children and youth development and animal-assisted social work, engaging in the practice and ethics of incorporating humane education into school-based settings.

List of Figures

List of Tables

CHAPTER 1

Introduction: The Case for Humane Education

Clair Linzey and Andrew Linzey

What wisdom can you find that is greater than kindness?
—Jean-Jacque Rousseau, *Emile*

1.1 I

In 1927, philosopher Bertrand Russell and his second wife, Dora, founded a school called Telegraph House on the South Downs in the United Kingdom. They wanted an ethos that offered an "unusual combination": freedom from "prudery and religious instruction," on one hand,

C. Linzey (✉)
Wycliffe Hall, University of Oxford, Oxford, UK
e-mail: depdirector@oxfordanimalethics.com

C. Linzey · A. Linzey
Oxford Centre for Animal Ethics, Oxford, UK
e-mail: director@oxfordanimalethics.com

C. Linzey and A. Linzey (eds.), *The Palgrave Companion to Humane Education*, The Palgrave Macmillan Animal Ethics Series,
https://doi.org/10.1007/978-3-032-19795-5_1

and more freedom than usual for the children, on the other, though they resisted a complete absence of discipline.[1] The experiment was not a success.

Russell recounts how perhaps the most serious cause of trouble was what he called "an undue proportion of problem children."[2] He laments,

> Whatever may have been the cause, many of the children were cruel and destructive. To let the children go free was to establish a reign of terror, in which the strong kept the weak trembling and miserable. A school is like the world: only government can prevent brutal violence. And so I found myself, when the children were not at lessons, obliged to supervise them continually to stop cruelty.[3]

He continues to narrate how, despite the adults' combined efforts, bullying was widespread. The school was divided into three groups—"bigs, middles, and smalls"—to help solve the problem. When Russell confronted one of the "middles" who constantly acted violently toward the "smalls," the child replied, "The bigs hit me, so I hit the smalls; that's fair."[4]

One of the troublesome incidents concerned animals. An unpopular child was given a pair of rabbits, and two other children tried to burn the poor creatures to death—and, "in the attempt, made a vast fire which blackened several acres and, but for a change of wind, might have burnt the house down."[5]

1.2 II

The Russells' experience invites many questions. What kind of discipline operated at the school? Was it too lax or, perhaps, nonexistent? What sanctions operated against violent behavior, and why were they not effective? Why, despite supervision and small numbers (only twenty children), did cruelty and bullying take place? Russell avers that the "problem children"

[1] Bertrand Russel, *The Autobiography of Bertrand Russel* (London: Unwin Books, 1975), 387.

[2] Russell, *Autobiography*, 389.

[3] Russell, *Autobiography*, 389.

[4] Russell, *Autobiography*, 389.

[5] Russell, *Autobiography*, 390.

were the fault of their parents and that the "ill-effects of their un-wisdom were renewed in each holiday."[6] But it seems too easy to simply blame parents. Doubtless, parents bear some responsibility, and parental supervision is essential, but it hardly suffices as a total explanation. After all, parents cannot be easily divorced or isolated from society and culture as a whole.

But there is even a bigger, deeper question that should be asked: Are children (and adults, for that matter) "naturally" violent? It certainly seems so. Even Michael Heseltine, the former deputy prime minister of the United Kingdom, reflected on his school days and judged that "no one will ever eliminate the instinct of little boys to behave as a predatory pack, picking on the weaker or more idiosyncratic members."[7] So the "bigs" hit the "middles," as the "middles" hit the "smalls." No wonder, then, that Heseltine's autobiography was titled *Life in the Jungle*. He was, of course, mainly reflecting on his time in politics (in that arena he was known as one of the "big beasts"), but it seems that the moral and political jungle may have begun for him years before when he was a schoolboy.

But violence and cruelty are not just modern problems. Philosopher John Locke, in the seventeenth century, remarked with distaste how he had observed children taking pleasure in ill-treating creatures in their power. This, he argues,

> should be watched in them, and if they incline to any such cruelty, they should be taught the contrary usage; for the custom of tormenting and killing beasts will, by degrees, harden their minds even towards men; and they who delight in the suffering and destruction of inferior creatures, will not be apt to be very compassionate or benign to those of their own kind.[8]

Cruelty, Locke argues, is "unnatural" and against the "natural temper of benignity."[9] But if such cruelty is so widespread—both then and now—how can it be "unnatural"?

[6] Russell, *Autobiography*, 389.

[7] Michael Heseltine, *Life in the Jungle: My Autobiography* (London: Hodder and Stoughton, 2000), 19.

[8] John Locke, "Some Thoughts concerning Education" (1693), in *The Works of John Locke in Ten Volumes*, 10th ed. (London, 1801), 9:112.

[9] Locke, "Some Thoughts," 112–15.

Of course, there is a long theological tradition that speaks of humans as "fallen," alienated from their true selves, even bearers of "original sin." This version of Christian anthropology has largely been lost nowadays but offers some important insights. If the notion of original sin sounds dated, even medieval (and of course, older than that), it is worth exploring how more recent theologians have interpreted it. Jonathan Edwards, for example, argues that original sin operates as a littleness of sensibility whereby human beings become progressively self-centered to the exclusion of sensitivity to other creatures. He writes:

> Immediately upon the fall, the mind of man shrank from its primitive greatness and expandedness, to an exceeding smallness and contractedness.... Before, his soul was under the government of that noble principle of divine love, whereby it was enlarged to the comprehension of all his fellow-creatures and their welfare.... Sin, like some powerful astringent, contracted his soul to the very small dimensions of selfishness; and God was forsaken, and fellow-creatures forsaken, and man retired within himself, and became totally governed by narrow and selfish principles and feelings.[10]

Whatever the truth of the theoretical basis, it does seem that Edwards has grasped a fundamental truth—namely, that there is a potential for all of us to become overwhelmingly selfish or, perhaps better, limited in our empathy.

Does this mean, then, that we should simply accept our condition as fallible, limited, selfish beings and realize that we cannot fundamentally change it? This view cannot be easily dismissed. It is understandable that people who in the last century lived through two world wars (and other wars since) are a little wary of schemes for moral self-improvement and grand utopias of both the political left and the right. It seems a truism that dictators are only too prepared to curtail our freedoms today in the hope of better society tomorrow. Grand schemes, then, are morally and politically problematic. Even the progressive Russell found in practice that the liberal notion of progress through education was not as he hoped. Progressive, liberal education—at least as shown in this instance—cannot, it seems, make us fundamentally altruistic.

[10] Jonathan Edwards, quoted and discussed in James M. Gustafson, *Theology and Ethics* (Oxford: Basil Blackwell, 1981), 304–5. For a general discussion of the possibility of moral progress, see A. G. B. Woollard, *Progress: A Christian Doctrine?* (London: SPCK, 1972).

Moreover, the worst is yet to come. The world, as we currently experience it, seems systemically hostile to peaceable living both within and between species. Overeager moralists should listen to the heartfelt and lyrical complaint of political theorist William Godwin, which is worth reproducing at length:

> Every animal preys upon his fellow. Every animal, however minute, has a curious and subtle structure, rendering him susceptible, as it should seem, of piercing anguish. We cannot move our foot without becoming the means of destruction. The wounds inflicted are of a hundred kinds. These petty animals are capable of palpitating for days in the agonies of death. It may be said with little licence of phraseology that all nature suffers.[11]

He continues:

> Let us turn our attention to our own species. Let us survey the poor; oppressed, hungry, naked, denied all the gratifications of life and all that nourishes the mind. They either are tormented with the injustice or chilled into lethargy. Let us view man writhing under the pangs of disease, or the fiercer tortures that are stored up for him by his brethren. Who is there that will look on and say "All this is well; there is no evil in the world?" Let us recollect the pains of the mind; the loss of friends, the rankling tooth of ingratitude, the unrelenting rage of tyranny, the slow progress of justice, the brave and honest consigned to the fate of guilt. Let us plunge into the depth of dungeons. Let us observe youth languishing in hopeless despair, and talents and virtue shrouded in eternal oblivion. The evil does not consist merely in the pain endured. It is the injustice that inflicts it that gives it its sharpest sting. Malignity, an unfeeling disposition, vengeance and cruelty, are inmates of every climate. As these are felt by the sufferer with peculiar acuteness, so they propagate themselves. Severity begets severity, and hatred engenders hate. The whole history of the human species, taken in one point of view, appears a vast abortion.[12]

Godwin is surely right to argue that a facile "creed of optimism has done much harm,"[13] and he is indubitably right that failing to grasp

[11] William Godwin, Enquiry concerning Political Justice and Its Influence on Modern Morals and Happiness (1798; London: J. Watson, 1842), 216–18.

[12] Godwin, *Enquiry Concerning*, 216–18.

[13] Godwin, *Enquiry Concerning*, 216–18.

the evils of the world is an exercise in self-delusion or at least radical insensitivity. Accepting that does not, however, mean that it is the last word. As Godwin acknowledges, while there is great evil—both moral and natural—in the world, there can be genuine moments of love, courage, fortitude, and generosity as well. As Russell wrote, "the pessimist has only half the truth, and to my mind the less important half."[14] The key point is this: even—and especially—if humans are only half-inclined to do good, then this makes humane education not less but more important.

1.3 III

This book largely takes a *qualified* optimistic view and, in doing so, reflects both the origins and the goals of the humanitarian and humane movements that have emerged in the United Kingdom, Europe, Russia, and the United States. Almost anyone who uses the word "humane" today is an heir to the intellectual heritage and social activism of the nineteenth century. The intellectual heritage is as articulated by Locke (and many before and since), who saw cruelty as endangering humanity itself. As we have already noted, Locke observed that "they who delight in the suffering and destruction of inferior creatures, will not be apt to be very compassionate or benign to those of their own kind."[15] This sense that not only animals but also humans would lose out if cruelty went unchecked has certainly been a long-standing idea in Western philosophy, but it is only in the nineteenth and twentieth centuries that it has become socially embodied in societies and institutions.

It is worth reminding ourselves of that movement and how late it emerged. In 1824, the first national Society for the Prevention of Cruelty to Animals (SPCA) was founded in the United Kingdom (only later, in 1840, to become a Royal SPCA and following an earlier Liverpool-based society as early as 1809). This, in turn, inspired the establishment of the American SPCA in 1866. A plethora of other societies followed. In a short period, many anticruelty societies were founded in Europe, including Stuttgart (1837), Dublin (1840), Berlin (1841), Vienna (1847), Basel

[14] Russell, *Autobiography*, 121.

[15] Locke, "Some Thoughts," 112–15.

(1849), Zurich (1856), and Rome (1873).[16] Russia also founded one of the earliest animal protection societies in 1865.[17] The Canadian Society for the Prevention of Cruelty to Animals was founded in 1869.[18] Contrary to common misconception, many of these societies also played significant roles in the emergence of the child protection movement. To give just two examples: RSPCA founders, including William Wilberforce, Fowell Buxton, and Lord Shaftesbury, also were active in movements for the abolition of slavery, better working conditions in factories, and child protection—indeed it was members of the RSPCA who helped fund and establish the National Society for the Prevention of Cruelty to Children in 1884. Similarly, in the United States, Henry Bergh (the founder of the ASPCA) helped form the Society for the Prevention of Cruelty to Children in 1874.[19]

The aim of almost all these organizations was not only the prevention of cruelty but also the promotion of kindness to all sentient beings—what we would now call the interspecies cultivation of empathy.[20] And education became the primary means through which these aims were to be achieved. Hence, the notion of humane education was born and quickly morphed into a wider movement under this banner. While often focused specifically on animals, humane education had both wider connotations and a wider reach. The American Humane Society (now American Humane) was founded in 1877, and the Humane Society of the United States in 1954. To the question "Why are you not more concerned about cruelty to humans?" Joyce Tischler of the Animal Legal Defense Fund responded that she was concerned about all suffering but was "working at the roots."[21]

[16] Sabrina Tonutti, "European Animal Protection," in *The Global Guide to Animal Protection*, ed. Andrew Linzey (Urbana: University of Illinois Press, 2013), 12.

[17] Irina Novozhilova, "The Emergence of Animal Protection in Russia," in Linzey, *Global Guide*, 10.

[18] Janet E. Baynger, "The Humane Movement in Canada," in Linzey, *Global Guide*, 14.

[19] Stephen L. Zawistowski, "Societies Against Cruelty in the United States," in Linzey, *Global Guide*, 16.

[20] Andrew Linzey, ed., *The Link Between Animal Abuse and Human Violence* (Brighton: Sussex Academic Press, 2009); see especially parts 1, 2, and 3.

[21] Joyce Tischer, cited and discussed in Linzey, *Global Guide*, 3.

That animal cruelty could lead to human violence has been accepted by many philosophers and theologians, not just Locke but also many others, including Pythagoras, Saint Thomas Aquinas, Alexander Pope, Arthur Schopenhauer, and Jeremy Bentham. But it is only in the last two centuries that this view has received a great deal of empirical support. Perhaps the most important breakthrough was the research by the Federal Bureau of Investigation in the 1970s. The FBI interviewed many violent offenders and discovered that a high proportion had histories of causing cruelty to animals. In the words of one FBI special agent, "by self-report, 36 percent described killing and torturing animals as children and 46 percent said they did this as adolescents."[22] This discovery, inter alia, led the FBI to add animal cruelty to its list of diagnostic criteria for assessing the dangerousness of actual or potential criminals.

Since the 1970s, a voluminous amount of scholarly literature has emerged on the link between animal abuse and human violence, almost all of it confirming the FBI's research and pushing the boundaries even further.[23] One of us has edited a volume on this very topic, bringing together most of the latest research.[24] Because the literature is so varied and multifaceted, we cannot begin to do it justice here, but one discovery by criminologists and sociologists Jack Levin and Arnold Arluke stands out especially. In a paper inauspiciously titled "Reducing the Link's False Positive Problem," Levin and Arluke address how it is that nonviolent individuals engage in animal abuse in childhood but never go on to commit violence to humans.[25] However, the authors indicate that their "results suggest that the false positive problem can be substantially reduced—if not almost eliminated—by limiting predictive acts of animal abuse to those in which dogs and cats are tortured in a hands-on manner."[26] The key point here, of course, is the important distinction between correlation and causation. That there is a link cannot now be doubted, but here in this paper is now the evidence that animal cruelty

[22] Alan C. Brantley, "An FBI Perspective on Animal Cruelty," interview by Randall Lockwood and Ann W. Church, in Linzey, *Link*, 224.

[23] See Sharon L. Nelson, "The Connection between Animal Abuse and Family Violence: A Selected Annotated Bibliography," *Animal Law Review* 17, no. 2 (2011): 369–414.

[24] Linzey, *Link*.

[25] Jack Levin and Arnold Arluke, "Reducing the Link's False Positive Problem," in Linzey, *Link*, 163–71.

[26] Levin and Arluke, "Reducing the Link," 69.

in early years can *predict* later egregious violence in the form of serial murder. This finding, in our view, has been insufficiently examined and its vital significance for human beings themselves hardly appreciated. Psychologist Eleanor Gullone's book on this matter is especially relevant in marshaling the latest evidence.[27]

We cannot move on without noting the significant and long-standing work by Phil Arkow and the National Link Coalition of the United States, who have explored and pioneered ways of understanding how animal abuse is related to a range of violent and harming activities to humans, including domestic abuse, elder abuse, spousal abuse, and child abuse.[28] Their work over decades has helped to inform and pioneer the need for remedial action, including cross-referring by professionals, which is so essential to ensure prevention through early intervention.

Most significantly of all, the United Nations elaborated in 2023 on its Convention on the Rights of the Child. In the General Comment (article 19) on the "right to freedom from all forms of violence," the United Nations asserts, "Children must be protected from all forms of physical and psychological violence and from exposure to violence, such as domestic violence *or violence inflicted on animals*."[29] This is confirmation by a major international body of the harmful nature of children's witnessing of animal abuse.

1.4 IV

Briefly summarized, these are the five key considerations that justify the case for humane education:

[27] Eleonora Gullone, *Animal Cruelty, Antisocial Behaviour, and Aggression: More than a Link* (Basingstoke: Palgrave Macmillan, 2012).

[28] See National Link Coalition, https://nationallinkcoalition.org/.

[29] UN Committee on the Rights of the Child, "General Comment No. 26 (2023) on Children's Rights and the Environment, with a Special Focus on Climate Change," United Nations Convention on the Rights of the Child (August 22, 2023), article 19, p. 7, our emphasis, https://www.right-to-education.org/sites/right-to-education.org/files/resource-attachments/CRC_General%20Comment%2026_2023_EN.pdf. We thank Louise van der Merwe for bringing this to our attention.

1. *Nature and Human Nature*. It is sometimes claimed that humans, along with nature itself, are inevitably violent and cruel. Well, it is certainly the case that there is (what we consider) violence in nature, though almost all of it is occasioned by necessity rather than deliberative action—for example, the need to eat in order to survive. Nature, however, is not a moral textbook, and acting morally (in the case of human beings) often requires acting *against* nature, or rather against natural impulses—for example, revenge or hatred. While many human beings are violent and cruel, that is not the last word. Humans are also capable of moral goodness, sacrifice, and love. Even if the pessimistic view is enjoined, this still makes the need for humane education not less but more important.
2. *Historical Connections*. While many philosophers and theologians have indicated the need to check acts of cruelty, the history of humane education can be located in the development of the humane movement, especially in the nineteenth century. While concerned with acts of cruelty to animals, humane societies as they grew also embraced children and vulnerable human beings. As the early anti-vivisectionists declared, "it is NOT a question of Your Dog or Your Baby, but rather of Your Dog AND Your Baby."[30] The point of issue is that experimenters performed on both children and animals, and anti-vivisectionists were opposed to both.
3. *Humane Goals*. The goal of humane educators was first described as the prevention of cruelty and the promotion of kindness, but it is nowadays more typified as the cultivation of interspecies empathy. Central to the development of such education has been the recognition that cruelty to animals can lead to acts of human violence. Indeed, this view has now received extensive empirical evidence through multidisciplinary research. It is now established that there is a correlation between acts of cruelty to animals and to human beings. Indeed, some criminologists argue that some cruelty to animals actually can predict the later murder of human beings. We

[30] Susan Lederer, *Subjected to Science: Human Experimentation in America Before the Second World War* (Baltimore: Johns Hopkins University Press, 1995), 101, original emphases. See also a discussion in "Normalizing the Unthinkable" in Andrew Linzey and Clair Linzey, eds., *The Ethical Case against Animal Experiments* (Urbana: University of Illinois Press, 2018); see especially the section "The Putative Justifications," 44–54.

now know that a world in which cruelty to animals goes unchecked is liable to be a less safe world for human beings.

4. *Interspecies Empathy*. The desirability of cultivating empathy as an educational goal is now widely accepted, but in many countries humane education has yet to establish a foothold among all the competing claims on the modern curriculum. But it remains a key tool that can be used to lessen violence and insensitivity to humans and animals alike. It is vital that the humane educational endeavor become institutionalized, so that it is normalized and regarded as a necessary feature of the potentially civilizing power of a moral and spiritual preparation for life.
5. *Normalizing Hope*. Humane education is an act of hope—what might be called hope in action. The importance of this consideration lies in the recognition that hope is the oxygen of moral effort. Even if humans are half-inclined to do evil, then the need for humane education becomes more, not less, acute. In many cases, we have left the task of humanizing humans to parents or often isolated enlightened individuals or communities, but what we now need is a worldwide program such that our "better selves" (for want of a better term) can at least be given full rein and realize their full potential.

1.5 V

This book is divided into three sections. Part I ("Traditions of Empathy") explores some of the origins of sensitivity and fellow feeling toward other creatures. Part I ("Making Progress") illustrates the attempts made internationally to put humane education on the educational agenda. And Part III ("Developing Practice") provides examples of how humane education can operate in schools and colleges and the challenges and difficulties that humane educators face. We move, as will be seen, through the historical to the contemporary and from the theoretical to the practical.

Our book opens with a chapter from Idan Breier exploring Sumerian fables in relation to animals ("Long before Aesop: Sumerian Animal Fables and Their Educational Messages"). Breier details how animals were used to convey human moral stories. While the stories offer little educational value in terms of our treatment of animals today, they may be considered a precursor to the development of humane thought. Miriam

Sampson continues the theme of literature as a form of humane education, by exploring the work of nineteenth-century female authors in England ("Compassion and Character: Animal Welfare in Fictional Narratives of Moral Improvement by Nineteenth-Century English Women Writers"). She examines how authors such as Anna Sewell and Anne Brontë believed that teaching children to be compassionate, especially to animals, was part of developing their Christian character and how this belief was reflected in their writings.

Scottish fantasy literature is the focus of Rebecca McLean's chapter ("The Human as Animal: Scottish Fantasy and the Importance of Imagination for Humane Empathy"). She explores novels, both Victorian and modern, that place humans in the position of being exploited as animals. The categories of horror and fantasy serve to elicit empathy in the reader as the humans are vivisected and farmed. The chapter demonstrates the importance of the imagination in developing empathy. The role of art in developing empathy is the subject of Mary Shannon Johnstone's chapter ("Empathy in Art"). Johnstone focuses on three aesthetic principles—perspective, gesture, and composition—and examines how they are used by photographers to evoke empathy in the viewer. Her hope is that illuminating how art creates empathy may encourage the production of more empathetic art for animals.

The work of novelist Richard Adams is the subject of Michael J. Gilmour's chapter ("In Defense of 'Sentimental Nonsense': Richard Adams's *The Plague Dogs* and Animal Compassion"). Gilmour analyzes how Adams used sentimentality and slippage between reality and fantasy to evoke compassion in his readers. Gilmour argues that Adams's work is an example of story-as-advocacy and demonstrates how narrative fiction can be used as a form of humane education. Empathy is again the focus of John Drew's chapter on *Charlotte's Web* ("Webs of Empathy: Animal Subjectivities and the Teaching of *Charlotte's Web*"). Drew argues that E. B. White's famous work can be used to foster interspecies empathy, by going beyond allegorical interpretations of animals in children's literature toward a recognition of animals as subjects. The appreciation of animal subjectivity is a key step in educating children in more ethical attitudes toward animals.

Matthew Webber's chapter ("A New Binary: Creator and Created") considers the role of Christian education in reinforcing an anthropocentric perspective on creation. He challenges the animal/human binary in Christian education and instead argues for replacing it with a Creator/

creature distinction instead. His chapter continues the discussion in this volume about the role of religious education in humane education. The first section of the book is rounded off by Kai Horsthemke's chapter, which argues for animal rights education ("Moral Education and Animal Rights"). He maintains that moral education that includes animals can help foster concern for social justice, moral feeling, and appropriate moral sentiments.

Kay Peggs and Barry Smart begin Part II, "Making Progress," by engaging in a critical reflection on the conceptualization of the words "humane" and "education" ("Concerning Critical Humane Education"). They argue for a critical pedagogical practice of education that challenges some of the anthropocentric biases found in humane education. Robyn Hederman's chapter explores how kindness to animals was used as a tool to educate children in nineteenth-century America ("The Kindness-to-Animals Ethic in Antebellum America"). She argues that the rapid changes taking place in America brought with them a concern for a decline in culture, which necessitated the teaching of kindness as a way to preserve and encourage moral sensitivity. Isabel Barber's chapter ("'Habits of Mercy': RSPCA Humane Education in Nineteenth-Century Britain") continues the focus on the nineteenth century, this time looking at moral reform in Victorian Britain. Her discussion highlights the RSPCA's program of humane education particularly in relation to school children.

The following three entries show the variability of progress in different international contexts. Bob Blackburn, Louisa McCune, and Nancy Snow's chapter ("Deep in the Heartland: Humane Education in Oklahoma") demonstrates how humane education was a foundational concept in the formation of Oklahoma's statehood. As one of the most conservative states in the US, fostering humane education in this state can have enormous impact for humans and animals. Specifically, the authors argue that rising industrial meat production in the state poses a significant challenge to maintaining humane attitudes toward animals. Shifting context, Low Wei Ling analyzes in her chapter ("Is the Singapore Education System Ready to Introduce Humane Education?") why humane education has not been on the agenda in Singapore, whether the time is right for the introduction of the concept, and how it might be introduced. Children's magazines are the focus in Boris Bakota and Lidija Bakota's chapter ("The Educational Role of Croatian Children's Magazines in Promoting Animal Rights"). The authors recount the significance of humane education in a context where there is little history of animal welfare or rights. They contend that children's magazines played a pivotal role in developing attitudes toward animals in Croatia by depicting and condemning animal abuse and cruelty.

Animal rescue as a form of humane education is the subject of Tarah Daly's thoughtful chapter ("Stories from an Animal Rescue: Changing Behavior through Indirect Education"). She argues that the indirect and unstructured humane education that volunteers encounter in animal sanctuary experiences can change participants' moral outlook as regards animals. Christina Nellist's chapter explores the significance of seminary education as a form of humane education ("The Importance of Seminary Education on 'Creation Care' and 'the Integrity of Creation'"). She details her work to try to incorporate creation care into theological courses, particularly in the Orthodox tradition, as a necessary step in addressing religious approaches to animals and the environment.

Part III, titled "Developing Practice," delves into different practical approaches toward humane education, considering the challenges and best practices. Magdalena Smrdelj, Amanda McKibbon, Elizabeth Davis, and Alyse Patterson begin by considering the importance of "OneKind" as a concept for humane education ("Building a Bright Future: Humane Education from Theory to Practice"). Their chapter reviews an educational resource called Animal Smart™ and considers its effectiveness as part of a school-based humane education program. Brynn M. Wiessner, Julia E. Senecal, Sarah M. Bexell, and Caroline A. Griffin further examine the importance of classroom-based humane education in their chapter ("'Cool to Be Kind:' An Evaluation of a Classroom-Based Humane Education Intervention"). The authors' study assesses the efficacy of humane education, finding that it has the potential to build empathy and interrupt patterns of violence directed at humans and animals.

Picture books are the educational focus in Donna Crawford and Melissa Logan's chapter ("Building Empathy through Picture Books: Grades 2 and 5 Teacher and Student Perspectives on an Animal-Themed Book Program"). Picking up on the literature themes from Part I, the authors argue that literature is an effective vehicle for fostering empathy. They examine the use of the AnimalTales book program in Alberta, Canada, in grade 2 and 5 classrooms as a humane education tool.

Julie O'Connor's first chapter in this book shifts our focus to consider the perceptions of teachers in humane education ("Can Humane Education Motivate Students to Learn? Teachers' Perceptions"). She presents the results of her phenomenological case study that explored teachers' perceptions of how learning about animal welfare motivates students. She concludes that all participants perceived humane education as a positive student experience. Louise van der Merwe's chapter ("The Five Freedoms for Animals as a Basic Learning Requirement for All Young Learners") focuses on a humane education pilot program in South Africa. She argues that previous attempts to incorporate humane education into South Africa's curriculum have led to little or no apparent improvement in the lives of companion animals. Instead, she advocates the introduction of the teaching of the Five Freedoms as a fundamental learning requirement in the South African National School Curriculum.

Laura Diamond's chapter ("An Unconscious Bias toward Non-Human Animal Exploitation in the British Schooling System") explores how the vegan child's perspective is unacknowledged in the British school system. She explores how education practices that are inhumane to animals, such as trips to zoos and "classroom pets," are widely accepted and detrimental to the vegan child. She calls on the British government to provide teacher training and policy implementation that addresses these concerns. Our penultimate chapter returns to North America, where Julie O'Connor and William Ellery Samuels consider humane education in relation to middle school students ("Effect of Humane Education on Middle School Reading Comprehension"). They consider a randomized control study of twelve- and thirteen-year-olds' reading comprehension and compare it with an experimental group who looked at animal welfare themes. They found a small, but significant, improvement in reading comprehension in the group whose reading focused on humane education. In our last chapter, Maureen O'Sullivan ("Raising Consciousness about Animals across the Curriculum: A Personal Journey") recounts her experience in trying to introduce animal-related issues into her Irish law classroom, where no distinct course on animal law existed at the time. The overall conclusion to the practical section of this volume supports the contention that including humane education in the curriculum helps instill attitudes of empathy and kindness to animals and also motivates students' learning.

1.6 VI

The purpose of this Companion volume is to inspire and facilitate the teaching of humane education. We think humane education should be on the agenda of every school and college in the world. Even if massively ambitious, the attempt is worthwhile. Perhaps nothing less will help humans become less cruel and violent.

This Companion volume is a project of the Oxford Centre for Animal Ethics, which pioneers ethical perspectives on animals through academic research, teaching, and publication. We are especially indebted to Louisa McCune and the Kirkpatrick Foundation for their generous support of this project. We are also grateful to all the contributors who have mapped out the area with such important and original research, including the many fellows of the Oxford Centre. This project would not have been possible without the great work of Stephanie Ernst and her careful and painstaking editing of the text.

All proceeds from the sale of this book will go to support the work of the Oxford Centre for Animal Ethics.

Oxford

February 2024

Bibliography

Baynger, Janet E. "The Humane Movement in Canada." In Linzey, *Global Guide*, 14.

Brantley, Alan C. "An FBI Perspective on Animal Cruelty." Interview by Randall Lockwood and Ann W. Church. In Linzey, *Link*, 224.

Godwin, William. *Enquiry Concerning Political Justice and Its Influence on Modern Morals and Happiness*. 1798. London: J. Watson, 1842.

Gullone, Eleonora. *Animal Cruelty, Antisocial Behaviour, and Aggression: More Than a Link.* Basingstoke: Palgrave Macmillan, 2012.

Gustafson, James M. *Theology and Ethics*. Oxford: Basil Blackwell, 1981.

Heseltine, Michael. *Life in the Jungle: My Autobiography*. London: Hodder and Stoughton, 2000.

Lederer, Susan. *Subjected to Science: Human Experimentation in America Before the Second World War*. Baltimore: Johns Hopkins University Press, 1995.

Levin, Jack, and Arnold Arluke. "Reducing the Link's False Positive Problem." In Linzey, *Link*, 163–71.

Linzey, Andrew, ed. *The Global Guide to Animal Protection*. Urbana: University of Illinois Press, 2013.

Linzey, Andrew, ed. *The Link Between Animal Abuse and Human Violence*. Brighton: Sussex Academic Press, 2009.

Linzey, Andrew, and Clair Linzey, eds. *The Ethical Case Against Animal Experiments*. Urbana: University of Illinois Press, 2018.

Locke, John. "Some Thoughts concerning Education" (1693). In *The Works of John Locke in Ten Volumes*, 10th ed., vol. 9. London, 1801.
Nelson, Sharon L. "The Connection between Animal Abuse and Family Violence: A Selected Annotated Bibliography." *Animal Law Review* 17, no. 2 (2011): 369–414.
Novozhilova, Irina. "The Emergence of Animal Protection in Russia." In Linzey, *Global Guide*, 10.
Russel, Bertrand. *The Autobiography of Bertrand Russel*. London: Unwin Books, 1975.
Tonutti, Sabrina. "European Animal Protection." In Linzey, *Global Guide*, 12.
UN Committee on the Rights of the Child. "General Comment No. 26 (2023) on Children's Rights and the Environment, with a Special Focus on Climate Change." United Nations Convention on the Rights of the Child (August 22, 2023), Article 19. https://www.right-to-education.org/sites/right-to-education.org/files/resource-attachments/CRC_General%20Comment%2026_2023_EN.pdf.
Woollard, A. G. B. *Progress: A Christian Doctrine?* London: SPCK, 1972.
Zawistowski, Stephen L. "Societies Against Cruelty in the United States." In Linzey, *Global Guide*, 16.

PART I

Traditions of Empathy

CHAPTER 2

Long Before Aesop: Sumerian Animal Fables and Their Educational Messages

Idan Breier

The *Oxford English Dictionary* defines "education" as "a process of teaching, training and learning, especially in schools, colleges or universities, to improve knowledge and develop skills."[1] Formal and informal education alike seek to inculcate habits, ways of thought and conduct, morals, character, and skills for specific purposes.[2] In ethical terms, the educational process is designed to correct human weaknesses and thus social failures, thereby promoting a society that it is more supportive

[1] See Dictionary.com, s.v. "education" https://www.oxfordlearnersdictionaries.com/definition/english/education?q=education (19.8.24). Cf. Cambridge Dictionary, s.v. "education," https://dictionary.cambridge.org/dictionary/english/education (19.8.24); Collins, s.v. "education," https://www.collinsdictionary.com/dictionary/english/education (19.8.24).

[2] Avraham Even-Shoshan, *The New Dictionary* [in Hebrew] (Jerusalem: Kiryat Sepher, 1969), 2:797.

I. Breier (✉)
The Israel and Golda Koschitzky Department of Jewish History and Contemporary Jewry, Bar-Ilan University, Ramat Gan, Israel
e-mail: idan.breier@biu.ac.il

C. Linzey and A. Linzey (eds.), *The Palgrave Companion to Humane Education*, The Palgrave Macmillan Animal Ethics Series,
https://doi.org/10.1007/978-3-032-19795-5_2

and moral.[3] From the earliest days of civilization in the ancient Near East, a system of formal education was created in which writing and a wealth of cultural knowledge were taught in schools. No less importantly, students were also trained in the ethical norms of society.[4] Among the sources employed in Mesopotamia were a collection of Sumerian proverbs and sayings. In this chapter, I shall look at these pithy aphorisms, which preceded Aesop's better-known Greek fables by around 1,500 years. To date, most scholars have paid attention only to the general Sumerian influence on the Greek counterparts.[5] As well as analyzing the Sumerian collection, I shall thus address the issue of the relationship between the two sets of literary texts.

Proverbs and aphorisms have formed an integral part of many cultures across the world from ancient to modern times.[6] In most cases, they originate in popular rather than sapiential circles. Reflecting the cultural and cognitive worldview of its creators,[7] the genre is known to have existed since the earliest Sumer dynasties—ED IIIa (2600–2500 BCE). During this period, mythic and epic literature began to be written down. Proverbs and sayings belong to didactic-wisdom literature, whose purpose is to guide people in how to behave properly in order to succeed in life. The Sumerian collections served as scribal-school material in Babylonia and

[3] Hanoch Ben-Pazi, *Emmanuel Levinas—Educational Contract: Responsibility, Hope, and Alliance* [in Hebrew] (Tel Aviv: Hakibbutz Hameuchad, 2016), 14.

[4] Nili Shupak, "The Father's Instruction in Ancient Egypt" [in Hebrew], in *Education and History: Cultural and Political Contexts*, ed. R. Feldahay and I. Etkes (Jerusalem: Zalman Shazar Centre, 1999), 13–21; Aaron Demsky, *Literacy in Ancient Israel* [in Hebrew] (Jerusalem: Bialik Institute, 2012), 68–71; Yoram Cohen, "The Scribal School at Emar" [in Hebrew], *Beit Mikra* 57, no. 1 (2012): 65–85.

[5] Francisco Rodrigues Adrados, *History of the Graeco-Roman Fable* (Leiden: Brill, 1999), 1:284, 291–93, 305–6; Niklas Holzberg, *The Ancient Fable: An Introduction*, trans. C. Jackson-Holzberg (Bloomington: Indiana University Press, 2002), 13, 15; Paula da Cunha Corrêa, "A Human Fable and the Justice of Beasts in Archilochus," in *Hesperos: Studies in Ancient Greek Poetry Presented to M. L. West on His Seventieth Birthday*, ed. P. J. Finglass, C. Collard, and N. J. Richardson (Oxford: Oxford Scholarship Online, 2007), 103–8.

[6] Edmund I. Gordon, *Sumerian Proverbs: Glimpses of Everyday Life in Ancient Mesopotamia* (New York: Greenwood, 1968), 1; Roland E. Murphy, *Proverbs* (Nashville: Thomas Nelson, 1998), xxviii.

[7] Rahamim Rajuan, *The Wisdom of Arabia: 1001 Proverbs, Sayings and Aphorisms* [in Hebrew] (Tel Aviv: Dvir, 1991), 7; Janice Raymond, *Mongolian Proverbs: A Window into Their World*, trans. O. Oidovdorj (Eugene, OR: Resource, 2010), vi.

Assyria for centuries and millennia after Sumerian had ceased to be a living language.[8] Most, although not all, are secular in nature, reflecting an agricultural society in which human beings and animals lived in close proximity to one another.[9] Some scholars maintain that the proverbs are late—belonging to the second millennium—and anachronistic.[10] Most, however, believe them to embody early Sumerian texts that were edited and canonized at a later stage.[11] Egyptian schools also copied and taught other early material.[12]

The collections being eclectic, aphorisms or sayings frequently assumed various forms in different collections.[13] Some later Babylonian examples, from the second half of the second millennium BCE, are longer. This

[8] Bendt Alster, "An Akkadian Animal Proverb and Assyrian Letter, ABL 555," *JCS* 41 (1989): 187–93; Bendt Alster, *Proverbs of Ancient Sumer: The World's Earliest Collections* (Bethesda MD: CDL Press, 1997), pp. 1: 18–19; William W. Hallo, "Proverbs Quoted in Epic," in *Lingering over Words: Studies in Ancient Near Eastern Literature in Honor of William L. Moran*, ed. T. Abrush, J. Huehnergard, and P. Steinkeller (Atlanta, GA: Scholars Press, 1990), 203–17.

[9] Jacob Klein, "Mesopotamian Literature: Genesis Traditions, Wisdom Literature, and Lamentations" [in Hebrew], in *The Literature of the Hebrew Bible: Introductions and Studies*, ed. Z. Talshir (Jerusalem: Yad Ben-Zvi, 2011), 2:550–56; Victor Avigdor Hurowitz, *Proverbs: Introduction and Commentary* [in Hebrew] (Tel Aviv: Am Oved; Jerusalem: Magnes, 2012), 1:55.

[10] Nicolas Christiaan Veldhuis, "Sumerian Proverbs in Their Curricular Context," *JAOS* 120, no. 3 (2000): 388–89; Jon Taylor, "The Sumerian Proverb Collections," *RA* 99, no. 1 (2005): 19–21.

[11] Herman L. J. Vanstiphout, "How Did They Learn Sumerian?," *JCS* 31 (1979): 118–26; Eleanor Robson, "The Tablet House: A Scribal School in Old Babylonian Nippur," *RA* 95 (2001): 40–48; Andrew R. George, "In Search of the É.DUB.BA.A: The Ancient Mesopotamian School in Literature and Reality," in *"An Experienced Scribe Who Neglects Nothing": Ancient Near Eastern Studies in Honor of Jacob Klein*, ed. Y. Sefati et al. (Bethesda, MD: CDL, 2005), 127–37; Bendt Alster and Takayoshi Oshima, "A Sumerian Proverb Tablet in Geneva with Some Thoughts on Sumerian Proverb Collections," *Orientalia* 75 (2006): 33–42; Jacob Klein and Nili Samet, "Ethics and Religion in Sumerian Proverb Literature" [in Hebrew], *Beit Mikra* 57, no. 2 (2012): 108–9.

[12] Stefan Wimmer, "Hieroglyphs: Writing and Literature," in *Egypt: The World of the Pharaohs*, ed. R. Schulz and M. Seidel (Cologne: Könemann, 1998), 348–53; Michael V. Fox, *Proverbs 1–9* (New York: Doubleday, 2000), 19.

[13] Bendt Alster, "Some New Sumerian Proverbs," in *Akkade Is King: A Collection of Papers by Friends and Colleges Presented to Aage Westenholtz*, ed. G. Barjmovich et al. (Leiden: Netherlands Instituut voor het Nabije Oosten, 2011), 9–10; Uri Gabbay, "Lamentful Proverbs or Proverbial Laments? Intertextual Connections between Sumerian Proverbs and Emesal Laments," *JCS* 63 (2011): 53.

genre did not become popular among Babylonian scholars, however, who appear to have dismissed it as a form of popular culture.[14]

Let us now look at the characteristics of the genre. Fables are brief tales in poetic verse or prose that frequently promote moral lessons by bestowing human characteristics on animals and, on occasion plants.[15] This artificial faunal society reflects the patterns and traits of its human counterpart.[16] Relating to everyday life, pithy sayings at times portray ironic or paradoxical situations.[17] In most cases, the animals adduced interact, thereby attracting the attention of their adult audiences and delivering their educational and moral points with greater punch.[18] In this way, they encourage the audience to ponder the fables' ramifications.[19]

From a technical perspective, fables customarily revolve around a single central plot. Because extraneous complexity would tend to blur their moral and educational message, they engage in no literary frills or flounces.[20] They embody a kernel of practical wisdom, presented via the drawing of analogies between the story and real life. Their particular form makes them universal and timeless, obscuring the boundaries between

[14] Wilfred G. Lambert, *Babylonian Wisdom Literature* (Oxford: Oxford University Press, 1969), 150, 151–212.

[15] Chris Baldick, *Oxford Concise Dictionary of Literary Terms* (Oxford: Oxford University Press, 2001), 26, 93.

[16] Tua Karhonen, "Anthropomorphism and the Aesopic Animal Fables," in *Animals and Their Relations to Gods, Humans and Things in the Ancient World*, ed. R. Mattila, S. Ito, and S. Fink (Wiesbaden: Harrassowitz, 2019), 211–12.

[17] Peter Childs and Roger Flower, *The Routledge Dictionary of Literary Terms* (London: Routledge, 2006), 82.

[18] Nicholas Howe, "Fabling Beasts: Traces in Memory," *Social Research* 62, no. 3 (1995): 642–43; Laura Gibbs, *Aesop's Fables: A New Translation* (Oxford: Oxford University Press, 2008), xviii.

[19] Heidi Northwood, "Making Music with Aesop Fables in the Phaedo," in *Plato's Animals: Gadflies, Horses, Swans, and Other Philosophical Beasts*, ed. J. Bell and M. Naas (Bloomington: Indiana University Press, 2015), 23.

[20] Shlomo Span, "Aesop and the Fable" [in Hebrew], in *Aesopi Fabulae*, 6th ed., trans. S. Span (Jerusalem: Bialik Institute, 2001), 175–76, 182.

humans and fauna.[21] Their universality indicates the existence of a principle espoused in classical Greece and later periods—namely, that human nature is the same across the world.[22]

The educational lesson the proverb seeks to deliver is not the high morality of the philosophers or prophets, which champions elevated virtues and going beyond the letter of the law, but the wisdom that brings practical benefits—friendship, loyalty, modesty, and so on. These are attributes that serve human beings in their relations with friends and enemies alike.[23]

Although fables are a well-known feature in biblical literature, herein they generally take the form of allegories—metaphors elaborated into very short stories[24] or parables—or epigrammatic sayings conveying a moral lesson.[25] The best-known biblical fables or fabulas are two that anthropomorphize plants—the trees in Jotham's parable[26] and the cedar and thorn bush in the account of Jehoash.[27]

[21] Azri'el Ukhmani, *Contents and Forms: Dictionary of Literary Terms* [in Hebrew] (Tel Aviv: Hapoalim Library, 1992), 2:67; Gibbs, *Aesop's Fables*, xv; Revital Refael-Vivante, *A Treasury of Fables: Isaac ibn Sahula's* Meshal Haqadmoni *(Castile, 1281)—Text and Subtext* [in Hebrew] (Ramat-Gan: Bar-Ilan University Press, 2017), 177; William W. Hallo, "Biblical and Ancient Near Eastern Literatures: A General Introduction" [in Hebrew], in *The Literature of the Hebrew Bible: Introductions and Studies*, ed. Z. Talshir (Jerusalem: Yad Ben-Zvi, 2011), 2:466.

[22] Thucydides, *The War of the Peloponnesians and Athenians*, trans. J. Mynott (Cambridge: Cambridge University Press, 2013), 3.82; David Asheri, *A Possession for All Times: Greek Historians and History Writing in Greece and Rome* [in Hebrew] (Jerusalem: Magnes, 2004), 71; Alexander Fuks, "Thucydides the Historian" [in Hebrew], in *Greek Historians and Greek Historiography*, ed. D. Asheri (Jerusalem: Bialik Institute, 2005), 142; Donald J. Puchala, *Theory and History in International Relations* (New York: Routledge, 2003), 37; Immanuel C. Y. Hsü, *The Rise of Modern China*, 6th ed. (New York: Oxford University Press, 2000), 360–61.

[23] Span, "Aesop and the Fable," 180–81.

[24] See Ezek. 17:3–12, 19:2–9, 24:3–11, 33:3–17, 34:2–31.

[25] 2 Sam. 12:1–4; Isa. 5:1–6.

[26] Judg. 9:8–15.

[27] 2 Kings 4:8–10; 2 Chron. 25:17–19; cf. Menahem Haran, "Mashal" [in Hebrew], in *Encyclopedia Biblica* (Jerusalem: Bialik Institute, 1978), 5:448–552. For early Egyptian animal anthropomorphization, see Nili Shupak, "The First Mickey Mouse: The Enigma of the Anthropomorphized Drawings in Ancient Egypt," in *Human Interaction with the Natural World in Wisdom Literature and Beyond: Essays in Honour of Tova L. Forti*, ed. M. Cogan, K. Dell, and D. Glatt-Gilad (London: T&T Clark, 2023), 33–44.

The first Greek proverb—"The Niggle and the Hawk"—appears to date to the eighth century BCE.[28] A didactic story relating to the struggle between justice and power, it teaches that one should not become involved with the wicked or strong.[29] Greek fables are most prominently associated with Aesop, however. According to Herodotus, Aesop was a slave who lived in the sixth century BCE.[30] Scholars dispute his place of origin, some arguing that he was born in Samos in the Aegean and others claiming that he came from Phrygia. His fables spread across Greece and the Roman Empire in both written and oral form. He composed numerous faunal fables, including those in which animals behave according to rather fixed stereotypes—the brave lion, cunning fox, cowardly hare, and so on. In some cases, however, he created paradoxes, the strong needing the assistance of the weak or the shrewd failing to deceive the innocent. In these fables, it can sometimes be difficult to discern the way in which the plot develops.[31] His fables also appear to have found their way to Socrates and Plato.[32]

Having briefly reviewed the theoretical background, let me now examine some of the animals who feature prominently in the Sumerian proverb collection. One of these is the arrogant fox, who refuses to acknowledge the limits of his capabilities:

> The fox said to his wife: "Come! Let us crush Uruk between our teeth like a leek; let us strap Kullab upon our feet like sandals." Hardly had they come within a distance of 600 nindan [100m] from the city, before the dogs began to howl from the city. "Slave girl from Tummal, slave girl from Tummal, come with me to your place! All kinds of evil howling from the city!"[33]

[28] Cf. Hesiod, *Op.* 202–12.

[29] Holzberg, *Ancient Fable*, 12–13; Edward Clayton, "Aesop, Aristotle, and Animals: The Role of Fables in Human Life," *Humanitas* 21, nos. 1–2 (2008): 180–82.

[30] Herodotus, *The Histories: A New Translation by Robin Waterfield* (Oxford: Oxford University Press, 2008), 2.134.

[31] Gibbs, *Aesop's Fables*, ix–xi, xix.

[32] Northwood, "Making Music," 13–26.

[33] SP 2.69, in Alster, Proverbs, 1:59.

Although he boasts that he can destroy the city of Uruk/Kullab, the fox in fact reveals himself to be a coward. The way he refers to his wife—"slave girl from Tummal"—also alludes to a Sumerian tradition regarding pusillanimity.[34] This fable was intended to teach the ancient Sumerians how to behave modestly, recognizing the boundaries of their abilities. Aesop represents the fox in a similar fashion.[35]

The same traits are attributed to the fox in extremely brief Sumerian sayings that closely resemble proverbs. In these, the fox utters a short sentence that conveys an educational-ethical message. Following is one example: "The fox, having urinated into the sea: 'All of the sea is my urine,' he said."[36] In two short epigrams, the fox attempts to provoke an ox or elephant, neither of whom he feels is relating sufficiently seriously to his silly antics.[37] This idea, of a small creature inciting a large one while completely ignoring the power relations between them, reaches the absurd in the following grotesque proverb:

> An elephant spoke to himself and said: "Among the wild creatures of Šakkan [the god of the wild animals of the steppe] there is no one comparable to me." The *altirigu*-bird [possibly a wren] answered him: "And yet, I, in my own proportion, I am equal to you," it said.[38]

This type of fable appears to warn against taking on those stronger or higher-ranked than oneself. A similar idea occurs in several of Aesop's fables.[39] In these fables, longer and more complex than the Sumerian proverbs, the weaker creature learns a hard lesson, only understanding his mistake when it is too late.[40]

The Sumerian fables also represent the fox as wicked and treacherous[41] and sometimes also as sarcastic and egotistical: "After a lion had fallen into a pit, a fox came to him and said: 'I am taking your sandals home to you

[34] Alster, *Proverbs*, 2:367.

[35] Gibbs, *Aesop's Fables*, fable 236, p. 116.

[36] SP 2.67, in Alster, *Proverbs*, 1:58; cf. SP 8 sec. B22.

[37] SP 2.65; 8 sec. B19, in Alster, *Proverbs*, 1:58, 1:169; cf. SP 8 sec. B34.

[38] SP 5.1, in Alster, *Proverbs*, 1:121, 2:400.

[39] Gibbs, *Aesop's Fables*, fables 228–36, pp. 113–16.

[40] Gibbs, *Aesop's Fables*, fables 265–66, pp. 128–29.

[41] SP 2.66, 8 sec. 33, in Alster, *Proverbs*, 1:58, 1:172.

on the other side.'"[42] Aesop likewise portrays the fox as cynical, mocking the hare rather than aiding the animal in his time of need.[43]

Another Sumerian fable relates to an old, worn-out ox. His powers failing him, he seeks to escape the fox who is harassing him.[44] Aesop also makes use of this motif, speaking of a wild boar, an ox, and even the harmless donkey tormenting a miserable old lion.[45] The poor state into which the strong animal falls in his old age, as an image of the person whose strength leaves him or her, appears to have been a prevalent concern in ancient civilizations.[46]

In the Sumerian fables, the fox is also impertinent and avaricious:

> Nine wolves having caught 10 sheep, there was one too many and they did not know how to share out the portions. A fox came along and said: "Let me allocate the portions for you! The nine of you take one sheep. I by myself shall take nine—this shall be my share!"[47]

This fable represents voracity not only as an immoral trait but also as ludicrous.[48] Aesop's fables too include a story about dividing the spoil among the lion, wolf, and fox. Here, however, the fox's cunningness rescues him from the wolf's fate of being eaten by the strong, ravenous lion.[49] Another Greek fable recounts how the fox takes all the prey hunted by a lion and bear, who have seriously injured one another in their fight over the meat.[50] The fox's avarice also is adduced in another Sumerian fable:

[42] SP 5.58, in Alster, *Proverbs*, 1:130.

[43] Gibbs, *Aesop's Fables*, fable 444, p. 205.

[44] SP 8 sec. B21, in Alster, *Proverbs*, 1:169–70; cf. SP YBC 7301.

[45] Gibbs, *Aesop's Fables*, fable 422, pp. 197–98; cf. fable 15, p. 12.

[46] Ps. 71:9; Vincent A. Tobin, trans., "The Maxims of Ptahhotep," in *The Literature of Ancient Egypt: An Anthology*, ed. W. K. Simpson (New Haven: Yale University Press, 2003), 130–31; Nili Shupak, *"No Man Is Born Wise": Ancient Egyptian Wisdom Literature and Its Contact with Biblical Literature* [in Hebrew] (Jerusalem: Bialik Institute, 2016), 67–68.

[47] SP 5 vers. A71–B74, in Alster, Proverbs, 1:132, 1:134.

[48] Alster, Proverbs, 2:415.

[49] Gibbs, *Aesop's Fables*, fable 15, p. 10.

[50] Gibbs, *Aesop's Fables*, fable 62, p. 33.

> A fox demanded of Enlil the horns of a wild bull. While it was wearing the horns of a wild bull, it started to rain, and they rose high before him, so he could not enter his hole. Until midnight the wind kept blowing, and the clouds gave rain. After it had stopped (raining) upon him, and he had become dry, he said: "Let me return the office to his lordship!"[51]

In yet another, the mice in Aesop's fable fall into a similar situation as the fox in the Sumerian saying:

> War had broken out between the mice and the weasels. The mice were inferior in strength and when they realized that their utter weakness and cowardice put them at a disadvantage, they elected satraps and generals who could be their leaders in war. The satraps wanted to be more remarkable and conspicuous than the other mice, so they put horns on the tops of their heads. Then the weasels attacked the mice once again and routed them completely. The other mice were able to scamper quickly and easily into the mouse holes which had been prepared for their concealment. The commanders, however, despite being the first to reach the holes in the retreat, were unable to get inside because of the horns on their heads. The weasels were thus able to seize the mouse generals and consign them to death.[52]

The "horn" fables appear to serve to inculcate the virtue of being content with little in both Sumer and ancient Greece.

The fox employs his attributes to find food. The fox frequently meets an ostensibly innocent creature who is ultimately revealed to be smarter and capable of escaping, however:

> A fox spoke to a goat: "Let me put my shoes in your house!" (The goat answered): "When the dog comes, let me hang them up on a nail!" (The fox answered): "If the dog stays like that in your house, bring me my shoes. Let me not stay till midnight!"[53]

As we saw previously—and as is indicated in another proverb—the fox ferociously fears the dog.[54] Aesop similarly portrays the fox as a cautious

[51] SP 8 sec. B20, in Alster, Proverbs, 1:169.

[52] Gibbs, *Aesop's Fables*, fable 455, p. 209.

[53] SP 8 sec. B28, in Alster, *Proverbs*, 1:171, 2:416.

[54] SP 8 sec. B29, in Alster, *Proverbs*, 1:171.

creature, who, while generally capable of protecting himself against predators or identifying imposters, on occasion gets into trouble because of his bravado.[55]

The goat who uses all her guile to outfox the fox also appears in a similar guise in another proverb:

> A lion had caught a helpless she-goat (and said), "Let me go! I will give you my fellow ewe in return." "If I let you go, tell me first your name!" The she-goat answered the lion: "You don't know my name? 'I Am Cleverer Than You' is my name!" After the lion had come to the sheepfold, he roared, "I released you!" She answered from the other side: "You released me, you were clever ... the sheep are not here."[56]

Aesop likewise presents the goat as a resourceful creature:

> There was a goat grazing up high on a cliff. At the bottom of the cliff there was a wolf who wanted to catch the goat and eat her. Since it was impossible for the wolf to climb up the cliff, he stood down below and said to the goat, "You poor creature! Why have you left the level plains and meadows in order to graze upon the cliff? Are you trying to tempt death from that height?" The goat said to the wolf in response, "I know how often I have managed to frustrate you! What makes you think that you can now get me to come down off this cliff so that you can eat me for dinner?"[57]

The motif of a weak animal using her wisdom to evade a predator is well known from other fables in Aesop's collection.[58] In some fables, however, the animal becomes a victim despite her acumen. These stories are designed to teach about the nature of the world.[59]

The dog also appears in a number of fables. Numerous Sumerian fables depict canines as wild and stupid.[60] In one of them, a dog is portrayed

[55] Gibbs, *Aesop's Fables*, fable 15, p. 12; fable 145, p. 75; fable 322, p. 154.

[56] SP 5.55, in Alster, *Proverbs*, 1:128, 2:416.

[57] Gibbs, *Aesop's Fables*, fable 100, p. 52.

[58] Gibbs, *Aesop's Fables*, fable 33, p. 20; fable 31, pp. 19–20; fable 117, p. 61; fable 312, p. 150; fable 356, p. 168.

[59] Gibbs, *Aesop's Fables*, fables 129–31, pp. 67–69.

[60] Idan Breier, "Sumer as the Cradle of History—Canine as Well: The Representation of the Dog in Sumerian Proverbial Literature," *Ma'arav* 21, nos. 1–2 (2014): 83–101.

as consulting with his tail—which is no wiser than himself.[61] *Inter alia*, the canine's idiocy derives from his voraciousness, as in the following very brief epigram: "The dog chewing some bones spoke to his anus, 'It will do harm to you,' they say."[62] Canines are also painted as failing to recognize social conventions: "A dog came to a brothel, and said, 'One does not see anything. Let me open this one!'"[63] On other occasions, the dog is described as ungrateful to his host.[64] The dog's insatiable appetite similarly points to his treachery toward others.[65]

The dog's forefather the wolf also is represented as dominated primarily by his stomach:

> After a wolf had taken a seat in a thorny bush, he spoke to Utu [the Sun-god], "When I come out, from now on, let me eat no more sheep. When I am hungry, the sheep I have taken, whatever you mention, what do they mean to me? I am bound by a just oath! Now, what shall I eat?," he said.[66]

With respect to food customs, we find the following brief sayings:

> When the lion caught a "bush-pig," he roared: "Until now your flesh has not filled my mouth, but your squeals have made me deaf!"[67]
> A pig which was being slaughtered by the pig-butcher squealed, "Along the road your ancestors and your uncles walked, you too are walking, [so why do you scream?]."[68]

These short epigrams teach that some things follow the way of the world and cannot be changed at all.[69] Aesop's collection also contains a

[61] SP 5.102, in Alster, *Proverbs*, 1:139, 2:407.

[62] SP 5.84, in Alster, *Proverbs*, 137.

[63] SP 5.77, in Alster, *Proverbs*, 1:135, 2:406.

[64] SP 5.116, in Alster, *Proverbs*, 1:142, 2:408.

[65] SP 5.42, in Alster, *Proverbs*, 1:126, 2:402.

[66] SP 5 vers. B72, in Alster, *Proverbs*, 1:133; cf. SP 5 vers. B71; SP 5 vers. A72.

[67] SP 5.57, in Alster, *Proverbs*, 1:129.

[68] SP 8 sec. A2, in Alster, *Proverbs*, 1:166.

[69] Alster, *Proverbs*, 2:414.

fable that presents a pig squealing on the way to be slaughtered in the city.[70]

The Sumerian aphorisms include a number of short sayings addressing the faunal love of freedom and release from human yoke. The horse, for example, complains against the burden laid upon him.[71] The donkey, in contrast, succeeds in throwing it off, contentedly returning to his pre-bondage state.[72] In another, a religious official attempts to persuade a lion to abandon his natural habitat and come and live with him in the city: "When the lamentation priest met a lion in the desert: 'Let him come to the town... to the gate of Inana, where the... dog is beaten with a stick. What is your brother doing in the desert?'"[73] A similar motif occurs in Aesop's fable about a wolf, a dog, and a collar:

> A comfortably plump dog happened to run into a wolf. The wolf asked the dog where he had been finding enough food to get so big and fat. "It is a man," said the dog, "who gives me all this food to eat." The wolf then asked him, "And what about that bare spot there on your neck?" The dog replied, "My skin has been rubbed bare by the iron collar which my master forged and placed upon my neck." The wolf then jeered at the dog and said, "Keep your luxury to yourself then! I don't want anything to do with it, if my neck will have to chafe against a chain of iron!"[74]

A short Sumerian fable also evinces the need to be sensitive to others' needs: "A dog spoke as fellows to his master: 'Indeed, what is pleasant to me does not exist to you. In your eyes it is not even there.'".[75]

As we have observed here, many of the Sumerian fables and Aesopian fables exhibit a close resemblance. Three Greek epigrams in particular appear to reflect the Sumerian Etana myth.[76] Herein, after an eagle makes a pact of friendship with a snake, the bird turns on the snake's offspring.[77]

[70] Gibbs, *Aesop's Fables*, fable 397, p. 185.

[71] SP 5.38.

[72] SP 5.39.

[73] SP 2.101; ETCSL translation t.6.1.02; cf. Alster, *Proverbs*, 1:65, 2:371. ETCSL = Proverbs: http://etcsl.orinst.ox.ac.uk/cgi-bin/etcsl.cgi?text=t.6.1.02#.

[74] Gibbs, *Aesop's Fables*, fable 3, p. 5; cf. fable 409, p. 191.

[75] SP 5.78, in Alster, *Proverbs*, 1:135, 2:406.

[76] Rodrigues Adrados, *History*, 1:305; Karhonen, "Anthropomorphism," 213.

[77] Tab. II 1'–72'.

As the snake bitterly mourns his young, the Sun-god Utu—also in charge of justice—tells him how he can capture and punish the eagle: "Cut its wings, feather and pinion, pluck it and throw it into a bottomless pit, let it die there of hunger and thirst!".[78]

Aesop's collection contains two passages from this myth. The first treats the alliance between the two animals—here an eagle and a fox—the attack on the young, and the retribution (again different in nature):

> The eagle befriended the fox but he later devoured the fox's pups. Since she had no power over the eagle, the fox prayed to the gods for justice. Then one day when a sacrifice was burning upon an altar, the eagle flew down and grabbed the sizzling meat to carry it off to his chicks. The meat was so hot that as soon as the chicks ate it, they died.[79]

The same story occurs around a century before Aesop in a poem by the Greek writer Archilochus.[80] Aristophanes also cites it a little later in *The Birds*.[81] The Sumerian motif of clipping the eagle's wings and the care humans devote to the eagle are known from another Aesopian fable. Here, a man crops the eagle's wings and keeps him shut up in a chicken coop, and another person releases the eagle and takes him under his own wing.[82] The third-millennium Sumerian fables, which appeared later in Mesopotamian collections (up to the eighth and seventh centuries BCE), thus appear to have found their way to Greece in oral form.

A brief examination of ancient Egyptian literature further substantiates this claim. Although very few written proverbs have survived from this corpus, paintings of animals—primarily from the Ramesside period (the last quarter of the second millennium BCE)—represent fauna in human

[78] Tab. II, 121–123; Stephanie Dalley, "Etana (1.131)," in *The Context of Scriptures*, ed. W. W. Hallo (Leiden: Brill, 2003), 1:455.

[79] Gibbs, *Aesop's Fables*, fable 155, pp. 80–81; cf. fable 154, p. 80.

[80] Frg. 174; Christopher G. Brown, "Iambus," in *A Companion to the Greek Lyric Poets*, ed. D. E. Gerber (Leiden: Brill, 1997), 59–62; da Cunha Corrêa, "Human Fable," 103–8; Renaud Gagné, "A Wolf at the Table: Sympotic Perjury in Archilochus," *TAPA* 139 (2009): 254–56.

[81] Ll. 652–53; Aristophanes, *The Birds*, trans. J. Henderson (Newburyport, MA: Focus, 1999), 49.

[82] Gibbs, *Aesop's Fables*, fable 83, p. 44.

situations and interaction.[83] Quite possibly, these represent oral traditions that were not written down. One of the few fables that have survived relates to a lion who catches a mouse. Pleading for his life, the latter promises to reward the lion at the first opportunity. Despite laughing in his face, the lion lets the mouse go. One day, the lion gets caught in a hunter's trap. The mouse finds him and gnaws through the net, thereby releasing him.[84]

Aesop's collection contains a very similar fable about a mouse who repays a favor after being trapped.[85] We may surmise that other Egyptian epigrams that have not survived in written form also influenced Aesop's. Dual Mesopotamian and Egyptian influence is evident in a number of biblical examples as well, in particular Proverbs.[86] Egypt's and Mesopotamia's ties with Greece are also well known, Gyges of Lydia in Western Anatolia being in contact first with Ashurbanipal II of Assyria (668–631 BCE)[87] and then Psamtek II of Egypt (664–610 BCE).[88]

2.1 Discussion

We may now address the question of whether the fables' educational message promotes an ethical attitude toward animals. Educational systems are unique to *Homo sapiens*, distinguishing human beings from the faunal kingdom.[89] Intended for the individual and collective alike, education

[83] Nili Shupak, "Father's Instruction," 19–24; Nili Shupak, "Ancient Egyptian Literature" [in Hebrew], in *The Literature of the Hebrew Bible: Introductions and Studies*, ed. Z. Talshir (Jerusalem: Yad Ben-Zvi, 2011), 621; Jennifer Miyuki Babcock, *Ancient Egyptian Animal Fables: Tree Climbing Hippos and Ennobled Mice* (Leiden: Brill, 2022).

[84] Mirian Lichtheim, *Ancient Egyptian Literature* (Berkeley: University of California Press, 2006), 3:158–59.

[85] Gibbs, *Aesop's Fables*, fable 70, pp. 37–38.

[86] Nili Shupak, *Olam Ha-Tanach: Sefer Mishley (Proverbs)* [in Hebrew] (Tel Aviv: Davidzon-Atai, 2007), 20–24; Hurowitz, *Proverbs*, 1:54–64.

[87] Mordechai Cogan and Hayim Tadmor, "Gyges and Ashurbanipal: A Study in Literary Transmission," *Orientalia* 46 (1977): 65–85.

[88] Herodotus, *Histories*, 2:154; Donald B. Redford, *Egypt, Canaan and Israel in Ancient Times* (Princeton: Princeton University Press, 1993), 434–35; Anthony J. Spalinger, "Pasammetichus King of Egypt," *JARCE* 15 (1978): 49–57.

[89] Immanuel Kant, "Lectures on Pedagogy," in *Immanuel Kant: Anthropology, History and Education*, ed. G. Zöller and R. B. Louden, trans. Mary Gregor (Cambridge: Cambridge University Press, 2007), 9.411, p. 43; cf. Manfred Kuehn, "Kant on

guides humans toward the good.[90] In mediating between young people and the dominant culture, it serves as a form of socialization.[91] Fostering skills, it greatly affects the human ability to become economically established and achieve power and social status.[92] As well as imparting both theoretical and practical knowledge, it yields ethical insights.[93] From an ethical standpoint, education is meant to develop personality, attributes, and a sense of responsibility for others,[94] which some scholars refer to as ethical intelligence.[95] Education thus aids in developing empathy, in providing people with the tools to acquire virtues and principles. While it also prompts them to employ the latter, the decision to do so resides in each person's own hands.[96] The task is not straightforward and depends on numerous factors, including family status and environment.[97]

Education, Anthropology, and Ethics," in *Kant and Education: Interpretations and Commentary*, ed. K. Roth and C. W. Suprenant (London: Routledge, 2012), 55–60.

90 Aristotle, *The Nicomachean Ethics*, trans. David Ross (Oxford: Oxford University Press, 2009), 5.1130b:24–29, pp. 83–84.

91 Peter Scott, "Education," in *Dictionary of Ethics, Theology and Sociology*, ed. P. B. Clarke and Andrew Linzey (London: Routledge, 1996), 273.

92 Peter Singer, *Practical Ethics*, 2nd ed. (Cambridge: Cambridge University Press, 1999), 45–46.

93 Scott, "Education," 273; Mike Morris, *Concise Dictionary of Social and Cultural Anthropology* (Oxford: Wiley-Blackwell, 2012), 78; Nimrod Aloni, *Good Education: Towards Meaningful Life, Moral Conduct, and Self-Realization* [in Hebrew] (Tel Aviv: Hakibbutz Hameuchad, 2013), 9.

94 Rosalind Hursthouse, *Ethics, Humans and Other Animals: An Introduction with Readings* (London: Routledge, 2000), 147; Peter Corruthers, "Animal Mentality: Its Character, Extent, and Moral Significance," in *The Oxford Handbook of Animal Ethics*, ed. T.L. Beauchamp, R.G. Frey (Oxford: Oxford University Press, 2011), 373–406; Nathan Rothenstreich, *The National and Individual Being* [in Hebrew] (Jerusalem: Magnes, 2014), 51–52; Jonathan Sacks, *Lessons in Leadership: A Weekly Reading of the Jewish Bible* (New Milford, CT: Orthodox Union/Jerusalem: Maggid, 2015), 10–11; Shalom Rosenberg, *The Varieties of Jewish Thought: From* The Kuzari *to Levinas* [in Hebrew] (Jerusalem: Koren, 2023), 17–18.

95 Avi Sagi and Daniel Statman, *Religion and Morality*, trans. B. Stein (Amsterdam: Rodopi, 1995), 45.

96 Yeshayahu Leibowitz, *Judaism, the Jewish People, and the State of Israel* [in Hebrew] (Jerusalem: Schocken, 1975), 59; Yeshayahu Leibowitz, *Faith, History, and Values* [in Hebrew] (Jerusalem: Academon, 2002), 237–38.

97 Aloni, *Good Education*, 105.

In human society, although theoretical and practical education is largely conducted in a formal institutional setting in schools, it also sometimes takes a more indirect, less institutional form.[98] Likewise, although it principally targets the young—infants through young adults—it can extend through adulthood.[99]

Like the ancient Egyptians and Israelites, the Sumerians and Babylonians regarded educating the young to develop virtues and good habits as a preeminent value.[100] In Mesopotamia and Egypt, only the upper classes could afford to send their children to school.[101] Large swathes of the populations therein thus acquired their education within the family or close environment. Herein, fables and proverbs played an important role, serving as learning aids for both youth and adults. As we saw previously, their purpose was, among other things, to equip students with social skills and moral principles.

With respect to human–animal relations, it must be remembered that human beings are not always compassionate, this virtue requiring prompting and cultivation.[102] In general, the empathy on which compassion rests is directed toward those for whom we care.[103] Human beings

[98] Charles Stafford, "Education," in *Encyclopedia of Social and Cultural Anthropology*, ed. A. Barnard and J. Spencer (London: Routledge, 2002), 241–42; Paul Waldau, "Animals as Legal Subjects," in *The Oxford Handbook of Animal Studies*, ed. L. Kalof (Oxford: Oxford University Press, 2017), 177.

[99] Morris, *Concise Dictionary*, 78.

[100] Samuel Noah Kramer, *The Sumerians: Their History, Culture, and Character* (Chicago: University of Chicago Press, 1963), 237–46; Nicolas Christiaan. Veldhuis, "Elementary Education at Nippur: The Lists of Trees and Wooden Objects" (PhD diss., University of Groningen, 1997), 27; Alexander Kleinerman, *Education in Early 2nd Millennium BC Babylonia: The Sumerian Epistolary Miscellany* (Leiden: Brill, 2011), 58n4, 100n23; Yoram Hazony, *The Philosophy of Hebrew Scripture* (Cambridge: Cambridge University Press, 2012), 112.

[101] Kramer, *The Sumerians*, 245; Eugen Strouhal, *Life of the Ancient Egyptians*, (Norman OK: University of Oklahoma Press, 1992), 31, 35; Mario Liverani, *The Ancient Near East: History, Society and Economy*, trans. S. Tabatabai (London: Routledge, 2014), 76.

[102] Andrew Linzey, *Creatures of the Same God: Explorations in Animal Theory* (New York: Lantern, 2009), 3.

[103] Khen Lampert, *A Language of Empathy* [in Hebrew] (Tel Aviv: Tel Aviv University Press, 2022), 8, 120.

are naturally concerned about their nearest and dearest.[104] This transpires to be an ethical truth independent of space and time.[105] Some moral principles being universal,[106] ethical education enables individuals to expand their circle of empathy and show compassion to those far away.[107] The majority draw a moral line between human beings and animals as lower species; however, pain (physical and mental) is pain, whether it is experienced by human beings or animals.[108]

Here, the question thus arises: how do fables of the type we have presented in this chapter induce empathy toward animals? In the main, the Sumerian and Greek examples reflect the mutual relationships within human society rather than addressing human–animal relations.[109] They thus virtually always relate to human beings, seeking to teach people practical lessons and arouse their moral sense.[110]

The fables present animals as rational beings with free will and the ability to speak. Bringing the two kingdoms into close proximity, this

104 Avishai Margalit, *The Ethics of the Memory* (Cambridge MA and London: Harvard University Press, 2004), 32–35; Julianna C. Oxley, *The Moral Dimensions of Empathy: Limits and Applications in Ethical Theory and Practice* (London and New York: Palgrave Macmillan, 2011), 71. Even those on the autism spectrum, who find it difficult to empathize, seek to avoid harming others.

105 Khen Lampert, *Traditions of Compassion: From Religions Duty to Social Activism* (London and New York: Palgrave Macmillan, 2005), vii.

106 Bernard Williams, *Ethics and the Limits of Philosophy* (London: Routledge, 2006), 100.

107 Mary Anne Warren, *Moral Status: Obligations and Other Living Things* (Oxford: Clarendon, 1997), 146–47.

108 Peter Singer, "Ethics beyond Species and beyond Instincts," in *Animal Rights: Current Debates and New Directions*, ed. C. R. Sunstein and M. C. Nussbaum (New York: Oxford University Press, 2004), 81–82.

109 Karhonen, "Anthropomorphism," 211–12.

110 Jacques Derrida, *The Animal That Therefore I Am*, trans. David Willis (New York: Fordham University Press, 2008); Gabriela Kompatscher, "Human-Animal Studies: Bridging the Lacuna between Academia and Society," in *Animals and Their Relation to God, Humans and Things in the Ancient World*, ed. R. Mattila, S. Ito, and S. Fink (Wiesbaden: Harrassowitz, 2019), 13.

On occasion, a physical behavioral resemblance exists between human beings and animals. Rob Parkinson, *Transforming Tales: How Stories Can Change People* (London: Jessica Kingsley, 2009), 161; Tyler J. Kasperbauer, *Subhuman: The Moral Psychology of Human Attitudes to Animals* (New York: Oxford University Press, 2018), 16.

anthropomorphic process reduces the human–animal divide.[111] Humanization of the "other" (primarily via education) often prompts caring and empathy, thereby improving ethical attitudes toward an individual or group.[112] The faunal anthropomorphization in the fables may thus achieve the same moral effect, in particular toward weaker animals who become prey for stronger ones or victims of their cunning. This is due to the way in which the stories present the faunal perspective, giving animals the ability to express themselves and their suffering.[113] While not always easy to do, "walking in their shoes" leads to humans' better treatment of animals,[114] with empathy and compassion demanding transcending one's own boundaries and understanding and experiencing the state of others—which in turn prompts action on their behalf.[115] Hearing or reading the fables thus helped cultivate the perception that animals are as much part of the world as human beings, all requiring proper care and concern.[116]

The use of fables to enhance moral human conduct has continued into the modern day, as demonstrated by George Orwell's *Animal Farm*, for example,[117] wherein Orwell forcefully portrays animal suffering,

[111] Edward Wheatley, *Mastering Aesop: Medieval Education, Chaucer and His Followers* (Gainesville: University Press of Florida, 2000), 114; Judith Still, *Derrida and Other Animals: The Boundaries of the Human* (Edinburgh: Edinburgh University Press, 2015), 5, 9, 73.

[112] Albert Bandera, "Selective Disengagement in the Exercise of Moral Agency," *Journal of Moral Education* 31, no. 2 (2002): 112; Michael Slote, *The Ethics of Core and Empathy* (London: Routledge, 2007), 30–31.

[113] Peter Singer, *Animal Liberation: A New Ethics for Our Treatment of Animals* (New York: Review Books, 1975), 15; Lori Gruen, *Ethics and Animals: An Introduction* (Cambridge: Cambridge University Press, 2011), pp. 24–25; Andrew Linzey, *Why Animal Suffering Matters: Philosophy, Theology*, and Practical Ethics (Oxford: Oxford University Press, 2013), 20, 42, 70, 80.

[114] Thomas Nagel, *Mortal Questions* (Cambridge: Cambridge University Press, 1991), 169–76; John Rawls, *A Theory of Justice* (Cambridge MA: Harvard University Press, 1999), 353.

[115] Oxley, *The Moral Dimensions*, 59; Susan Lanzoni, *Empathy: A History* (New Haven: Yale University Press, 2018), 280; Avi Sagi, *Living with the Other: The Ethic of the Inner Retreat*, trans. B. Stein (Cham Switzerland: Springer Nature, 2018), 12–13; Lampert, 2022, 121, 127, 132.

[116] Williams, *Ethics and the Limits of Philosophy*, 111; cf. Tom Regan, *The Case for Animal Rights* (Berkeley CA: University of California Press, 1983), 239, 242.

[117] Parkinson, *Transforming Tales*, 198.

awakening the reader's compassion toward not only those under cruel tyrannical rule but also the faunal world.[118]

2.2 Conclusion

Herein, I have examined some of the fables in the Sumerian proverb collection in relation to the animals to which they refer. Dating to the third millennium BCE, these fables served as study material in scribal schools in Babylonia and Assyria for thousands of years. Although belonging to the genre of Mesopotamian wisdom, they are unique in deriving from popular rather than sapiential circles. Designed to function as a tool for acquiring practical wisdom and inculcating morality, the messages they convey were intended to guide the ancient Sumerians in their behavior and make them decent people, skilled in avoiding mistakes and successful and prosperous in life. The short, pithy epigrams delivered ethico-educational content in a way that was easily heard and understood, attracting the attention of the audience and on occasion even arousing antagonism because of their moralization. As we have seen, the fauna adduced possess particular traits—arrogance, cunning, profound insight, and so on. Some of the fables are humorous, and in others the conclusion is unexpected. The traditional correspondence between the human and animal worlds teaches both to acknowledge the other, recognizing their shared habits. The "human" behavior attributed to the animal world appears to be based primarily on the original features of each species, which are anthropomorphized in the fables. The fox is thus portrayed as cunning but not always wise, the goat as smart (at least in comparison with the sheep), the dog as not always capable of understanding human social conventions, and so on. Nor do human

[118] Dictatorships tend to dehumanize people and groups in order to avoid recognizing them as moral agents, thereby making them into objects that can be abused and oppressed without conscience. See Bandera, "Selective Disengagement," 109; cf. Janice T. Gibson and Mika Haritos-Fatouros, "The Education of a Torturer," *Psychology Today* 20, no. 11 (1986): 52. Frequently, they harness educational systems and state propaganda to this purpose, as in Nazism. Tamar Ketko, *Ice Creatures: The Nazi Educational System* [in Hebrew] (Tel Aviv: Resling, 2016), 16; Pnina Langberg, *The Commanding Voice of Auschwitz* [in Hebrew] (Tel Aviv: Resling, 2023), 52; David Clark, "On Being 'The Last Kantian in Nazi Germany': Dwelling with Animals after Levinas," in *Animal Acts: Configuring the Human in Western History*, ed. J. Ham and M. Senior (London: Routledge, 1997), 165–98.

beings always comprehend the needs and distress suffered by animals. The lessons cover various areas: knowledge of one's limitations, avoidance of bragging or provoking someone stronger or more powerful, extending a hand rather than mocking, eschewal of voraciousness and rapaciousness, honesty without naïveté, dodging danger, and more. This practical wisdom was meant to aid the ancient Sumerians not only in succeeding in life but also in creating an ethical society.

Fables generally being stories with a universal message, they are not necessarily dependent on the time or place in which they originated. They thus appear to have spread across civilizations marked by the cultural interaction that frequently accompanies political and economic ties. Analyzing the affinities between the Sumerian fables and Aesop's fables, we have seen how, where influence is palpable, the later Greek fables developed and elaborated their earlier Sumerian counterparts in many instances. Our study of Aesop's collection indicates that they also exhibit the influence of Sumerian myth and Egyptian fables. From the dawn of history, animals thus appear to have served ancient civilizations as a useful tool for inculcating moral traits in human society, thus migrating one culture to culture.

Although as educational tools fables seek to influence human interrelations, the anthropomorphization process bestows rationality, free will, and human speech on the faunal world. In thus approximating the two categories of species, they present the perspective of animals rather than human beings, thereby arousing compassion and empathy toward the faunal kingdom.

Bibliography

Aloni, Nimrod. *Good Education: Towards Meaningful Life, Moral Conduct, and Self-Realization.* [In Hebrew.] Tel Aviv: Hakibbutz Hameuchad, 2013.

Alster, Bendt. "An Akkadian Animal Proverb and Assyrian Letter, ABL 555." *JCS* 41 (1989): 187–93.

Alster, Bendt. *Proverbs of Ancient Sumer: The World's Earliest Proverb Collections.* Bethesda, MD: CDL, 1997.

Alster, Bendt. "Some New Sumerian Proverbs." In *Akkade Is King: A Collection of Papers by Friends and Colleges Presented to Aage Westenholtz*, edited by G. Barjmovich et al., 17–35. Leiden: Netherlands Instituut voor het Nabije Oosten, 2011.

Alster, Bendt, and Takayoshi, Oshima. "A Sumerian Proverb Tablet in Geneva with Some Thoughts on Sumerian Proverb Collections." *Orientalia* 75 (2006): 33–42.

Aristophanes. *The Birds*. Translated by J. Henderson. Newburyport, MA: Focus, 1999.

Aristotle. *The Nicomachean Ethics*. Translated by David Ross. Oxford: Oxford University Press, 2009.

Asheri, David. *A Possession for All Times: Greek Historians and History Writing in Greece and Rome*. [In Hebrew.] Jerusalem: Magnes, 2004.

Baldick, Chris. *Oxford Concise Dictionary of Literary Terms*. Oxford: Oxford University Press, 2001.

Bandera, Albert. "Selective Disengagement in the Exercise of Moral Agency." *Journal of Moral Education* 31, no. 2 (2002): 101–19.

Ben-Pazi, Hanoch. *Emmanuel Levinas—Educational Contract: Responsibility, Hope, and Alliance*. [In Hebrew.] Tel Aviv: Hakibbutz Hameuchad, 2016.

Breier, Idan. "Sumer as the Cradle of History—Canine as Well: The Representation of the Dog in Sumerian Proverbial Literature." *Ma'arav* 21, nos. 1–2 (2014): 83–101.

Brown, Christopher G. "Iambus." In *A Companion to the Greek Lyric Poets*, edited by D. E. Gerber, 11–88. Leiden: Brill, 1997.

Childs, Peter, and Roger Flower. *The Routledge Dictionary of Literary Terms*. London: Routledge, 2006.

Clark, David. "On Being 'The Last Kantian in Nazi Germany': Dwelling with Animals After Levinas." In *Animal Acts: Configuring the Human in Western History*, edited by J. Ham and M. Senior, 165–98. London: Routledge, 1997.

Clayton, Edward. "Aesop, Aristotle, and Animals: The Role of Fables in Human Life." *Humanitas* 21, nos. 1–2 (2008): 179–200.

Cogan, Mordechai, and Hayim Tadmor. "Gyges and Ashurbanipal: A Study in Literary Transmission." *Orientalia* 46 (1977): 65–85.

Cohen, Yoram. "The Scribal School at Emar." [In Hebrew.] *Beit Mikra* 57, no. 1 (2012): 65–85.

Corruthers, Peter. "Animal Mentality: Its Character, Extent, and Moral Significance." In *The Oxford Handbook of Animal Ethics*, edited by T. L. Beauchamp and R. G. Frey, 373–406. Oxford: Oxford University Press, 2011.

da Cunha Corrêa, Paula. "A Human Fable and the Justice of Beasts in Archilochus." In *Hesperos: Studies in Ancient Greek Poetry Presented to M. L. West on His Seventieth Birthday*, edited by P. J. Finglass, C. Collard, and N. J. Richardson, 101–17. Oxford: Oxford Scholarship Online, 2007.

Dalley, Stephanie. "Etana (1.131)." In *The Context of Scriptures*, edited by W. W. Hallo, 1:453–57. Leiden: Brill, 2003.

Demsky, Aaron. *Literacy in Ancient Israel*. [In Hebrew.] Jerusalem: Bialik Institute, 2012.

Derrida, Jacques. *The Animal That Therefore I Am*. Translated by David Willis. New York: Fordham University Press, 2008.
Even-Shoshan, Avraham. *The New Dictionary*. [In Hebrew.] Jerusalem: Kiryat Sepher, 1969.
Fox, Michael V. *Proverbs 1–9*. New York: Doubleday, 2000.
Fuks, Alexander. "Thucydides the Historian." [In Hebrew.] In *Greek Historians and Greek Historiography*, edited by D. Asheri, 111–43. Jerusalem: Bialik Institute, 2005.
Gabbay, Uri. "Lamentful Proverbs or Proverbial Laments? Intertextual Connections Between Sumerian Proverbs and Emesal Laments." *JCS* 63 (2011): 51–64.
Gagné, Renaud. "A Wolf at the Table: Sympotic Perjury in Archilochus." *TAPA* 139 (2009): 251–74.
George, Andrew R. "In Search of the É.DUB.BA.A: The Ancient Mesopotamian School in Literature and Reality." In *"An Experienced Scribe Who Neglects Nothing": Ancient Near Eastern Studies in Honor of Jacob Klein*, edited by Y. Sefati et al., 127–37. Bethesda, MD: CDL, 2005.
Gibbs, Laura. *Aesop's Fables: A New Translation*. Oxford: Oxford University Press, 2008.
Gibson, Janice T., and Mika, Haritos-Fatouros. "The Education of a Torturer." *Psychology Today* 20, no. 11 (1986): 50–52, 56–58.
Gordon, Edmund I. *Sumerian Proverbs: Glimpses of Everyday Life in Ancient Mesopotamia*. New York: Greenwood, 1968.
Gruen, Lori. *Ethics and Animals: An Introduction*. Cambridge: Cambridge University Press, 2011.
Hallo, William W. "Biblical and Ancient Near Eastern Literatures: A General Introduction." [In Hebrew.] In *The Literature of the Hebrew Bible: Introductions and Studies*, edited by Z. Talshir, 2:457–70. Jerusalem: Yad Ben-Zvi, 2011.
Hallo, William W. "Proverbs Quoted in Epic." In *Lingering over Words: Studies in Ancient Near Eastern Literature in Honor of William L. Moran*, edited by T. Abrush, J. Huehnergard, and P. Steinkeller, 203–17. Atlanta, GA: Scholars Press, 1990.
Haran, Menahem. "Mashal." [In Hebrew.] In *Encyclopedia Biblica*, vol. 5, cols. 548–53. Jerusalem: Bialik Institute, 1978.
Hazony, Yoram. *The Philosophy of Hebrew Scripture*. Cambridge: Cambridge University Press, 2012.
Herodotus. *The Histories: A New Translation by Robin Waterfield*. Oxford: Oxford University Press, 2008.
Holzberg, Niklas. *The Ancient Fable: An Introduction*. Translated by C. Jackson-Holzberg. Bloomington: Indiana University Press, 2002.

Howe, Nicholas. "Fabling Beasts: Traces in Memory." *Social Research* 62, no. 3 (1995): 641–59.

Hsü, Immanuel C. Y. *The Rise of Modern China*. 6th ed. New York: Oxford University Press, 2000.

Hurowitz, Victor Avigdor. *Proverbs: Introduction and Commentary*. [In Hebrew.] 2 vols. Tel Aviv: Am Oved; Jerusalem: Magnes, 2012.

Hursthouse, Rosalind. *Ethics, Humans and Other Animals: An Introduction with Readings*. London: Routledge, 2000.

Kant, Immanuel. "Lectures on Pedagogy." In *Immanuel Kant: Anthropology, History and Education*, edited by G. Zöller and R. B. Louden, 437–85, translated by Mary Gregor. Cambridge: Cambridge University Press, 2007.

Karhonen, Tua. "Anthropomorphism and the Aesopic Animal Fables." In *Animals and Their Relations to Gods, Humans and Things in the Ancient World*, edited by R. Mattila, S. Ito, and S. Fink, 211–31. Wiesbaden: Harrassowitz, 2019.

Kasperbauer, Tyler J. *Subhuman: The Moral Psychology of Human Attitudes to Animals*. New York: Oxford University Press, 2018.

Ketko, Tamar. *Ice Creatures: The Nazi Educational System*. [In Hebrew.] Tel Aviv: Resling, 2016.

Klein, Jacob. "Mesopotamian Literature: Genesis Traditions, Wisdom Literature, and Lamentations." [In Hebrew.] In *The Literature of the Hebrew Bible: Introductions and Studies*, edited by Z. Talshir, 2:522–79. Jerusalem: Yad Ben-Zvi, 2011.

Klein, Jacob, and Nili Samet. "Ethics and Religion in Sumerian Proverb Literature." [In Hebrew.] *Beit Mikra* 57, no. 2 (2012): 106–31.

Kleinerman, Alexander. *Education in Early 2nd Millennium BC Babylonia: The Sumerian Epistolary Miscellany*. Leiden: Brill, 2011.

Kompatscher, Gabriela. "Human-Animal Studies: Bridging the Lacuna Between Academia and Society." In *Animals and Their Relation to God, Humans and Things in the Ancient World*, edited by R. Mattila, S. Ito, and S. Fink, 11–22. Wiesbaden: Harrassowitz, 2019.

Kramer, Samuel Noah. *The Sumerians: Their History, Culture, and Character*. Chicago: University of Chicago Press, 1963.

Kuehn, Manfred. "Kant on Education, Anthropology, and Ethics." In *Kant and Education: Interpretations and Commentary*, edited by K. Roth and C. W. Suprenant, 55–68. London: Routledge, 2012.

Lambert, Wilfred G. *Babylonian Wisdom Literature*. Oxford: Oxford University Press, 1969.

Lampert, Khen. *Traditions of Compassion: From Religions Duty to Social Activism*. London and New York: Palgrave Macmillan, 2005.

Lampert, Khen. *A Language of Empathy*. [In Hebrew.] Tel Aviv: Tel Aviv University Press, 2022.

Langberg, Pnina. *The Commanding Voice of Auschwitz.* [In Hebrew.] Tel Aviv: Resling, 2023.
Lanzoni, Susan. *Empathy: A History*. New Haven: Yale University Press, 2018.
Leibowitz, Yeshayahu. *Faith, History, and Values.* [In Hebrew.] Jerusalem: Academon, 2002.
Leibowitz, Yeshayahu. *Judaism, the Jewish People, and the State of Israel.* [In Hebrew.] Jerusalem: Schocken, 1975.
Lichtheim, Miriam. *Ancient Egyptian Literature*. 3 vols. Berkeley: University of California Press, 2006.
Linzey, Andrew. *Creatures of the Same God: Explorations in Animal Theory.* New York: Lantern, 2009.
Linzey, Andrew. *Why Animal Suffering Matters: Philosophy, Theology, and Practical Ethics*. Oxford: Oxford university Press, 2013.
Liverani, Mario. *The Ancient Near East: History, Society and Economy.* Translated by S. Tabatabai. London: Routledge, 2014.
Lorberbaum, Menachem. *I Seek Thy Face.* [In Hebrew]. Jerusalem: Carmel, 2018.
Margalit, Avishai. *The Ethics of the Memory.* Cambridge MA and London: Harvard University Press, 2004.
Miyuki Babcock, Jennifer. *Ancient Egyptian Animal Fables: Tree Climbing Hippos and Ennobled Mice*. Leiden: Brill, 2022.
Morris, Mike. *Concise Dictionary of Social and Cultural Anthropology.* Oxford: Wiley-Blackwell, 2012.
Murphy, Roland E. *Proverbs*. Nashville: Thomas Nelson, 1998.
Nagel, Thomas. *Mortal Questions*. Cambridge: Cambridge University Press, 1991.
Northwood, Heidi. "Making Music with Aesop Fables in the Phaedo." In *Plato's Animals: Gadflies, Horses, Swans, and Other Philosophical Beasts*, edited by J. Bell and M. Naas, 13–16. Bloomington: Indiana University Press, 2015.
Oxley, Julianna C. *The Moral Dimensions of Empathy: Limits and Applications in Ethical Theory and Practice*. London and New York: Palgrave Macmillan, 2011.
Parkinson, Rob. *Transforming Tales: How Stories Can Change People*. London: Jessica Kingsley, 2009.
Puchala, Donald J. *Theory and History in International Relations*. New York: Routledge, 2003.
Rajuan, Rahamim. *The Wisdom of Arabia: 1001 Proverbs, Sayings and Aphorisms*. [In Hebrew.] Tel Aviv: Dvir, 1991.
Rawls, John. *A Theory of Justice*. Cambridge MA: Harvard University Press, 1999.
Raymond, Janice. *Mongolian Proverbs: A Window into Their World.* Translated by O. Oidovdorj. Eugene, OR: Resource, 2010.

Redford, Donald. B. *Egypt, Canaan and Israel in Ancient Times*. Princeton: Princeton University Press, 1993.
Refael-Vivante, Revital. *A Treasury of Fables: Isaac ibn Sahula's* Meshal Haqadmoni *(Castile, 1281)—Text and Subtext*. [In Hebrew.] Ramat-Gan: Bar-Ilan University Press, 2017.
Regan, Tom. *The Case for Animal Rights*. Berkeley CA: University of California Press, 1983.
Robson, Eleanor. "The Tablet House: A Scribal School in Old Babylonian Nippur." *RA* 95 (2001): 39–66.
Rodrigues Adrados, Francisco. *History of the Graeco-Roman Fable*. 2 vols. Leiden: Brill, 1999.
Rosenberg, Shalom. *The Varieties of Jewish Thought: From* The Kuzari *to Levinas*. [In Hebrew.] Jerusalem: Koren, 2023.
Rothenstreich, Nathan. *The National and Individual Being*. [In Hebrew.] Jerusalem: Magnes, 2014.
Sacks, Jonathan. *Lessons in Leadership: A Weekly Reading of the Jewish Bible*. New Milford, CT: Orthodox Union/Jerusalem: Maggid, 2015.
Sagi, Avi. *Living with the Other: The Ethic of the Inner Retreat*. Translated by B. Stein. Cham Switzerland: Springer Nature, 2018.
Sagi, Avi., and Daniel Statman. *Religion and Morality*. Translated by B. Stein. Amsterdam: Rodopi, 1995.
Scott, Peter. "Education." In *Dictionary of Ethics, Theology and Sociology*, edited by P. B. Clarke and A. Linzey, 273–76. London: Routledge, 1996.
Shupak, Nili. "The Father's Instruction in Ancient Egypt." [In Hebrew.] In *Education and History: Cultural and Political Contexts*, edited by R. Feldahay and I. Etkes, 13–21. Jerusalem: Zalman Shazar Centre, 1999.
Shupak, Nili. *Practical Ethics*. 2nd ed. Cambridge: Cambridge University Press, 1999.
Shupak, Nili. "Ethics Beyond Species and Beyond Instincts." In *Animal Rights: Current Debates and New Directions*, edited by C. R. Sunstein and M. C. Nussbaum, 78–92. New York: Oxford University Press, 2004.
Shupak, Nili. *Olam Ha-Tanach: Sefer Mishley (Proverbs)*. [In Hebrew.] Tel Aviv: Davidzon-Atai, 2007.
Shupak, Nili. "Ancient Egyptian Literature." [In Hebrew.] In *The Literature of the Hebrew Bible: Introductions and Studies*, edited by Z. Talshir, 2:605–56. Jerusalem: Yad Ben-Zvi, 2011.
Shupak, Nili. *"No Man Is Born Wise": Ancient Egyptian Wisdom Literature and Its Contact with Biblical Literature*. [In Hebrew.] Jerusalem: Bialik Institute, 2016.
Shupak, Nili. "The First Mickey Mouse: The Enigma of the Anthropomorphized Drawings in Ancient Egypt." In *Human Interaction with the Natural World*

in Wisdom Literature and Beyond: Essays in Honour of Tova L. Forti, edited by M. Cogan, K. Dell, and D. Glatt-Gilad, 33–44. London: T&T Clark, 2023.
Singer, Peter. *Animal Liberation: A New Ethics for Our Treatment of Animals*. New York: Review Books, 1975.
Slote, Michael. *The Ethics of Core and Empathy*. London: Routledge, 2007.
Spalinger, Anthony J. "Pasammetichus King of Egypt." *JARCE* 15 (1978): 49–57.
Span, Shlomo. "Aesop and the Fable." [In Hebrew.] In *Aesopi Fabulae*, 6th ed., translated by S. Span, 170–98. Jerusalem: Bialik Institute, 2001.
Stafford, Charles. "Education." In *Encyclopedia of Social and Cultural Anthropology*, edited by A. Barnard and J. Spencer, 271–75. London: Routledge, 2002.
Still, Judith. *Derrida and Other Animals: The Boundaries of the Human*. Edinburgh: Edinburgh University Press, 2015.
Strouhal, Eugen. *Life of the Ancient Egyptians*. Norman OK: University of Oklahoma Press, 1992.
Taylor, Jon. "The Sumerian Proverb Collections." *RA* 99, no. 1 (2005): 13–38.
Thucydides. *The War of the Peloponnesians and Athenians*. Translated by J. Mynott. Cambridge: Cambridge University Press, 2013.
Tobin, Vincent A., trans. "The Maxims of Ptahhotep." In *The Literature of Ancient Egypt: An Anthology*, edited by W. K. Simpson, 129–48. New Haven: Yale University Press, 2003.
Ukhmani, Azri'el. *Contents and Forms: Dictionary of Literary Terms*. [In Hebrew.] 2 vols. Tel Aviv: Hapoalim Library, 1992.
Vanstiphout, Herman L. J. "How Did They Learn Sumerian?" *JCS* 31 (1979): 118–26.
Veldhuis, Nicolas Christiaan. "Elementary Education at Nippur: The Lists of Trees and Wooden Objects." PhD diss., University of Groningen, 1997.
Veldhuis, Nicolas Christiaan. "Sumerian Proverbs in Their Curricular Context." *JAOS* 120, no. 3 (2000): 388–99.
Waldau, Paul. "Animals as Legal Subjects." In *The Oxford Handbook of Animal Studies*, edited by L. Kalof, 167–88. Oxford: Oxford University Press, 2017.
Warren, Mary Anne. *Moral Status: Obligations and Other Living Things*. Oxford: Clarendon, 1997.
Wheatley, Edward. *Mastering Aesop: Medieval Education, Chaucer and His Followers*. Gainesville: University Press of Florida, 2000.
Williams, Bernard. *Ethics and the Limits of Philosophy*. London: Routledge, 2006.
Wimmer, Stefan. "Hieroglyphs: Writing and Literature." In *Egypt: The World of the Pharaohs*, edited by R. Schulz and M. Seidel, 348–53. Cologne: Könemann, 1998.

CHAPTER 3

Compassion and Character: Animal Welfare in Fictional Narratives of Moral Improvement by Nineteenth-Century English Women Writers

Miriam Sampson

In the autumn of 1877, a novel was published in London that took the world by storm. Its author was a single woman who had been disabled for much of her life and virtually bedridden for the last eight years. It was the only book she wrote, and she lived just long enough to become aware of the favorable reception it received. By the end of April 1878, she was dead. The book was *Black Beauty, His Grooms and Companions*[1]; its author, Anna Sewell. It told the story of the life of a horse, with the horse himself as narrator, recounting "conversations" between the horses

[1] Anna Sewell, *Black Beauty, His Grooms and Companions: The Autobiography of a Horse* (London: Jarrold and Son, 1877); Sewell, *Black Beauty* (Ware, UK: Wordsworth Editions, 1993).

M. Sampson (✉)
Portsmouth, UK
e-mail: sampsons@dsl.pipex.com

C. Linzey and A. Linzey (eds.), *The Palgrave Companion to Humane Education*, The Palgrave Macmillan Animal Ethics Series,
https://doi.org/10.1007/978-3-032-19795-5_3

in the story. It highlighted many cruelties inflicted on horses in Victorian society.

Within a year, *Black Beauty* sold 12,000 copies; within twelve years, 100,000 had been sold in England, and it subsequently sold millions across the world. Written in a simple style, to appeal to people who actually looked after horses, it was marketed as an accessible horse-care manual, a work of humane protest and an exciting story featuring an empathetic hero. Reviewers praised the narrative as full of variety and interest and commended it as "especially valuable for the lessons of humanity which it inculcates, not only towards horses, but towards all living creatures."[2]

Black Beauty was also widely read by children. Early advertising promoted it particularly for boys, since many boys and men looked after horses, and when the book was published, there was widespread amazement that it could have been written by a woman, since it was so full of accurate detail relating to horse care. One hundred years after its publication, it was rated the favorite book of ten-year-olds in a major survey of British children's reading and had become particularly popular with girls.[3]

Victorian society depended heavily on horses for transport and labor. *Black Beauty* raised awareness of the ways that horses were ill-used, whether through ignorance or through deliberate cruelty. The Royal Society for the Prevention of Cruelty to Animals (RSPCA) was campaigning for better treatment of cab horses, and Edward Fordham Flower was leading a campaign against the fashionable but cruel bearing rein, which forced horses to hold their heads in an unnaturally high position. *Black Beauty* gave an enormous impetus to both causes. When it was published in the United States in 1890, it was explicitly used as a campaigning document by George Thorndike Angell, the founder of the American Humane Education Society, who saw its educational potential

[2] Review in the *Eastern Daily Press*, December 22, 1877, quoted in Adrienne E. Gavin, introduction to *Black Beauty*, by Anna Sewell, Oxford World Classics ed. (Oxford: Oxford University Press, 2012), xiv.

[3] Adrienne E. Gavin, *Dark Horse: A Life of Anna Sewell* (Stroud, UK: Sutton Publishing, 2004), 205–6.

as "the best book ever written teaching kindness to the horse."[4] He aimed to "put a copy... in every home in America."[5]

There is no doubting the impact of Anna Sewell's *Black Beauty*, but where did her passion for kindness come from? What was the background from which Anna Sewell's novel emerged, and how did it relate to the thinking and writing of other women over the previous century? Sewell was not the first woman writer in the nineteenth century to address cruelty to animals, and her book was part of a wider discourse on moral improvement. Her narrative contained a variety of moral messages "not only about cruelty to horses and other animals, but also against drinking, hunting, irresponsible smoking, war, ignorance, blindly following party politics, and not keeping Sunday as a day of rest."[6] Sewell saw this moral framework as a coherent whole; for example, she saw drunkenness as contributing to cruelty to horses and believed that horses and cab men deserved a day of rest on the Sabbath like anyone else. Where were these values rooted, and what did Anna Sewell have in common with other women who explored these themes in their writing?

3.1 An Impetus for Moral Improvement

The underlying values and educational motivation that informed Anna Sewell's narrative can be traced back to the mid-eighteenth century, a time of considerable social change. The Industrial Revolution was underway, with workers leaving the countryside to work in factories, eroding traditional social structures.[7] Poverty was widespread, child labor was the norm,[8] and the "gin craze" had had a detrimental effect on families and fueled a rise in crime. Contemporary writers bemoaned the barbaric state

[4] George T. Angell, *Autobiographical Sketches and Personal Recollections* (Boston: American Humane Society, 1898), 94, quoted in Gavin, "Dark Horse," 187.

[5] Angell, "Autobiographical Sketches," 95, quoted in Gavin, "Dark Horse," 188. As a result of Angell's highly successful promotion, many people think that *Black Beauty* was written by an American author.

[6] Gavin, "Dark Horse," 179.

[7] Manchester, the center of the textile industry in the north of England, experienced a sixfold increase in its population between 1771 and 1831.

[8] Douglas A. Galbi, "Child Labour and the Division of Labour in the Early English Cotton Mills," *Journal of Population Economics* 10 (1997): 357–75.

of society, often indicting not only the behavior of the laboring classes but also that of the upper classes.[9]

One aspect of this "barbarism" was the popularity of cruel sports involving animals. In the early eighteenth century, activities such as bull-baiting, cockfighting, and dogfighting were widespread, popular with both the gentry and the laboring classes. But as the century progressed, critical voices began to be heard.[10]

Those critical voices frequently came from figures involved with the Evangelical Revival, which gained influence from the 1740s onward. John Wesley's Methodism was one important strand, but as the century progressed, evangelicalism grew within the Anglican Church too. The Clapham Sect was particularly influential: so-called because it gathered around the antislavery campaigner and member of Parliament William Wilberforce, who lived in Clapham, South London, and worshiped at Holy Trinity Clapham.[11]

The Clapham Sect belonged to the upper echelons of society, but the impact of the Evangelical Revival was felt much more widely. Members of the Clapham Sect were acutely aware that others did not share their social privileges; they campaigned and took direct action against a range of social evils: slavery, the condition of the poor, and also cruelty to animals.[12] The impact of this commitment to social and moral improvement can be seen particularly clearly in an educational innovation that had a remarkable impact in the late eighteenth and early nineteenth centuries: the Sunday school. Through the Sunday school a direct link can be traced to Anna Sewell and *Black Beauty*, as Sewell herself was active in the Sunday

[9] See, for example, Hannah More, *Thoughts on the Importance of the Manners of the Great to General Society* (London: T. Cadell, 1788).

[10] See Robert W. Williamson, *Popular Recreations in English Society, 1700–1850* (Cambridge: Cambridge University Press, 1973).

[11] Two prominent men in the Evangelical movement were incumbents at Holy Trinity, Clapham. Henry Venn was curate from 1754 to 1759. His son John was rector from 1792 till 1813.

[12] Wilberforce attended the founding meeting of the Society for the Prevention of Cruelty to Animals (later RSPCA) in 1824 and supported animal welfare legislation in the nineteenth century. For a full discussion of pre-twentieth-century Evangelical theology and animal welfare initiatives, see Philip J. Sampson, *Animal Ethics and the Nonconformist Conscience* (Basingstoke: Palgrave Macmillan, 2018).

school movement. Sunday schools were unknown before the late eighteenth century, and they emerged from a context of wider educational innovation.

3.2 Educational Innovation and Moral Improvement

Educational innovation was another aspect of rapid social change in England during the eighteenth century. Numerous books of educational theory explored how a child's character was formed and how educational approaches could be suited to a child's age and capabilities.[13] The most influential on the early Sunday schools was Isaac Watts's *Treatise on the Education of Children and Youth*.[14]

Watts was an evangelical Congregationalist minister, now best known as a hymn writer. He worked as a tutor in a nonconformist family and wrote from his own teaching experience. Watts based his approach on a statement in the book of Proverbs: "train up a child in the way he should go, and when he is old he will not depart from it."[15] He believed that educating children well was the best way to establish them in virtue and piety in adulthood and that education should enable children to improve their natural powers of body and mind. Education in religion and the development of moral capacities should begin as soon as a child was old enough for a basic understanding.

For Watts, learning to read was a vital skill: "it is a pity that any children should be born and brought up in Great Britain without the skill of reading."[16] He favored education for girls as well as boys.[17] Watts

[13] The growing middle class was interested in gaining a practical education for their children, and works of educational theory by philosophers were being widely read. John Locke's treatise *Some Thoughts concerning Education* (published 1693) had become a bestseller, and later Rousseau's *Emile* (published in France in 1762; English translation, 1763) attracted widespread attention. See W. A. C. Stewart and W. P. McCann, *The Educational Innovators 1750–1880* (London: Macmillan, 1967).

[14] Originally published as an appendix to the second part of Isaac Watts, *The Improvement of the Mind* (London: James Brackstone, 1741), but later presented separately. *A Treatise on the Education of Children and Youth*, 2nd ed. (London: Buckland, Longman, Dilly, and Field, 1769).

[15] Prov. 22:6.

[16] Watts, "A Treatise on Education," 56.

[17] Watts, "A Treatise on Education," 59.

also recognized that recreation was important for children, but he was adamant that games and pastimes should not involve cruelty to animals.[18]

While older boys were educated in schools where men taught, younger children, both boys and girls, were educated in the home, largely supervised by women. From the mid-eighteenth century, educational publications began to appear, putting into practice the more "child-centered" ideas of contemporary writers, including Watts.[19] These included reading primers and books of "dialogues" between mother and child or teacher and pupil, which encouraged children to observe the world around them and take note of the habits of the domesticated animals, plus the birds and insects, whom they encountered every day: not just to read about them but also to learn how to treat them.[20]

One particularly influential publication was *The First Principles of Religion, and the Existence of a Deity, Explained in a Series of Dialogues, Adapted to the Capacity of the Infant Mind*. Written by Dorothy Kilner, it was then taken up by Sarah Trimmer, a pioneer in the Sunday school movement, who published an abridged edition titled *Sunday School Dialogues*.[21] It consists of a dialogue between a young girl and her teacher. This extract comes from their conversation about how small creatures should be treated[22]:

> *Mary*: Is it wicked to kill wasps and such kinds of things madam?
> *Teacher*: Not if we kill them without tormenting them. But it is very wrong to pull them to pieces, and hurt them.... There is a reason why we kill wasps and that is, because they bite and sting us; ... but

[18] Watts, " A Treatise on Education," 110–11.

[19] Other widely read eighteenth-century writers on education included Anna-Laetitia Barbauld and Richard and Maria Edgeworth. Barbauld's *Lessons for Children* (1778–79) was the earliest book of educational dialogues.

[20] Anna Sewell's mother, Mary, wrote a book of this type, *Walks with Mama,* which was published in 1820. It was based on the educational approach she was using with Anna and her brother.

[21] Dorothy Kilner published the original version under the pseudonym M. P. The original publication date is not clear; the copy in the British Library catalogue is dated 1795, but Sarah Trimmer's abridged version was probably published in 1790.

[22] Dialogue 20 in Sarah Trimmer, ed., *Sunday School Dialogues; Being an Abridgement of a Work, by M. P. Entitled "The First Principles of Religion and the Existence of a Deity Explained in a Series of Dialogues, Adapted to the Infant Mind"* (London: John Marshall, [1790?]), 58.

we should be very careful when we do kill them to do it at once, as quickly as possible, that they may not suffer for any time. It is very cruel to torment anything....

Mary: Do dumb creatures feel as much as we do, madam?

Teacher: There is reason to think they do; and it is not unlikely, that when hurt many of them suffer *more* than we, because their bodies are so small and delicate.

Mary: Poor things! They do not cry out.

Teacher: We do not know that, Mary; perhaps they do, but our ears cannot hear them, because their cries are not loud enough: but God can hear them; at least he knows when we are barbarous to them and certainly will punish us for being so; for God is a friend to all creatures that are unjustly treated....

And you would not like to be cut in two, to have a spit run through your body, to have your legs and arms pulled off, or to be shut up in a box so you could not breathe, to be kept without victuals in a cage, to have a string tied to your leg or arm, and to be pulled about so your limbs were broken, should you?

Mary: No, to be sure, madam.

Teacher: Then I beg you will never use flies, birds or butterflies in that manner. How common is it to see children doing those things without any thought of what the poor creatures suffer.

Mary: I never will do anything cruel as long as I live.

This child may sound like a bit of a prig, but she is learning an important lesson, and the teacher pulls no punches in describing the kinds of cruelties to which small creatures might be subjected. Kindness to animals is an important theme running through these dialogues; although they emphasize the *differences* between the children and the animals they encounter—focusing on the rationality of the child and the possession of a "soul"—they also stress the *similarities*, since animals can feel pain and should not be made to suffer. The children are taught that human beings have a God-given responsibility to care for animals.

The eighteenth-century educational dialogues were originally written for middle-class parents to use with their children, and Sarah Trimmer also wrote a fictional work, *Fabulous Histories, Designed for the Instruction of Children Respecting Their Treatment of Animals,* featuring a human family in comfortable circumstances and a family of robins.[23] This became

[23] Sarah Trimmer, *Fabulous Histories, Designed for the Instruction of Children, Respecting Their Treatment of Animals* (London, 1785).While emphasizing kindness to

very popular, but through the *Sunday School Dialogues* Trimmer reached a much wider audience in different social circumstances.

3.3 Origins of the Sunday School Movement

Robert Raikes was a key figure in the early Sunday school movement.[24] As managing editor of the *Gloucester Journal*,[25] he was a leading voice in campaigns against a number of social evils, including cruel sports, but Raikes soon moved beyond campaigning.

Raikes had been visiting the local prison and became convinced that vice would be better prevented than cured. He saw schooling as the best intervention. The best available time was Sunday since children were often working on the other six days. The best available teachers were laypeople. The textbook was the Bible, and the curriculum started with learning to read.

Raikes started his Sunday school in Gloucester in 1780, with a local woman as teacher. Within two years, he opened several more. He published an account in his paper in 1783, and word began to spread. Sarah Trimmer, Hannah More (a personal friend of Raikes), and others started Sunday schools in poor communities. Numbers rose rapidly. By 1851, nearly two and a half million children—approximately 13 percent of the population—attended weekly across the country.[26]

Sunday schools were controversial. They were led by laypeople, rather than clergy, and most teachers were women. The schools were suspected of educating working people above their station and fomenting revolution. This was a particular fear at the end of the eighteenth century, with the French Revolution across the Channel causing alarm. However,

animals, Trimmer also asserted the importance of feeding hungry people before animals. *Fabulous Histories* remained in print until the early twentieth century. Mary Wollstonecraft refers to it favorably in Chapter 2 of her book of educational dialogues: *Original Stories, from Real Life: With Conversations Calculated to Regulate the Affections and Form the Mind to Truth and Goodness* (London: Johnson, 1788).

[24] The first Sunday school was probably started by a Methodist woman called Hannah Ball in 1769, but Raikes played a key role in initiating a Sunday school movement.

[25] The *Gloucester Journal* was established in 1722. It espoused social causes and pursued campaigns against the inhumane treatment of debtors and criminals and against cock-fighting. Robert took over as proprietor and editor on his father's death in 1757. Guy Kendall, *Robert Raikes: A Critical Study* (London: Nicholson and Watson, 1939), 29.

[26] Kendall, *Robert Raikes*, 148.

most Sunday schools were led by members of the Anglican Church,[27] who were very committed to upholding the social order. But it was the evangelical wing of the Anglican Church that took the lead, and many people involved in the Sunday school movement were also active in the antislavery campaign.[28]

Sunday schools aimed not only to increase religious knowledge but also, crucially, to change behavior. A key behavior that they wanted to change was the way people treated animals. They also wanted to reduce drunkenness and violence and increase diligence and Sabbath observance.[29] These aims related to their opposition to cruel sports, since cockfights, dogfights, and bullbaiting often took place around ale houses on Sundays and public holidays.

The central commitment of all Sunday schools was teaching children to read, following the recommendation given by Watts. Previously, religious instruction had primarily involved catechizing: learning to answer a set of questions by rote. Reading gave more agency to the pupils; they could read the Bible and other texts and find things out for themselves. But once the poorer classes had learned to read, the question arose: what reading material was available for them? The Sunday school pioneers took on this challenge.

3.4 Literature of Moral Improvement—*Cheap Repository Tracts*

In the eighteenth century a variety of cheap literature aimed at working-class readers, such as chapbooks and broadside ballads, was sold by traveling salesmen in enormous quantities, but the themes were often crude and the illustrations vulgar. Then in the 1790s, as the French

[27] The founder of Methodism, John Wesley, commented approvingly of Sunday schools that he observed in action as he traveled around the country, but there is little evidence that he started many such schools himself. Kendall, *Robert Raikes*, 141–43.

[28] More had been strongly influenced by John Newton, the London preacher and former slave trader who became a leading voice in the antislavery campaign. More and her sister started their Sunday schools in the Cheddar villages with funding from William Wilberforce, who saw the poverty there while visiting the More sisters in Bristol.

[29] In many respects the curriculum in the early Sunday schools was socially conservative. For example, More did not allow the teaching of writing in her schools and prepared young people for domestic service.

Revolution took hold on the other side of the English Channel, radical political writings began to circulate, causing alarm among the ruling classes in England, who felt steps were needed to counteract revolutionary sentiments. In the 1790s the bishop of London, a reformer who actively opposed the slave trade but wanted to avoid revolution, prevailed upon Hannah More to turn her hand to writing popular literature aimed at combating revolutionary tendencies and promoting moral improvement.[30]

More had taught at a girls' school in Bristol, where she wrote dialogues and dramas to enliven the students' education. She had then been feted on the London literary scene in the 1770s.[31] Her poems were widely read and her plays staged to great acclaim. But by the 1780s, More had embraced an evangelical faith and become associated with Wilberforce and the Clapham Sect. More moved back to Bristol, and at Wilberforce's instigation, she and her sister started Sunday schools in the impoverished villages around Cheddar in North Somerset. By the mid-1790s, they were supervising twelve thriving schools in the area, and More was writing short "tracts" expounding the Bible in a lively style, to be used as instructional material.[32] In 1792, in response to the bishop's request, she produced a booklet, *Village Politics*,[33] affirming the values of the British constitution. But this was just the start.

From 1794, More embarked on a series known as *Cheap Repository Tracts*, produced at the rate of three a month and marketed quite skillfully.[34] They were a phenomenal success, selling over two million copies in two years. They featured homely characters in familiar settings and promoted frugality, sobriety, industry, Sabbath observance, and kindness

[30] A full account of Hannah More's life and work can be found in Karen Swallow Prior, *Fierce Convictions: The Extraordinary Life of Hannah More—Poet, Reformer, Abolitionist* (Nashville, TN: Nelson, 2014).

[31] More had been friendly with Samuel Johnson and the actor and theater producer David Garrick, who had produced one of her plays.

[32] Martha More, *Mendip Annals* (London, James Nisbet, 1859), introduction, 9.

[33] More published this pseudonymously under the name Will Chip.

[34] The tracts were priced very competitively, and More used traveling salesmen and sympathetic contacts among the privileged classes to distribute them, since most poor people would never enter a bookshop. She understood the need to appeal to the popular imagination, choosing recognizable formats and including woodcut illustrations. She enlisted several collaborators and instructed them to write in a similar way.

to animals. But through Hannah More's lively writing style, they managed to avoid merely didactic moralizing and were enjoyable to read.

The History of Tom White, the Postilion, published in 1795, illustrates how More associated swearing and drunkenness with the ill-treatment of animals. Tom is a laborer's son. His father has many mouths to feed, so as soon as Tom is old enough, he goes to work for a local farmer, who trains him to drive a horse-drawn wagon and ensures that he attends Sunday school each week, where he learns to read his Bible. But Tom wants to see the world, so he heads off to town and gets work looking after horses at a local inn.

> Tom's good education stood by him for some time. At first he was frightened to hear the oaths and wicked words which are too often uttered in a stable-yard ... By degrees he began to think it manly, and a mark of spirit in others to swear; ... at first when he swore it was with fear and in a low voice. But he was soon laughed out of his sheepishness, as they called it; ... he would too often use a dreadful bad word when he was in a passion with his horses. And here I cannot but drop a hint on the great folly as well as wickedness of being in a great rage with poor beasts, who not having the gift of reason, cannot be moved like human creatures, with all the wicked words that are said to them; but who, unhappily, having the gift of feeling, suffer as much as human creatures can do, at the cruel and unnecessary beatings given them.[35]

Tom engages in an ill-advised race with another coach driver when they are both drunk, resulting in a dreadful accident in which he and his horses are injured. While recovering in hospital, he resolves to change his ways. He gives his horses time to rest on Sundays, taking the opportunity to go to church. He reads in the Bible that he should give consideration to animals, and he puts this into practice:

> Tom indeed was very careful ... in never giving his horses too much water when they were hot; nor, whatever was his haste, would he ever gallop them up hill, (or) strike them across the head ... What helped to cure Tom of these bad practices, was that remark he met with in the Bible, that a good man is merciful to his beast.[36]

[35] Hannah More, *The History of Tom White, the Postilion, Part 1* (London: J. Marshall and R. White, 1795), 4.

[36] More, *Tom White*, 14. The Bible verse cited is Prov. 12:10.

Through his hard work and care for his horses, Tom prospers and ends up as a tenant farmer.

3.5 Popular Ballads

More wrote not only in prose but also in verse, recognizing that rhythm and rhyme helped to keep words in the memory. An example is *The Hackney Coachman; or, The Way to Get a Good Fare*, a ballad about a sober and hardworking Hackney cab driver:

> Though my beasts should be dull, yet I don't use them ill;
> Though they stumble I swear not, nor cut them up hill;
> For I firmly believe there's no charm in an oath
> That can make a nag trot, when to walk he is loath.

Hannah More was not the only female writer to turn her hand to popular ballads for the new working-class readers. Anna Sewell's own mother, Mary, was prolific in this genre, and during her lifetime Mary was far better known as an author than her daughter was.[37]

Mary had grown up a Quaker, but in her middle years she and Anna came into contact with evangelical doctrines and started worshiping in the Anglican Church. Both women were very active in Sunday schools, particularly during the late 1850s, when they lived in a poor Gloucestershire village. They embarked on charitable works in what they considered to be a "thoroughly neglected parish"[38] where there was a Sunday school, but few helpers. Here Mary started to write her ballads, verses "designed for the moral and practical edification of the working classes and children."[39] They condemned cruelty, unkindness, sloth, alcohol, and wastefulness and promoted sound moral behavior, charity, and love for one's neighbor. For women they gave practical instruction in housekeeping and parenting. The ballads were printed in hundreds of thousands, widely distributed and read.

[37] Her most popular ballad was translated into several languages and sold over a million copies throughout the world. Mary Sewell, *Mother's Last Words* (London: Jarrold & Son, 1860).

[38] Gavin, "Dark Horse," 127.

[39] Gavin, "Dark Horse," 132.

Kindness to animals was an integral theme. For example, *Abel Howard and His Family*[40] features Tom, Abel's son, who is being considered for a job working with horses. He is recommended to a potential employer because he is gentle and kind to animals:

> I often see him with his father's team;
> the very horses are so fond of him.
> I've marked it scores of times, and found it suit,
> and judge a boy as he will treat a brute.
> Suppose I see him speak a pony fair,
> rubbing his nose, and stroking down his hair,
> driving the sheep, and keeping back the dog,
> feeling a man although he does not flog,...
> I'd trust that boy to have a kindly heart,
> and that of men, I think the better part.[41]

Anna acted as editor for her mother's work. Both were acutely aware of the miseries of others and were committed to remedying them. While living in the village, they started a lending library, taught in both the Sunday school and the day school, and founded a Temperance Brotherhood.[42] They held "mothers' meetings" for women, teaching them to read and giving simple explanations of Bible verses, alongside practical instruction in cooking, housework, health care, childcare, and "how to maintain a moral household and deal with a drunken husband."[43]

At this time a family friend, Mary Bayly, introduced Anna to the work of an American evangelical, Horace Bushnell, who proposed a God-given order embracing humans and animals, arguing that "according to God's will, animals would be very happy serving humans provided they were well treated; it was therefore incumbent on people to treat them kindly."[44] When writing *Black Beauty*, Anna told Bayly, "The thoughts you gave me from Horace Bushnell years ago have followed me entirely through

[40] Published in Mary Sewell's collection *Homely Ballads for the Working Man's Fireside* (London: Smith & Elder, 1858).

[41] M. Sewell, *Homely Ballads*, 36.

[42] *Black Beauty* includes a strong temperance message, highlighting the damage done to horses by drunken riders and coachmen.

[43] Gavin, "Dark Horse," 145.

[44] Gavin, "Dark Horse," 142.

the writing of my book, and have... helped me to feel it was worth a great effort to try, at least, to bring the thoughts of men more in harmony with the purposes of God on this subject."[45]

3.6 Compassion and Character—A Critique of Masculinity

Why did these women writers involved in popular education place so much emphasis on treating animals kindly?[46] To address this question, we turn to an example of literary fiction, rather than the educational texts and pamphlets discussed so far: *Agnes Grey* by Anne Brontë.

Anne was the youngest of the three Brontë sisters and the least known. She taught in her father's Sunday school and later became a governess. Her first novel, *Agnes Grey* (1847),[47] reflects her own teaching experience, recounting many examples of cruelty to animals that she witnessed in that role. As a governess, Agnes is not included in the social circle of her employers, and neither does she belong with the servants; she is a "hapless dependent,"[48] giving her an affinity with the animals within the household—Snap the dog, the cat, and the pony. Her isolated and dependent position also gives her sympathy for free-living animals whom the male characters enjoy abusing.

Her first pupil, Tom, delights in whipping his rocking horse, boasting that he would do the same to a real pony, and threatens his sister with his whip. He proclaims his delight in trapping and torturing small animals and is about to tear the wings and heads off fledglings in a nest before Agnes prevents him by dropping a large stone onto the nest to put the birds out of their misery quickly. Tom is following the example of his father and uncle, who commend his masculine spirit and joke that he is "beyond petticoat government already"[49]; in other words, at a young age he is already dominating women. Agnes's plea on behalf of the animals,

[45] Letter from Anna Sewell to Mary Bayly in 1877 or 1878, quoted in Gavin, "Dark Horse," 142.

[46] This emphasis is not present in the works of Dickens, for example, who addressed many social evils in his fiction but said little about cruelty to animals.

[47] Anne Brontë, *Agnes Grey* (1847; Ware, UK: Wordsworth Editions, 1998).

[48] Maggie Berg, "'Hapless Dependents'": Women and Animals in Anne Bronte's 'Agnes Grey,'" *Studies in the Novel* 34, no. 2 (2002), http://www.jstor.org/stable/29533502.

[49] Anne Brontë, *Agnes Grey*, 37.

quoting the injunction from Proverbs that "the righteous man shows mercy to his beast,"[50] is considered to be evidence of feminine weakness.

Like More and Sewell, Brontë links drunkenness with cruelty. However, there is one man in the book whose behavior is very different: the curate, Mr. Weston. We never see Mr. Weston drunk; he is kind to animals and considerate to women in socially inferior positions. He is thoughtful to Agnes; he visits a poor elderly villager, Nancy, and is kind to her cat. In contrast, the vicar treats Nancy's cat roughly. Weston also rescues Agnes's little dog, Snap, who had been sold to a cruel rat catcher, and restores Snap to her at the end of the book, when he proposes marriage, and Agnes accepts.

Brontë offers a trenchant critique of a masculinity that uses power to oppress animals and women. She proposes a model of masculine character associated with compassion and dependability, illustrated through kindness to animals and consideration for people in positions of weakness.

This critique was also implied in the popular tracts and ballads of More and Sewell. Mary Sewell's characterization of young Tom Howard as "feeling a man although he does not flog" and her comment "I'd trust that boy to have a kindly heart, and that of men I think the better part"[51] make it quite explicit. And More and Brontë both quote the same Bible verse in support—Proverbs 12:10. Such a critique is also evident throughout *Black Beauty*.

All these women taught children in Sunday schools and other settings. They all witnessed the consequences of male violence and overbearing control toward women and children and also toward animals. They understood how weaker members of society depended on the stronger, and they recognized how abuse took place. They saw that drunkenness could lead to cruelty and the exploitation of women and of animals. From these experiences they started to construct a critique of masculinity that applied as much to the upper classes as the lower. As they taught and wrote about kindness to animals, they were implementing that critique and advocating an alternative model of masculine character, embodying compassion, not violence.

[50] Prov. 12:10; Brontë, *Agnes Grey*, 38.

[51] M. Sewell, *Homely Ballads*, 36. For context, see the longer quote cited previously.

Bibliography

Berg, Maggie. "'Hapless Dependents': Women and Animals in Anne Brontë's 'Agnes Grey.'" *Studies in the Novel* 34, no. 2 (2002): 177–97. http://www.jstor.org/stable/29533502.
Brontë, Anne. *Agnes Grey*. 1847. Ware, UK: Wordsworth Editions, 1998.
Galbi, Douglas A. "Child Labour and the Division of Labour in the Early English Cotton Mills." *Journal of Population Economics* 10 (1997): 357–75.
Gavin, Adrienne E. *Dark Horse: A Life of Anna Sewell*. Stroud, UK: Sutton Publishing, 2004.
Gavin, Adrienne E. Introduction to *Black Beauty*, by Anna Sewell. Oxford World Classics ed. Oxford: Oxford University Press, 2012.
Kendall, Guy. *Robert Raikes: A Critical Study*. London: Nicholson and Watson, 1939.
More, Hannah. *The Hackney Coachman; or, the Way to Get a Good Fare*. London: John Marshall, 1796.
More, Hannah. *The History of Tom White, the Postilion, Part 1*. London: J. Marshall and R. White, 1795.
More, Hannah. *Thoughts on the Importance of the Manners of the Great to General Society*. London: T. Cadell, 1788.
More, Hannah. *Village Politics: Addressed to All the Mechanics, Journeymen, and Day-Labourers in Great Britain*. London: F. & C. Rivington, 1792.
More, Martha. *Mendip Annals*. London: James Nisbet, 1859.
Sampson, Philip J. *Animal Ethics and the Nonconformist Conscience*. Basingstoke: Palgrave Macmillan, 2018.
Sewell, Anna. *Black Beauty*. 1877. Ware, UK: Wordsworth Editions, 1993.
Sewell, Mary. *Homely Ballads for the Working Man's Fireside*. London: Smith & Elder, 1858.
Sewell, Mary. *Mother's Last Words*. London: Jarrold & Son, 1860.
Stewart, W. A. C., and W. P. McCann. *The Educational Innovators 1750–1880*. London: Macmillan, 1967.
Swallow Prior, Karen. *Fierce Convictions: The Extraordinary Life of Hannah More—Poet, Reformer, Abolitionist*. Nashville, TN: Nelson, 2014.
Trimmer, Sarah. *Fabulous Histories, Designed for the Instruction of Children, Respecting Their Treatment of Animals*. London, 1785.
Trimmer, Sarah, ed. *Sunday School Dialogues; Being an Abridgment of a Work, by M. P. Entitled "The First Principles of Religion and the Existence of a Deity Explained in a Series of Dialogues, Adapted to the Infant Mind."* London: John Marshall [1790?].
Watts, Isaac. *A Treatise on the Education of Children and Youth*. 2nd ed. London: Buckland, Longman, Dilly, and Field, 1769.
Williamson, Robert W. *Popular Recreations in English Society, 1700–1850*. Cambridge: Cambridge University Press, 1973.

Wollstonecraft, Mary. *Original Stories, from Real Life: With Conversations Calculated to Regulate the Affections and Form the Mind to Truth and Goodness.* London: Johnson, 1788.

CHAPTER 4

The Human as Animal: Scottish Fantasy and the Importance of Imagination for Humane Empathy

Rebecca McLean

4.1 The Human Animal: Victorian Science and Popular Understanding

The scientific upheavals of the Victorian era led to a reconceptualization of humankind's understanding of creation and humans' place within it. More specifically, this chapter focuses on the advent of evolutionary biology and its appearance within the popular consciousness. Although most commonly associated with the works of Charles Darwin, this destabilization began earlier, with Charles Lyell's *The Principles of Geology* (1830–1833). Lyell's work made reference to the fossil record to demonstrate that the earth was more than the 6,000 years old that had been calculated to accord with a biblical creation narrative. Lyell argued the earth was much older and had been subject to geological

R. McLean (✉)
London, UK
e-mail: rebeccamclean2022@pm.me

C. Linzey and A. Linzey (eds.), *The Palgrave Companion to Humane Education*, The Palgrave Macmillan Animal Ethics Series,
https://doi.org/10.1007/978-3-032-19795-5_4

change over previously inconceivable lengths of time. Of course, the crisis of human identity is most commonly associated with the publication of Charles Darwin's *On the Origins of Species* (1859). However, of even greater significance when examining the relationship between humans and animals are Darwin's *The Descent of Man* (1871) and *The Expression of the Emotions in Man and Animals* (1872). These two works clearly and directly place humans in the animal world, through evolutionary processes. *The Descent of Man* explicitly applies the principles of evolution to humans. Having compared a human fetus and monkey fetus, Darwin writes, "Consequently we ought frankly to admit their community of descent: to take any other view, is to admit that our own structure and that of all animals around us, is a mere snare laid to entrap our judgment."[1] *The Expression of Emotions in Man and Animals* focuses on how human emotions attach physical, recordable movements to psychological states and how these emotions can then be universally understood. The human, on an emotional, non-linguistic level, communicates like other animals, selects mates like other animals, and has evolved just as other species do. What these works do is destabilize the notion that humanity is unique within creation and as such is entirely separate from and holds dominion over other life forms. The responses of Christians to this destabilization were varied. Charles Kingsley embraced the ideas and found them compatible with Christian belief systems,[2] while figures such as Samuel Wilberforce argued that "the principle of natural selection is absolutely incompatible with the word of God."[3] What ultimately emerged from these discussions is the idea of humans as partially animal, and this idea can be seen in a wide range of Victorian literature, as has been discussed by scholars such as Gillian Beer.[4] I argue here that the association of humans with animals led to an enduring shift in our perception of both humans and animals that is reflected in a range of Scottish fantasy texts, aimed at an adult readership, that depict human subjects in the

[1] Charles Darwin, *The Descent of Man* (London: John Murray, 1871), 32–33.

[2] Will Abberley provides a detailed discussion of Kingsley's views in "Animal Cunning: Deceptive Nature and Truthful Science in Charles Kingsley's Natural Theology," *Victorian Studies* 58, no. 1 (2016): 34–56.

[3] Samuel Wilberforce, "Darwin's Origin of Species," *Quarterly Review* 108 (July 1860): 225–64.

[4] See Gillian Beer, *Darwin's Plots: Evolutionary Narrative in Darwin, George Eliot and Nineteenth-Century Fiction*, 3rd ed. (Cambridge: Cambridge University Press, 2009).

role of animals. These texts develop the reader's empathy toward animals through placing humans in the role of animals.

Literature assists in developing support for animal advocacy by encouraging humans to develop an empathetic response to animal suffering at our hands. I should say from the outset that humans are exceptionally bad at expressing empathy for nonhumans or rather are very good at coming up with reasons to excuse or support the practices that lead to animals' suffering. The ability of texts to circumvent this is evident in Scottish fantasy in the Victorian period. There followed a fallow period in the production of Scottish fantasy and an intense focus on Scottish identity in the texts produced, leading to a sustained focus on humans rather than animals by midcentury. In his 1934 essay "The Land," Lewis Grassic Gibbon argues, "To me it is inconceivable that sincere and honest men would go outside the range of their own species with gifts of pity and angry compassion and rage when there is horror and dread among humankind. I am unreasonably and mulishly prejudiced in favour of my own biological species."[5] However, the literary representation of humans in the role of the nonhuman animal reemerged toward the end of the twenty-first century.[6] Here the reader gains empathy toward animals through the substitution of the nonhuman animal in the place of the animal. In developing empathy toward the nonhuman animal, these works of fantasy place the human in the position of an animal, provoking horror, providing readers with a moment of what the theorist Julia Kristeva terms "abjection." Kristeva says that abjection is a response to a threatened breakdown in the established meaning or categorization of subject and object—in this case the human and nonhuman. Kristeva's work *Powers of Horror* focuses on moments where we are confronted with the physicality of death: "The corpse, seen without God and outside of science, is the utmost of abjection. It is death infecting life. Abject."[7] Moments

[5] Lewis Grassic Gibbon, "The Land," in *Smeddum: A Lewis Grassic Gibbon Anthology*, ed. Valentina Bold (Edinburgh: Canongate, 2001), 95. Grassic Gibbon's quote and the marginalization of the nonhuman animal in post-2000 Scottish literature are discussed by Timothy C. Baker in "Perpetual Vanishing: Animal Lives in Contemporary Scottish Fiction," *Humanities* 8, no. 1 (2019): 12.

[6] Timothy C. Baker provides a detailed examination of this in *Writing Animals: Language, Suffering, and Animality in Twenty-First-Century Fiction* (London: Palgrave MacMillian, 2019).

[7] Julia Kristeva, *Powers of Horror: An Essay on Abjection* (New York: Columbia University Press, 1982), 4.

where a human is tortured (as an animal may well be in the real world) beyond the point of real-world death provide a sense of horror for the reader. However, it is the humanity of the subject that elicits an empathetic response to these scenes. The horror of depicting the human in the position of the animal allows a state of abjection to develop in the mind of the reader, and this abject horror, when retained beyond the text, leads us to reconsider what previously have been socially acceptable modes of interacting between humans and animals. When works of literature demonstrate the importance of extending empathy toward the nonhuman animal by placing a human in the role of animal, further questions arise for the reader regarding the accepted treatment of animals in the world beyond the text.

4.2 Late Victorian Responses: *Lilith* and "The Land of Darkness"

So let us begin by examining some examples of how Victorian literature began to reassess the relationship between humans and animals. George MacDonald's 1895 novel *Lilith* is a seminal work of adult fantasy revolving around the adventures of a young man, Mr. Vane, as he traverses a fantasy world called "the region of the seven dimensions."[8] These dimensions effectively function as a purgatorial space, where souls are shown trapped in various forms of torment, and mythological and biblical figures act as guides, including Adam, Eve, and Lilith. As this description suggests, the text has strongly religious themes. But prefacing this is a scientific frame to the tale. Mr. Vane is first encountered in the real world, having just completed his studies at Oxford. He is described as highly bookish, with a specific interest in "the physical sciences,"[9] and a sample of his holiday reading is provided:

> In the great room I mainly spent my time, reading books of science, old as well as new; … Ptolemy, Dante, the two Bacons and Boyle were even more to me than Darwin or Maxwell.[10]

[8] George MacDonald, *Lilith: A Romance* (London: Chatto & Windus, 1986), 25.

[9] MacDonald, *Lilith*, 1.

[10] MacDonald, *Lilith*, 2.

When Mr. Vane follows first a ghost and later a raven through a mirror into the secondary world, he also discovers that the raven transforms into a man, who in due course we discover to be the biblical Adam. With his scientific training and interests—key proponents of evolution are included in his reading list—Mr. Vane challenges Adam's shape-shifting ability, especially when, in his raven form, Adam turns worms into butterflies. Mr. Vane argues that this ability to change will lead to individuals becoming "proud, and cease to recognise [their] superiors."[11] Adam's response is to conflate human and animals, thereby rejecting the idea of superiority of one species over another:

> Upon occasion, ... it is more convenient to put one's bird-self in front. Everyone, as you ought to know, has a beast-self—and a bird-self, and a stupid fish-self, ay, and a creeping serpent-self too—which it takes a deal of crushing to kill! In truth he has also a tree-self and a crystal-self, and I don't know how many more selves more—all to get into harmony. You can tell what sort a man is by his creature that comes oftenest to the front.[12]

This melding not just of species,[13] but of all living things, still allows for prejudice to attach to specific species, with fish being called stupid and a serpent-self requiring crushing.[14] However, because these are all present within the same individual alongside many other selves, any prejudice becomes a form of self-criticism. The human is also animal. MacDonald negates any attempt at alterity, where something is defined against the "other," between human and nonhuman life. While this revelation is not

[11] MacDonald, *Lilith*, 23. See also Fernando Soto, "The Worm as Metaphor in Lilith," *North Wind: A Journal of George MacDonald Studies* 25, no. 1 (2006): 111–23.

[12] MacDonald, *Lilith*, 37.

[13] MacDonald's inclusion of crystals in his list of life-forms indicates his understanding of contemporary efforts toward classification of life-forms. For something to be classified as living, it must have the following characteristics: order, sensitivity or response to the environment, reproduction, growth and development, regulation, homeostasis, and energy processing. Crystals do fit within the majority but not all of these categories, which led many to argue that crystals were a not yet fully understood type of life. By 1900 Nikola Tesla argued in *The Century Magazine* that "in a crystal, we have clear evidence of the existence of a formative life principle, and though we cannot understand the life for a crystal, it is no the less a living being." Nikola Tesla, "The Problem of Increasing Human Energy," *The Century Magazine* 60, no. 2 (June 1900): 180.

[14] See Roland Hein, *Doors In: The Fairy Tale World of George MacDonald* (Eugene, OR: Cascade Books, 2018), 110.

developed much beyond this conversation in the early stages of *Lilith*'s narrative, it is deeply significant since it provides an exaggerated fantasy version of Darwin's suggestion that humanity shares common traits with animals.

George MacDonald's suggestion in *Lilith* highlights an inherent animality in all humans, while a less overt but equally significant representation of the human–animal relationship can be found in Margaret Oliphant's 1889 novella *The Land of Darkness*.

The tale follows an unnamed narrator as he arrives in a dystopic hellscape, where he has no needs—no need to sleep, eat, or drink and no requirement for money. He also cannot die and will heal from any injury. And so our nameless narrator begins wandering through the landscape, entering a variety of towns. In the first, the citizens focus on what they can gain for themselves to the detriment of any social cohesion. After being captured and enslaved in a gold mine in the wilderness between cities, the narrator enters a second city, which is controlled by strict rules and in which his existence becomes purposeless. He moves to a third city where the denizens are constantly feasting, dancing, and drinking, which becomes exhausting. So the narrator moves to the fourth and final city, a dystopic industrial landscape where all human purpose is replaced by robots. At this point he discovers that these locations are the extent of hell, and though he has the option to attempt to climb the surrounding mountains and enter heaven, this effort causes such pain that he returns to moving between the cities.

Oliphant provides two instances of public experimentation that drive the narrator out of both the first and second cities. In the first city, the protagonist joins a crowd in the town square, and though he acknowledges that there is a living being onstage, he initially is not disturbed by this or by the lecture on how to use the nerves to increase pain in the subject. However, nonchalant curiosity rapidly changes: "I with consternation made a step forward, crying with horror. The object on the table, writhing, twitching, to get free, but bound down by every limb, was a living man."[15] The humanity of the test subject appears to be the reason for the narrator's horror, and he identifies strongly with the individual as the vivisection begins: "I shrieked out, unable to restrain myself. The sight of the deliberate wound, the blood, the cry of agony that came from

[15] Margaret Oliphant, "The Land of Darkness," in *A Beleaguered City and Other Tales of the Seen and Unseen*, ed. Jenni Calder (Edinburgh: Canongate, 2000), 327.

the victim, the calmness of all the lookers-on, filled me with horror and rage indescribable."[16] This moment of empathy is promptly challenged when in the course of defending the test subject, the narrator is betrayed by the subject, who shouts to the crowd, "Take him instead; curse him! take him instead."[17] The narrator has to escape the town to avoid taking the man's place as the subject of public vivisection.

In the second experiment, Oliphant develops from this first incident. This time the reader takes the place of the narrator, acting as witness to the experiment, while the narrator takes the place of the test subject. In the following extract the narrator attempts to show the suffering of the test subject through detailed description.

> I was caught by every joint, sustained, supported, exposed to the gaze of what seemed a world of upturned faces ... Above my head was a strong light, more brilliant than anything I had ever seen, and which blazed upon my brain till the hair seemed to singe and the skin shrink. I hope I may never feel such a sensation again. The pitiless light went into me like a knife; but even my cries were stopped by the framework in which I was bound. I could breathe and suffer, but that was all.... Then he [the scientist] pulled over me a great movable lens as of a microscope, which concentrated the insupportable light.... I was let down among the crowd, and exhibited to them, every secret movement of my being, by some awful process which I have never fathomed. A burning fire was in my brain, flame seemed to run along all my nerves, speechless, horrible, incommunicable fury raged in my soul. But I was like a child—nay, like an image of wood or wax in the pitiless hands that held me.[18]

This process of removing the subject's skin with some unknown technological apparatus is here described in highly emotive terms by the narrator. Through this description, he appeals directly to the readers, seeking to get them to empathize with and be revolted by his experiences. We should remember that the audience in hell, those viewing the experiment firsthand, are amused by the spectacle and display no empathy for the subject.

[16] Oliphant, "Land of Darkness," 328.

[17] Oliphant, "Land of Darkness," 328.

[18] Oliphant, "Land of Darkness," 342.

The comparison between the first and second vivisection is directly referred to by the narrator, who questions, "What was the cut of a surgeon's knife to this? And I had thought *that* cruel! And I was powerless, and could do nothing."[19] This recalls his initial horror roused by the initial cut of the subject at the first vivisection, thereby encouraging his readers to not only share his emotional response to the spectacle but also empathize with him directly as the victim of the second experiment. Oliphant uses humans in the place of animals to increase the horror of these scenes and develop the depravity of hell. While nonhuman animals are not depicted at any point in this narrative, it is by no means suggested that humans are seen as in any way different from the animal subjects of vivisection; indeed, it is suggested that the narrator must "contribute [his] quota to our enlightenment."[20] Whereas my first example, George MacDonald's *Lilith*, shows human and animals as part of the same being, Oliphant relies on the humanity of her test subjects to provoke a response from her readers. The horror attached to the human vivisection subject therefore allows readers to reconceive of the ethical implications of animal vivisection in the real world, as they have now gained insight into how such practices may feel to the subject.

4.3 Modern Representations of the Human Animal

Michel Faber's novel *Under the Skin* provides an example of so-called best-practice free-range farming, but as these processes are applied to human subjects, the reader encounters a tale of body horror that raises questions about any form of meat farming. It is through an appeal to the reader's imagination and empathy with the human-animal subject that new perspectives regarding the human–nonhuman animal divide are developed.

Faber's tale is told primarily through the perspective of Isserley, an alien who has undergone what she terms "mutilation" to make her look human. She is tasked with capturing male hitchhikers on the A9 and knocking them out with a drug called "icpathua." The men are then brought to a rural farm to be processed for their meat, an expensive

[19] Oliphant, "Land of Darkness," 342.

[20] Oliphant, "Land of Darkness," 341.

delicacy on her home planet. The aliens call themselves "human" and view what we would consider the humans of the tale as animals named "vodsels." I will return to this blurring of human and animal or "other" identity in a moment. But first I want to briefly set out the similarities between the treatment of the vodsels and the treatment of farmed pigs.

First, the environment they are kept in is described as follows:

> The celling was less than seven feet high … The vodsel enclosures, a corona of linked pens all along the walls, took up almost the entire floor space; there was just enough room left down the middle for a walkway. In the cages to the left, the monthlings, to the right the transitionals; at the deep end, against the far wall facing the lift, the new arrivals.[21]

This system of pens and designation of individuals to specific stages is echoed in intensive pig farming, where "the majority of growing and finishing pigs are housed indoors and grouped into pens."[22] The fact that the vodsels are finished over the course of a month, via alien technology, allows them to live the majority of their lives free-range on earth. The point at which the vodsels are kept inside during their final month is shown to be in their best interests; at one point in the narrative, four break free, and after only a few hours they have frostbitten extremities and are injured because of their modification. This process of modification loosely follows the preparation of a pig for slaughter: the vodsels are "shaved, castrated, fattened, intestinally modified, [and] chemically purified"[23] and, perhaps most significantly in terms of this discussion, have their tongues docked.[24] By offering detailed descriptions of the putrid conditions, the vodsels are kept in and the grotesque changes to their bodies, Faber leaves his readers feeling a sense of horror that they would not necessarily feel for a nonhuman animal in a similar free-range, supposedly ethical farming scenario. Readers are encouraged to empathize with the vodsels as they understand them to be a representative of themselves

[21] Michel Faber, *Under the Skin* (Edinburgh: Canongate, 2014), 168.

[22] Dale Arey and Phil Brooke, *Animal Welfare Aspects of Good Agricultural Practice: Pig Production* (Compassion in World Farming, 2006), 40, https://www.ciwf.org.uk/media/5492194/gap_pig_book_full.pdf.

[23] Faber, *Under the Skin*, 97.

[24] Faber, *Under the Skin*, 100.

in the novel. The abjection or moment where we see them as other relative to the reader is not in a moment of death (the packaged meat and corpses of the vodsels are not dwelled upon) but is in their reduction to the status of farmed animals, or "livestock."

As promised, I now return to the blurring of distinctions between human and animal in the novel. This is best approached through the idea of sentience. We as readers know that humans (vodsels) are sentient and can communicate, but in the text it appears that only Isserley is aware of this; she converses with her victims in the car, but their tongues are removed before any other farmworker is aware of their ability to speak. During a visit from Amlis Vess, the farm owner's son, this is challenged:

> Isserley watched, disturbed, as the vodsel scrawled a five-letter word with great deliberation …
> "No one told me they have language," marvelled Amlis, too impressed, it seemed, to be angry. "My father always describes them as vegetables on legs."
> "It depends what you classify as language, I guess," said Isserley dismissively …
> "But what does in *mean*?" persisted Amlis.
> Isserley considered the message, which was M E R C Y.… the word was untranslatable in her own tongue; it was a concept that just didn't exist.[25]

The denial of language and the removal of the vodsels' tongues becomes a denial of their sentience. This insistence becomes increasingly important in the pages immediately following this scene, where Isserley considers the implications of vodsel sentience. When denying the word means anything, she says to Amlis, "I'm a human being, not a vodsel."[26] Michel Faber's decision to use the term "human" for the alien visitors (in their natural form, seven-foot-long dog-type creatures) and to label what the reader identifies as humans as "vodsels" in alien terms ensures that there is a sense of alienation between the readers and their closest counterpart in the tale. The figure who comes closest to expressing the readers' concerns and horror at the treatment of the vodsels is, oddly, the entirely alien Amlis. In addition to noticing the attempts of his livestock to communicate, Amlis refuses to eat vodsel meat, criticizing those who do—"That meat

[25] Faber, *Under the Skin*, 171.

[26] Faber, *Under the Skin*, 172.

you're eating… is the body of a creature that lived and breathed"[27]—and facilitates the vodsels' escape, believing they should be returned to their natural environment.[28]

Amlis's empathetic response to the processing of vodsels is contrasted by Isserley's refusal to see the people she collects as similar to her, despite her being the only person who interacts with them in their natural environment. At one point, when troubled by their apparent sentience, Isserley even lists all the alien qualities and abilities that these vodsels lack.[29] The empathy Amlis shows is criticized because it privileges a species other than his own. When he takes Isserley to view the "livestock," he interprets her reaction as disgust at the conditions, not her sense of personal discomfort, and she reflects that Amlis is "so obsessed with his own idealism he was incapable of feeling empathy for a human being suffering right under his nose."[30] Isserley's critique here reaches to the heart of my argument. She shows how easy it is for individuals to refuse to engage with the suffering of animals and to rely on the idea that their species is distinct and special in some way. Amlis challenges this but ultimately fails to change Isserley's mind because she fails to empathize or imagine herself in the place of the creatures she works with on a daily basis. By centering this debate around the perspective of two aliens, who term themselves human, Michel Faber directly challenges his readers to think about what it means to classify oneself as human. By placing what we conventionally think of as human subjects in a scenario that is considered acceptable and even humane by a majority of the population—that of an industrialized meat production system—under the control of a new, alien category of "human," Faber forces the reader to reconsider our current treatment of animals within the industrial farming process.

Humane empathy becomes central to our reaction, as readers, to Faber's text. However, such empathy is clearly developed further than in my earlier examples. While George MacDonald's *Lilith* recategorizes humans as containing various other life-forms, Faber's work simply changes the being attached to the label of "human" and by so doing places what we consider to be human into the position of animal.

[27] Faber, *Under the Skin*, 163.

[28] Faber, *Under the Skin*, 115.

[29] Faber, *Under the Skin*, 174.

[30] Faber, *Under the Skin*, 167.

Margaret Oliphant's novella *The Land of Darkness* places a human subject in the position of an animal one, just as in *Under the Skin* the vodsel pens situate humans in the place of farmed animals. For Oliphant the purpose of this is to allow the readers a moment of horror in her hellscape and to highlight the lack of empathy shown by its inhabitants. Isserley's lack of empathy is contrasted with Amlis's possession of it, and this only serves to heighten the abjection and horror of the text as the reader is drawn to continue sympathizing with the plight of the vodsels, despite Isserley's thoughts dominating the narrative. By not just suggesting that humans are inherently animals, but by starkly placing them in the role of animal and using arguments we as society may make to support our choices to intensively farm animals and consume meat to support what is tantamount to the torture of humans, *Under the Skin* forces readers to reconsider our treatment of animals. We as readers are led into accepting a framework of empathy by applying it first to our own species and then extending this to others as we read Faber's work.

4.4 Conclusion

All three of the literary examples discussed here allow human protagonists to inhabit the role or category of the nonhuman. MacDonald's conflation of human and nonhuman life forms is merely unsettling. Oliphant and Faber replace animal test subjects or farmed animals with humans in a way that initially evokes horror in their readers. What all three works do is develop their readers' empathy toward the nonhuman other. It is very difficult to promote the ethical treatment of animals without at some point imagining or empathizing with the suffering animal. These three examples of Scottish fantasy literature all present this idea in a nonconfrontational way, allowing readers to fully consider the implications of their choices in a safe, fictional environment. The texts provide a conceptual space for us to work through difficult ideas and their implications, especially the human–animal relationship.

Bibliography

Abberley, Will. "Animal Cunning: Deceptive Nature and Truthful Science in Charles Kingsley's Natural Theology." *Victorian Studies* 58, no. 1 (2016): 34–56.

Arey, Dale, and Phil Brooke. *Animal Welfare Aspects of Good Agricultural Practice: Pig Production*. Compassion in World Farming, 2006. https://www.ciwf.org.uk/media/5492194/gap_pig_book_full.pdf.

Baker, Timothy C. "Perpetual Vanishing: Animal Lives in Contemporary Scottish Fiction." *Humanities* 8, no. 1 (2019): 12. https://doi.org/10.3390/H8010012.

Baker, Timothy C. *Writing Animals: Language, Suffering, and Animality in Twenty-First-Century Fiction*. London: Palgrave MacMillian, 2019.

Beer, Gillian. *Darwin's Plots: Evolutionary Narrative in Darwin, George Eliot and Nineteenth-Century Fiction*, 3rd ed. Cambridge: Cambridge University Press, 2009.

Darwin, Charles. *The Descent of Man*. London: John Murray, 1871.

Faber, Michel. *Under the Skin*. Edinburgh: Canongate, 2014.

Grassic Gibbon, Lewis. "The Land." In *Smeddum: A Lewis Grassic Gibbon Anthology*, edited by Valentina Bold, 81–97. Edinburgh: Canongate, 2001.

Hein, Roland. *Doors In: The Fairy Tale World of George MacDonald*. Eugene, OR: Cascade Books, 2018.

Kristeva, Julia. *Powers of Horror: An Essay on Abjection*. New York: Columbia University Press, 1982.

MacDonald, George. *Lilith: A Romance*. London: Chatto & Windus, 1986.

Oliphant, Margaret. "The Land of Darkness." In *A Beleaguered City and Other Tales of the Seen and Unseen*, edited by Jenni Calder, 313–62. Edinburgh: Canongate, 2000.

Soto, Fernando. "The Worm as Metaphor in Lilith." *North Wind: A Journal of George MacDonald Studies* 25, no. 1 (2006): 111–23.

Tesla, Nicola. "The Problem of Increasing Human Energy." *The Century Magazine* 60, no. 2 (June 1900): 175–210.

Wilberforce, Samuel. "Darwin's Origin of Species." *The Quarterly Review* 108 (July 1860): 225–64.

CHAPTER 5

Empathy in Art

Mary Shannon Johnstone

5.1 Introduction

How is empathy created in art? This chapter defines three visual principles used to create empathy (perspective, gesture, and composition) and examines how each evokes empathy in artwork. In particular, this discussion will look at examples of how these principles are used by contemporary photographers and visual artists working with nonhumans. This essay examines how empathy is communicated and what concepts are shared with the viewer as a result.

This chapter includes three sections:

1. Definitions.
2. Two case studies of photographic historical examples.
3. Examples of each principle through the lens of contemporary artists featuring nonhumans.

M. S. Johnstone (✉)
Art Department, Meredith College, Raleigh, NC, USA
e-mail: johnsmar@meredith.edu

C. Linzey and A. Linzey (eds.), *The Palgrave Companion to Humane Education*, The Palgrave Macmillan Animal Ethics Series,
https://doi.org/10.1007/978-3-032-19795-5_5

While there is a plethora of resources dedicated to teaching principles of art and design, it is a struggle to find any that address how empathy is evoked. The three visual principles of empathy defined here arose from studying activist photography with messages of social justice.[1,2] Most of the examples in this chapter are photographic, but these principles can be applied to any kind of visual art. It is worth noting that this list of principles is in no way exhaustive. In fact, artists make many visual choices to evoke empathy. However, through my naming and defining three umbrella principles, I hope young artists can gain control over them and perhaps become more effective in creating their own empathetic art.

5.2 Definitions

Empathy is the ability to share feelings and thoughts with another.[3,4] This is different from compassion or sympathy which engage in feeling *for* another, without necessarily feeling *with* them.[5] Empathy can occur spontaneously and involuntarily, as is the case a person yawns after seeing another yawn. This type of empathy is what psychologists call "mimicry"[6] or "mirror-matching"[7] and a person does not have to be aware they are engaging in it, nor do they have to be physically present to experience

[1] Michelle Bogre, *Photography as Activism: Images for Social Change, 2nd Edition* (Routledge, 2025).

[2] Robert Hirsch, *Seizing the Light: A Social & Aesthetic History of Photography, 4th Edition* (Routledge, 2024).

[3] Amy Coplan, "Understanding Empathy: Its Features and Effects," in *Empathy: Philosophical and Psychological Perspectives*, ed. Amy Coplan and Peter Goldie (Oxford University Press, 2011), 17.

[4] Lauren Wispé. "The Distinction Between Sympathy and Empathy: To Call Forth a Concept, a Word Is Needed," *Journal of Personality and Social Psychology* 50, no. 2 (1986), 316.

[5] Martha Nussbaum, "Compassion: Human and Animal," in *Species Matters: Humane Advocacy and Cultural Theory*, ed. Marianne DeKoven and Michael Lundblad (Columbia University Press, 2012), 146.

[6] Martin Hoffmann, "Empathy, Justice, and the Law," in *Empathy: Philosophical and Psychological Perspectives*, ed. Amy Coplan and Peter Goldie (Oxford University Press, 2011), 233.

[7] Alessio Avenanti, Ilaria Minio-Paluello, Ilaria Bufalari, Salvatore M. Aglioti, "The pain of a model in the personality of an onlooker: Influence of state-reactivity and personality traits on embodied empathy for pain," *NeuroImage* 44, Issue 1 (2009): 275–283. https://doi.org/10.1016/j.neuroimage.2008.08.001.

this type of empathy. In a 2009 study, neuropsychologists found mirror-matching occurred in brain and muscle responses in the same area when observers were shown video clips of "flesh to bone" injections of a needle piercing a hand, although the degree of mimicry varied between subjects.[8] This research can have implications for art and photography because it suggests that empathy can be evoked in humans through visual representation of another's pain or suffering. For artists interested in creating social change, evoking empathy in the viewer can sometimes inspire action on behalf of a subject that the viewer has never met, as demonstrated in the two case studies discussed in Sect. 5.3 in this chapter.

Another type of empathy that involves thought (unlike mimicry) is called "perspective-taking."[9] This type of empathy involves the person thinking about another's situation, and then experiencing their emotions vicariously.[10] It is important to note that experiencing another's emotions is not always a positive thing and does not always inspire altruistic behavior, nor is it what empathy critic Paul Bloom calls a "magic bullet of morality."[11] As an example of how empathy in photography might negatively impact someone, psychologist Martin Hoffmann discusses the empathetic distress caused by photographs depicting trauma.[12] Empathetic distress is a phenomenon that can occur when a viewer looks at photographs of violence and/or suffering inflicted on another, imagines what it feels like to be in that situation, and then feels the trauma themselves. Hoffmann calls this "vicarious trauma" and cautions that it can have a negative impact because the viewer's reaction can be so strong that they become immobilized and/or shut down.[13] In this chapter, I purposely chose examples of art and photography that avoid graphic

[8] Ibid.

[9] Hoffmann, "Empathy, Justice, and the Law," 233.

[10] Coplan, "Understanding Empathy," 7, 17. In her essay, Coplan discusses three different types of perspective-taking which include self oriented—imagining one's self in another's situation; other oriented—thinking about what another might be experiencing given their unique perspective; and self-other oriented—empathy that allows us to experience another's emotions and situation while not let it eclipse our own thoughts and feelings (9–17). These are important distinctions for recognizing how empathy functions, but for the purpose of this chapter, I do not distinguish between them.

[11] Paul Bloom, *Against Empathy: A Case for Rational Compassion* (Ecco, 2016), 11.

[12] Hoffmann, "Empathy, Justice, and the Law," 236.

[13] Hoffmann, "Empathy, Justice, and the Law," 251.

violence for two reasons. First, by foregrounding imagery without graphic violence, I hope to navigate away from some of these unintended consequences such as emotional shutdown and vicarious trauma. Second, I want to demonstrate how art and photography can strategically communicate empathy through three visual principles that center the animal as an active subject worthy of our moral consideration. These three principles are perspective, gesture, and composition.

Perspective

In the context of this chapter, perspective refers to the angle the artist takes toward the subject. This can be literal or metaphorical, or both. For example, if a photographer wants to evoke empathy for a small dog, the photographer could lie down on the floor and photograph from the dog's perspective. In this example, the photographer is literally taking the point of view, or perspective, of the dog and offering what the world looks like from the dog's vantage point. This is an example of a literal kind of perspective-taking that can lead the viewer to an empathic response. Perspective can also include a sense of scale for the subject, allowing the viewer to experience what the world might look like from the subject's unique perspective. When working with smaller subjects, this means the photographer lowering their angle of view to meet the subject on their terms. However, perspective does not always have to give the subject's literal point of view. Perspective can also include a conceptual approach, where the artist depicts the subject from a metaphorical standpoint. This is examined more closely in the photographic works depicting nonhumans in Sect. 5.4 later in this chapter.

Gesture

In defining gesture as a principle of empathy, there are two distinct categories to explain: the subject's gesture and the artist's gesture.

When examining empathy derived from the subject's gesture, the subject's action, expression, and body language is analyzed, followed by a discussion on what is communicated through that particular gesture. Key questions include: What is the subject doing? What is suggested by this action or expression? And what is being shared with the viewer as a result?

An artist's gesture refers to the rendering, style, or artistic choices made by an artist. This can include things like hand drawn lines, particular camera controls, film speed, or the artist's interactions with the subject that remain visible to the viewer within the art. As an example, in photography, an artistic gesture might be the use of shallow depth of field, where just one subject is in focus and the rest of the image appears blurry. With this example, we would then analyze how this choice of focusing on just one encourages the viewer to attune to just one subject in a particular way and how that choice can elicit empathy.

Composition

In the context of this chapter, composition refers to both the overall design and the placement of the subject(s) within the frame. Empathy can be communicated through the way the viewer's eye is encouraged to move through the art through leading lines, by juxtaposing the subject in relation to something else or to edges of the frame, or by the composition itself being a symbol or metaphor. Each of these will be examined in detail through the photographic examples of nonhumans in Sect. 5.4 later in this chapter. But for now I would like to discuss these principles of empathy through two historical photographs that resulted in social change.

5.3 Two Historical Case Studies

Case Study 1: Dorothea Lange

Migrant Mother[14] (Fig. 5.1) from 1936 by Dorothea Lange is often cited as one of the most influential photographs in American history.[15] It was even reproduced on a commemorative US stamp from 1998, as part of the "Celebrate the Century" series. *Migrant Mother* has been widely written about and has been the center of two ethical controversies. The first involves Florence Owens Thompson, the woman in the photograph. She claimed she never saw the photograph until the 1970s and says she

[14] The title and caption Lange provided for this photograph is "Destitute pea pickers in California. Mother of seven children. Age thirty-two. Nipomo, California." However the photograph became colloquially known as "Migrant Mother." Neither title names subject. Both titles appear within the Library of Congress.

[15] Time Inc., "*Migrant Mother*, Dorothea Lange 1936," *100 Photographs: The Most Influential Images of All Time* (Time Inc., 2016), accessed March 18, 2019, http://100photos.time.com/.

never gave Lange permission to circulate her likeness.[16] Second, Lange either misunderstood or embellished a few of the published details about Thompson's life and situation.[17] There are many more interesting facts and tangential stories about this photograph, but the following will focus on a discussion of this photograph through the lens of the three principles of empathy.

Fig. 5.1 Dorothea Lange, *Destitute pea pickers in California. Mother of seven children. Age thirty-two. Nipomo, California,* (also known as *Migrant Mother*), 1936, nitrate negative, 5 × 4″, Library of Congress, Washington, DC, public domain

[16] Jeff Jardine, "Dust Bowl Part IV: Symbol of an Era," *The Modesto Bee*, September 17, 2008, https://www.modbee.com/latest-news/article3114135.html.

[17] Richard Stockton, "The True Story of the Iconic 'Migrant Mother' Photograph," *ATI: All That's Interesting*, April 26, 2017, last modified March 19, 2018, allthatsinteresting.com/migrant-mother-photo.

Perspective

The perspective of this photograph is eye level with the seated woman. The viewer is neither looking down nor up at her. Her eyes are slightly north of the center of the photograph, and situated in the same approximation as if one was looking in a mirror. This eye-level point of view provides an equality between the viewer and the woman. In other words, the perspective Lange took invites the viewer to look, listen, and attune with her—actions that are cornerstones for empathy.[18]

Gesture

The photograph depicts a woman seated and surrounded by three children; two who hang on her shoulders, and a nursing baby in her arms. The children are turned away so that the only facial expression visible to the viewer is the worry on the woman's face. This is accentuated through sharp focus. Additionally, the woman's hand is cupped just below her chin—a gesture that denotes thinking. She looks beyond the camera suggesting there is even more outside of the frame to be concerned about. In this way, the subject's gesture and gaze invites viewers to feel the weight of her worry and perhaps even feel the weight of the children who hang onto her.

Composition

Dorothea Lange took five–seven photographs of this family.[19,20] Looking at four other images taken that day (Figs. 5.2, 5.3, 5.4, and 5.5) reveal alternate Lange's choices and offer insight into what makes *Migrant*

[18] Lori Gruen, *Entangled Empathy: An Alternative Ethic For Our Relationship With Animals* (Lantern Books, 2015), 27.

[19] This number of photographs Lange took at this camp is debated. Sarah Meister, curator of photography at the Museum of Modern Art, claims there are seven, but notes that only five were sent to Washington D.C. Since the images in this essay come from the Library of Congress, I am only discussing Lange's five images that reside within their archive.

[20] Sarah Meister, "Piecing Together Dorothea Lange's Migrant Mother," *Museum of Modern Art Magazine,* February 6, 2020, https://www.moma.org/magazine/articles/233.

Mother a more empathetic photograph. First, the elimination of the background forces us to look at details—the wrinkles in the woman's face, her tattered clothing, and the dirt and holes in her children's clothes. This up-close focused composition creates an intimate space and invites the viewer to feel along with this mother in need.

Fig. 5.2 Dorothea Lange, *Migrant agricultural worker's family, Seven hungry children. Mother aged thirty-two. Father is native Californian. Nipomo California*, 1936, nitrate negative, 5 × 4″, Library of Congress, Washington, DC, public domain

Fig. 5.3 Dorothea Lange, *Migrant agricultural worker's family, Seven hungry children. Mother aged thirty-two. Father is native Californian. Destitute in pea picker's camp, Nipomo, California, because of the failure of the early pea crop. These people had just sold their tent in order to buy food. Of twenty-five hundred people in this camp most of them were destitute*, 1936, nitrate negative, 5 × 4″, Library of Congress, Washington, DC, public domain

Fig. 5.4 Dorothea Lange, *Migrant agricultural worker's family, Seven children without food, Mother aged thirty-two. Father is native Californian. Nipomo California.* 1936, nitrate negative, 5 × 4″, Library of Congress, Washington, DC, public domain

Fig. 5.5 Dorothea Lange, *Migrant agricultural worker's family. Seven hungry children. Mother aged thirty-two. Father is native Californian. Nipomo California,* 1936, nitrate negative, 5 × 4″, Library of Congress, Washington, DC, public domain

Second, this tight crop creates a triangular shape for the viewer's eyes to move around in. This is important because the photographer creates a hierarchy within the frame which directs the viewer to a more controlled scene. In this case, the woman is at the apex of the triangle. She is responsible for everything below. She is the center of the photograph's attention, and the viewer's emotion is directed toward her.

Outcomes

Initially the *San Francisco News* ran two of Lange's photographs (Figs. 5.3 and 5.4), but not *Migrant Mother*.[21] However, a few days later *Migrant Mother* was distributed via the news agency United Press (now UPI) and the government sprang into action and sent large quantities of food to the migrant camp.[22] Although Thompson had already moved on from the camp by the time the food arrived and did not directly benefit from the photograph,[23] *Migrant Mother* remains an example of an empathetic photograph that spurred the distribution of humanitarian aid.

Case Study 2: Lewis Hine

Lewis Hine was a sociologist and photographer who worked with the National Child Labor Committee to expose child labor practices and to reform laws in the early 1900s. One 1908 photograph from this series titled *Rhodes Mfg. Co., Lincolnton, N.C. Spinner. A moments glimpse of the outer world Said she was 10 years old. Been working over a year. Location: Lincolnton, North Carolina* (Fig. 5.6), provides a strong example for examining the principles of empathy in art.

Perspective

In this photograph (Fig. 5.6), Hine, like Lange, uses an eye-level camera angle to connect the viewer with the girl. This eye-level perspective can be a powerful tool in creating empathy. As Michelle Bogre writes, "this point of view forces us to look directly into the child's eyes, catapulting us into their lives, challenging us to consider how the children's health, education, and futures were being sacrificed so the middle class could have less expensive clothing, food, or products."[24] In other words, Hine's choice to photograph the girl at eye level near her work equipment allows the viewer to look at her from her height and compare her size to the size of the industrial loom in the background. This perspective in both placing

[21] Hirsch, *Seizing the Light*, 347.

[22] Ibid.

[23] Bogre, *Photography As Activism*, 71.

[24] Bogre, *Photography as Activism*, 65.

Fig. 5.6 Lewis Hine, *Rhodes Mfg. Co., Lincolnton, N.C. Spinner. A moments glimpse of the outer world Said she was 10 years old. Been working over a year. Location: Lincolnton, North Carolina*, 1908, black-and-white photograph from negative, Library of Congress, Washington, DC, public domain

the subject at eye level with the viewer and making sure the mechanical equipment was visible for a sense of scale were strategic choices that Hine made several photographs.[25] These choices frame the child empathetically because the viewer is given her perspective with the environment and loom appearing to overwhelm her size.

Gesture

In terms of the subject's gesture, the photograph depicts a girl with an apron looking out a window. Although what she is looking at is not visible, the world beyond the window has her attention. She stands alone in a factory with her back turned to a loom that overpowers her. Her

[25] Hirsch, *Seizing the Light*, 353.

shoes are too big for her. In other words, the photograph shows a girl waiting alone in a dangerous world that does not fit her, while she longs to be somewhere else. Hine emphasizes this gesture by his use of shallow depth of field calling the viewer's attention and focus only to the area where the girl is standing. The horizontal lines of the loom and the lines of the wall and floor all converge to an ominous out-of-focus space just beyond the girl. In other words, the viewer's attention is directed to the girl in a dismal environment who longs to be elsewhere. These choices ask the viewer to engage in empathy through perspective-taking by imagining what it might feel like to be in her oversized shoes.

Composition

The composition of this photograph is mostly symmetrical and balanced. This means that the photograph's composition is mirrored in weight from left to right and top to bottom. The leading lines (implied lines formed by elements in the composition that lead the viewer's eye throughout the frame) are formed by the edges of the wall and loom converge toward the center of the image just beyond where the girl is standing. This creates a triangular shape on either side of the girl with the apex of each triangle pointing toward the girl, and leading the viewer to the singular area of focus within the image—the girl. Encouraging viewers to focus on individuals is part of Hine's empathetic approach. This technique shifts the viewer's attention from a general problem impacting many, to a specific person impacted by child labor.[26] Focusing on individuals is one of the ways psychologists suggest empathy can be fostered and cultivated and then applied to larger groups of people.[27] In fact, a 2002 study found that empathy for an individual within a stigmatized group led to readiness to help the entire group even if that individual didn't directly benefit.[28] Although this research came along almost 100 years later, Hine used this approach of highlighting an individual in order to question the role of

[26] Hirsch, *Seizing the Light*, 330.

[27] Hoffmann, "Empathy, Justice, and the Law," 236.

[28] C. Daniel Batson, Johee Chang, Ryan Orr, and Jennifer Rowland, "Empathy, Attitudes, and Actions: Can Feeling for a Member of a Stigmatized Group Motivate One to Help the Group?" *Society for Personality and Social Psychology Bulletin: Sage Publications* 28, no. 12 (2002): 1657, https://doi.org/10.1177/01461670223764.

children in the workforce, asking the viewer to consider how the overpowering, desolate industrialized labor environment is harmful to children's physical or psychological needs. The perspective-taking empathy in this composition is derived from these compositional choices and encourages the viewer to look at this situation from the girl's perspective.

Outcomes

Hine considered himself an "interpretive photographer" meaning that he was aware that his photographic choices could persuade viewers.[29] In a 1909 speech, Hine said, "With a picture thus sympathetically interpreted, what a lever we have for the social uplift."[30] And in fact, Hine's child labor photographs were so persuasive that by 1916 Congress passed the Keating–Owens Act, which restricted the age and hours that children could work.[31] This is another example of empathetic photographs inciting social change.

5.4 Empathy for Nonhumans Through the Lens of Contemporary Artists

The last section of this chapter examines the three principles of empathy in art in images made by contemporary artists and photographers working with nonhumans.

Perspective

Two photographers who practice an empathic perspective by photographing at eye level with their subjects are Janet Holmes (Fig. 5.7) and Isa Leshko (Fig. 5.8).

[29] Hirsch, *Seizing the Light*, 333.

[30] Lewis W. Hine, "Social Photography: How the Camera May Help in the Social Uplift," in *Proceedings of the National Conference of Charities and Correction at the Thirty-Sixth Annual Session Held in the City of Buffalo, New York, June 9–16, 1909*, ed. Alexander Johnson (Fort Wayne, IN: Press of Fort Wayne, 1909), 356. This quote also appears in Hirsch, *Seizing the Light*, 333.

[31] Bogre, *Photography as Activism*, 68.

Fig. 5.7 Janet Holmes, *Jon Snow (Heartwood Haven Animal Sanctuary)*, 2017. Reprinted by permission of the artist

Fig. 5.8 Isa Leshko, *Bessie*, from *Allowed to Grow Old*, published by University of Chicago Press, 2019, p. 69. Reprinted by permission of the artist

With Janet Holmes's *Jon Snow (Heartwood Haven Animal Sanctuary)*[32] (Fig. 5.7), Holmes photographs the rescued chicken, Jon Snow, at (human) eye level while he is looking out a window on a kitchen counter. In this case, bringing the subject to eye level with the viewer creates connection and a sense of intimacy. Jon Snow is no longer part of the agricultural industry, and has not become a piece of meat on a kitchen counter like so many others. Instead, Holmes photographs Jon Snow as a family member enjoying a view within a kitchen. Holmes' eye-level perspective amplifies this point.

Isa Leshko's *Bessie, Holstein Cow, Age 20* (Fig. 5.8) depicts a 20 year-old cow who was rescued on her way to the slaughterhouse after spending years producing milk for humans.[33] This photograph is part of Leshko's book *Allowed to Grow Old* which features black and white portraits of elderly farm animals who now live in sanctuaries. Like the other portraits in this book, instead of depicting the abuse or scars, Leshko photographs Bessie with the reverence of bright sunlight as she lies down in soft hay in a moment of rest. In order to photograph Bessie, Leshko has also dropped to the ground making sure that the viewer will be looking directly into Bessie's eyes. This perspective eliminates photographic distortion and creates equality since neither the subject nor the viewer are looking up or down at each other. Each animal is depicted as an individual with unique needs, desires, and histories, rather than a stand-in for their species. In addition, this perspective and attention given to Bessie underscores Leshko's larger point about how unique it is to even be in the presence of an elderly farm animal since most are slaughtered before the age of two. Leshko's choices frame the rescued farm animals with dignity and encourages empathy by asking the viewer to attune with the subjects on their own terms.

Entangled Empathy

In her 2015 book, *Entangled Empathy*, ecofeminist and moral philosopher Lori Gruen writes about how humans can improve their relationships with nonhumans through a type of empathetic perspective-taking she

[32] Janet Holmes, *Nest: Rescued Chickens at Home* (Kehrer Verlang, 2020), 5.

[33] Isa Leshko, *Allowed To Grow Old: Portraits of Elderly Animals from Farm Sanctuaries* (University of Chicago Press, 2019), 27.

calls "entangled empathy."[34] Because humans often experience the world differently from nonhumans, entangled empathy begins with the human learning something specific about how the nonhuman experiences their environment and community, and then using that information to imagine "what the world seems, feels, smells, and looks like from their situated position."[35] This involves both a blend of cognition and affect, and the ability to distinguish the self and others.[36] Entangled empathy offers artists a route to ethically depicting nonhumans, while encouraging a connection between the subject and the viewer. For instance, one way artists can engage in entangled empathy is to first research their subject and try to understand how they experience the world biologically, socially, and perhaps emotionally, and then use that information as part of their depiction. Two examples of artists who use this approach are Chris Jordan and Lee Deigaard.

In his 2017 film *Albatross* (Fig. 5.9), Chris Jordan examines the beauty and tragedy of where human and albatross intersect, and bears witness to the dying albatross chicks on Midway Island—an unintended consequence of human mass consumption.[37] In order to introduce the audience to the albatross, Jordan makes sure he is eye level with the albatross when they are on the ground. However, Jordan goes a step further. At one point in the film, Jordan discusses how albatross process information at a much quicker rate than humans do, which means albatross experience time differently than humans do.[38] So in order to show the beauty and synchronicity of the albatross mating dance, Jordan slows the frame rate down to show how this dance is experienced by the albatross. The result is, as Jordan says, "a series of graceful and sensuous gestures."[39] Being able to appreciate this beautiful dance the way an albatross does gives the viewer a perspective that we could not otherwise see and, therefore, a new perspective with greater empathy for the albatross.

[34] Gruen, *Entangled Empathy*, 3.

[35] Gruen, *Entangled Empathy*, 66.

[36] Ibid.

[37] Chris Jordan, "Albatross," 2017, 97 min., 43 sec., https://www.albatrossthefilm.com/watch-albatross.

[38] Chris Jordan, "Albatross," 23:55–24:33.

[39] Chris Jordan, "Albatross," 24:50–24:53.

Fig. 5.9 Chris Jordan, *Albatross* (film still), 2017. [Film 1:37:43, www.albatr ossthefilm.com.watch-albatross.] Reprinted by permission of the artist

Another example of an artist using entangled empathy is Lee Deigaard's photographic series *Equuleus* (Fig. 5.10). In this work, Deigaard examines human–horse relationships. Deigaard has spent many years working with horses and studying their ability to gain knowledge through a variety of senses including vision, touch, smell, taste, body awareness, and sound.[40] For instance, she writes "Each whisker on a horse's chin corresponds to a discrete site in the horse's brain. Through pallesthesia, the horse feels the firmament, its subtle signatures and unique vibrations, through his vascularized hooves and through his sinuses."[41] Using specific knowledge about horses' navigate their world and relate to her presence as a human, Deigaard created a set of portraits called *Equuleus*.

In the photograph, *Copper Valley,* the perspective is an up-close, tightly cropped view of a horse's back. The angle of view is so close that the horse's back might be mistaken for a landscape with a small tree growing

[40] Lee Deigaard, "Horse/Horizon: On Photography and Multispecies Bonding," *Humanimalia* 14, no. 2 (2024), https://doi.org/10.52537/humanimalia.15525.

[41] Ibid, paragraph 21.

Fig. 5.10 Lee Deigaard, *Copper Valley*, from *Equuleus* series, 2015. Reprinted by permission of the artist

on a hill or mountain. Deigaard's abstraction of the horse's back is two two-fold. First, even though this photograph represents a human perspective, Deigaard's choice to reference a horizontal landscape is a nod to the horse's rectangularly shaped pupils that continually orient them toward the horizon line.[42] Second, Deigaard is referencing the horse's strong sense of proprioception and the trust that comes with it if one is allowed in such proximity. In other words, the fact that the horse has allowed the photographer to get so close without touching or restraint implies a relationship of trust. The implied landscape becomes a metaphor for this human-horse relationship. In reflecting on her photographs, Deigaard poetically writes "To deeply know another, horse or human, is a wide-open frontier. The horizon, the convergence of earth and the sensory with the aspirational and ethereal sky, draws us and eludes us in equal

[42] Ibid, paragraph.

measure. This is the meeting of species, of two, in a congruence of the granular and celestial."[43] With *Copper Valley* Deigaard engages in a kind of entangled empathy because she has identified something specific and particular about horses' biologies, and then used that information as part of the way she pictures them and her ongoing relationship with them, which offers the viewer a new and unexpected perspective.

Gesture

Two photographers that empathetically depict gestures of nonhumans that may result in a type of mimicry in the viewer are Annie Marie Musselman and Jo-Anne McArthur. I would like to discuss an example from each.

Annie Marie Musselman's 2009 photograph, *Eeya* (Fig. 5.11), of a rehabilitated owl leaving the cage speaks about recovery and freedom. The photograph depicts the legs and flutter of the tail as the owl takes off into the upper right corner of the frame. The cage occupies the lower right corner and the viewer is left frozen in this moment of takeoff, and perhaps can feel the anticipation of independence that the bird is about to experience.

In Jo-Anne McArthur's 2011 photograph, *A Monkey Protecting His Blind Cage Mate at a Macaque Breeding Facility, Laos* (Fig. 5.12), two macaques comfort each other in a place of despair. The gray manmade floor and walls are devoid of anything natural. The two macaques hold each other as they huddle in fear in the left half of the frame. Seeing this gesture and the conditions these monkeys are living in may produce a sense of fear and desperation in the viewer as well.

Julia Oldham's short animated film *Laika's Lullaby* (Fig. 5.13) represents an empathetic artistic gesture by the way Oldham has drawn Laika, the stray Soviet dog who was launched into space and orbited Earth with no plan for her recovery.[44] In *Laika's Lullaby*, Oldham chronicles Laika's terrifying journey into space and subsequent death due to overheating inside the space capsule. The word "terrifying" is not conjecture. Laika wore a space suit with sensors that monitored her breath and

[43] Ibid, paragraph 10.

[44] Julia Oldham, "Laika's Lullaby (2015)," Artist website, http://www.juliaoldham.com/index.php/portfolio/laikas-lullaby/.

Fig. 5.11 Annie Marie Musselman, *Eeya*, from *Finding Trust* series, 2007. Reprinted by permission of the artist

heartrate, and when the rocket took off her heartbeat tripled and breath rate increased by four times.[45] In order to visualize this fear, Oldham drew Laika's body, face, and eyes with anxious and frenetic lines that quiver. This artistic gesture conveys a feeling of panic and confusion, thereby evoking empathy for what Laika must have experienced in her last moments.

Additionally, in order to depict Laika's vulnerability as she orbited in space, Oldham drew her body curved into a fetal position (Fig. 5.14). This rendering depicts Laika as defenseless and innocent, and perhaps evokes a sense of helplessness in the viewer as they imagine what Laika must have experienced in space.

Emma Kisiel's series *At Rest* depicts photographs of memorials built by Kisiel for animals who were killed by vehicles while crossing the road.[46] Although the photographs are taken with a camera angle that positions the viewer above and looking down on the animal, it is Kisiel's artistic

[45] Alice George, "The Sad, Sad Story of Laika, the Space Dog, and Her One-Way Trip Into Orbit," *Smithsonian Magazine*, April 11, 2018, https://www.smithsonianmag.com/smithsonian-institution/sad-story-laika-space-dog-and-her-one-way-trip-orbit-1-180968728/.

[46] Emma Kisiel, "At Rest," Artist website, https://www.emmakisiel.com/at-rest.

Fig. 5.12 Jo-Anne McArthur, *A macaque protectively holds his blind cage mate at a macaque breeding facility. Laos*, 2011, We Animals Media. Reprinted by permission of the artist (Description: These photos and videos of captive long-tailed macaque monkeys in Laos, shot for the Canadian documentary "Maximum Tolerated Dose," resulted from an investigation into animals used in research. During the investigation, three macaque breeding facilities were visited, all of which housed wild-caught and purpose-bred monkeys destined for international shipment for use in medical research or entertainment)

gesture of adorning the area around the bodies with flowers, leaves, and rocks that communicates empathy. For example, in *Coyote* (Fig. 5.15), a coyote's body lays motionless on a dirt ground. The snow surrounding her body has melted everywhere except for her shadow, suggesting this photograph was taken in the morning and that she has been dead for a while. The coyote is surrounded by a ring of bright blue flowers laid out by the artist. With this gesture, Kisiel asks the viewer to pay respects and give this slain coyote a moment of pause, thought, and attention. And "empathetic attunement" or attention is part of Gruen's notion of entangled empathy.[47] In other words, instead of documenting the world as is—perhaps depicting a forgotten creature who has been left for dead on the side of the road—Kisiel intervenes, altering the site with adornment

[47] Gruen, *Entangled Empathy*, 46.

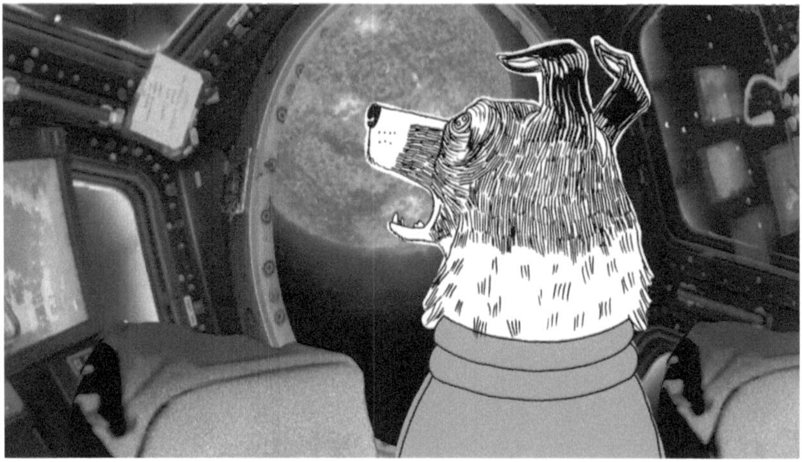

Fig. 5.13 Julia Oldham, *Laika's Lullaby* (film still), 2015. [Film 6:29 min., https://vimeo.com/123684913.] Reprinted by permission of the artist

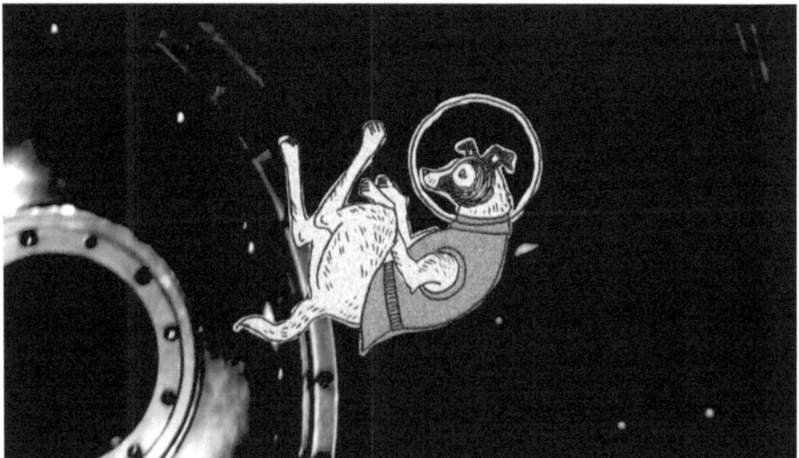

Fig. 5.14 Julia Oldham, *Laika's Lullaby* (film still), 2015. [Film 6:29 min., https://vimeo.com/123684913.] Reprinted by permission of the artist

Fig. 5.15 Emma Kisiel, *Coyote*, from *At Rest* series, 2011. Reprinted by permission of the artist

and reverence, thereby increasing the viewer's sensitivity to a subject they might have otherwise overlooked.

Composition

As noted above, evoking empathy through composition can take on different forms. In this section, I will examine photographs that encourage empathy through compositions in three different ways—leading lines that form geometric shapes, framing the subject metaphorically, and juxtaposing the subject with the edges of the frame.

Traer Scott's *Dutch* (Fig. 5.16)[48] uses a traditional apex portrait, much like the *Migrant Mother* photograph by Dorothea Lange. This means that Dutch is not only eye level with the viewer but a triangular shape is formed by the leading lines of his body making him the sole focus

[48] Traer Scott, *Finding Home:Shelter Dogs & Their Stories* (Princeton Architectural Press, 2015), 42.

of attention. Empathy is encouraged through the attunement of looking directly into the eyes of this overlooked shelter dog.

Sometimes composition itself can serve as a symbol or metaphor for something else, like human–nonhuman relationships. Reexamining the human–nonhuman relationship is a theme in another body of work by artist Lee Deigaard, *Unbidden II*.

In *Connect the Dots* (Fig. 5.17), Deigaard photographs wild nonhumans at night. The photographs are circular, referencing a mediated gaze of humans looking at wild nonhumans. Deigaard is speaking about a particular type of gaze through a lens—a camera, binoculars, a microscope, or even a rifle scope, where the human is in control and the animal is unaware and/or unable to give any kind of meaningful consent. In *Connect the Dots*, the deer's tapetum lucidum appears to glow while they are temporarily stunned by the infrared light.[49] Here the circle becomes

Fig. 5.16 Traer Scott, *Dutch*, from *Finding Home* series, 2015. Reprinted by permission of the artist

[49] "Curated Conversations With Revelations Photographer, Lee Deigaard Part II: Featuring Photographs From the Unbidden Series," Ogden Museum of Southern Art, June 3, 2020, https://ogdenmuseum.org/curated-conversation-with-revelations-photographer-lee-deigaard-part-ii/.

a symbol of objectification and control with the viewer as a voyeur. Through this framing Deigaard offers empathy for the creatures who are caught within a scope.

In Chris Jordan's series *Midway: Message from the Gyre* (Fig. 5.18), he photographs young albatrosses of Midway Island who have died after ingesting bits of plastic disguised in algae. Before photographing, Jordan removes the breastplate of the baby albatross and positions his camera above the deceased chick. The composition is always centered and the undigested plastic waste in the belly of the albatross appears midway through each photograph. With this work, the midway point of the photograph is a metaphor; Jordan says humankind is at a "midway," where we can hopefully arrest climate emergency.[50] The centering of the composition provides empathy for the innocent creatures who have died

Fig. 5.17 Lee Deigaard, *Connect the Dots*, from *Unbidden II* series, 2011. Reprinted by permission of the artist

[50] Chris Jordan, personal interview via Skype, June 4, 2019.

from unintentionally consuming our waste and perhaps leaves the viewer with a sense of the grief for what is lost through climate emergency.

Sometimes empathy is created by juxtaposition or placement of elements within the frame. For instance, Rebecca Norris Webb's *Buffalo Roundup* (Fig. 5.19), depicts a green railing of a cattle pen with a smattering of red paint. The paint is strategically cropped directly above the body of the buffalo, suggesting blood or violence. The composition suggests a foreshadowing, privileging the viewer to information the buffalo might be unaware of. This careful attention to composition may evoke a sense of vulnerability and empathy for the unsuspecting buffalo.

Kelly Guerin's 2020 photograph "Rescued, dying chicks outside a Perdue chicken farm" (Fig. 5.20) depicts an outstretched human hand presenting two weary chicks both with their eyes closed. They are the only subjects in the frame in sharp focus, which amplifies their fluffy yellow feathers and the feces that caked to one of the chick's beaks. The rich color contrasting the yellow chicks and the green grass outside the buildings provides a visceral contrast to the gray, sterile windowless buildings. In the background, there are three long rows of factory farm buildings.

Fig. 5.18 Chris Jordan, *Midway, Unaltered Remains of a Laysan Albatross Fledgling, Midway Island*, 2009. Reprinted by permission of the artist

Fig. 5.19 Rebecca Norris Webb, *Buffalo Roundup*, from *My Dakota*, published by Radius Books, 2012. Reprinted by permission of the artist

The leading lines of the left and right buildings converge at a midpoint in the center building just beyond the chicks head. The center building has a series of industrial circular fans while the roof forms a triangular shape just above the chicks heads, visually framing them like a house without windows. All the leading lines in the composition point to these two chicks who are so weak that they cannot open their eyes or stand on their own, suggesting there are thousands more living in bleak, horrendous conditions inside the rows of buildings in the background. The empathy in the photograph is a type of attunement, perspective-taking, and asking the viewer to pay attention to these two victims of industrialized farming, and then think about the billions of others who live in similar conditions and whose only experience with sunlight comes during a trip to the slaughterhouse.

Fig. 5.20 Kelly Guerin. *Rescued, dying chicks outside a Perdue chicken farm*, North Carolina, USA, 2020, We Animals Media. Reprinted by permission of the artist (Description: While our team was documenting conditions inside a Perdue Farms growing facility, the farmer demonstrated the industry standard culling practice on a sick, young chick by grabbing her head in one hand, her body in the other, and pulling them apart in rapid decapitation. When he picked up the next one, I asked whether it would be alright if I took him with me instead of letting him be killed. He agreed and set the chick aside. Later, while documenting the other chicks in the barn, I noticed a chick far smaller than all the others, stumbling around with dried feces caked to her beak. She was too short to reach the water droppers, and was at risk of dying of dehydration. The farmer agreed to let her go with me as well. The first chick passed away on the two-hour drive home. The second was revived with watered-down Gatorade and the next day began her new life at Piedmont Farm Animal Refuge in Pittsboro, NC)

5.5 Conclusion with Humor

I would like to conclude with a slightly different tone: humor. Dan Piraro is the genius behind the cartoon series *Bizarro*. *Bizarro* has been published every day since 1985 and has a tremendous, wide-reaching audience because it is featured in a few hundred newspapers throughout

North and South America, Europe, and Asia.[51] While Piraro's cartoons employ punchy color and witty graphics, it is the pithy humor that evokes emotion for the subject. Often viewers are caught off guard and find themselves laughing at something unexpected. For instance, consider this piece on animal cruelty, published May 31, 2008 (Fig. 5.21).

Fig. 5.21 Dan Piraro, from *Bizarro* cartoon series, 2008, published May 31, 2008. Reprinted by permission of the artist

[51] Dan Piraro, "25 Years of Scribbles," *BizarroBlog*, January 22, 2010, http://bizarrocomic.blogspot.com/2010/01/25-years-of-scribbles.html.

Looking at this piece through the lens of the principles of empathy, Piraro has taken the perspective of an omniscient observer, giving the viewer a wide-angle view of the scene. Additionally, he uses an anthropomorphic voice, which could be problematic due the fact that the human words are being put in the elephant's mouth.[52] However, anthropomorphism here is used as satire and as a method to realign the viewer's perspective with the elephant providing a moral compass for the situation, leading the viewer to an empathetic response to the elephant's situation.

In terms of gesture, the elephant is depicted in chains and drawn with heavier lines, while the two children are enjoying the frivolity of balloons and cotton candy and are depicted with more colors such as green and purple. The children are turned away, but the elephant is facing the viewer and therefore calls for the viewer's attention and emotion. The artist's gesture is also visible in Piraro's rendering of bright color and sharp wit.

The composition provides leading lines that encourage a triangular eye movement for the viewer, starting at the top of the elephant, down to the children, and back to the chain. The entire composition is weighted toward the elephant, with the elephant occupying more than half of the composition. All of these artistic choices lead the viewer to the importance of the elephant and the unfairness of the situation she is in. In his own words, Piraro says, "My intention was, in fact, to educate readers and create some empathy for the animals we use commercially in hopes that when activists manage to get initiatives on ballots, voters will already be aware that there are problems to be solved in these areas and be less likely to ignore them as fringe beliefs that need not be taken seriously."[53] Engaging in empathy alters our perceptions, and with that there is hope for more understanding and perhaps greater sensitivity in the world. This is because, as Gruen writes, "in the process of coming to share the perspective of another— figuring out what she cares about, what worries her, and even getting a glimpse of how she sees you— we are changed."[54] Art offers us this change in perspective, and by examining three visual principles of empathy my hope is this will lead to greater control over

[52] Eva Meijer, *When Animals Speak: Toward an Interspecies Democracy* (New York University Press: 2019), 65.

[53] Dan Piraro, personal interview via email, May 11, 2019.

[54] Gruen, *Entangled Empathy*, 75.

images and image-making. and with that perhaps greater sensitivity to the world and all who live here.

Bibliography

Avenanti, Alessio, and Ilaria Minio-Paluello, Ilaria Bufalari, Salvatore M. Aglioti. "The Pain of a Model in the Personality of an Onlooker: Influence of State-Reactivity and Personality Traits on Embodied Empathy for Pain." *NeuroImage* 44, Issue 1 (2009): 275–283. https://doi.org/10.1016/j.neuroimage.2008.08.001.

Batson, Daniel C, Johee Chang, Ryan Orr, and Jennifer Rowland. "Empathy, Attitudes, and Actions: Can Feeling for a Member of a Stigmatized Group Motivate One to Help the Group?" *Society for Personality and Social Psychology Bulletin: Sage Publications* 28, no. 12 (2002): 1656–1666. https://doi.org/10.1177/014616702237647.

Bloom, Paul. *Against Empathy: A Case for Rational Compassion,* Ecco, 2016.

Bogre, Michelle. *Photography As Activism: Images for Social Change, 2nd Edition.* Routledge, 2025.

Coplan, Amy. "Understanding Empathy: Its Features and Effects." In *Empathy: Philosophical and Psychological Perspectives,* edited by Amy Coplan and Peter Goldie. Oxford University Press, 2011.

Deigaard, Lee. "Horse/Horizon: On Photography and Multispecies Bonding." *Humanimalia* 14, no. 2 (2024): https://doi.org/10.52537/humanimalia.15525.

George, Alice. "The Sad, Sad Story of Laika, the Space Dog, and Her One-Way Trip Into Orbit." *Smithsonian Magazine*, April 11, 2018. https://www.smithsonianmag.com/smithsonian-institution/sad-story-laika-space-dog-and-her-one-way-trip-orbit-1-180968728/.

Gruen, Lori. *Entangled Empathy: An Alternative Ethic for Our Relationships with Animals.* Lantern Books, 2015.

Hine, Lewis. "Social Photography: How the Camera May Help in the Social Uplift." In *Proceedings of the National Conference of Charities and Correction at the Thirty-Sixth Annual Session Held in the City of Buffalo, New York, June 9–16, 1909*, edited by Alexander Johnson, 356. Fort Wayne, IN: Press of Fort Wayne, 1909.

Hirsch, Robert. *Seizing the Light: A Social History of Photography, 4th Edition.* Routledge, 2024.

Hoffmann, Martin. "Empathy, Justice, and the Law." In *Empathy: Philosophical and Psychological Perspectives,* edited by Amy Coplan and Peter Goldie. Oxford University Press, 2011.

Holmes, Janet. *Nest: Rescued Chickens at Home.* Kehrer Verlang, 2020.

Jardine, Jeff. "Dust Bowl Part IV: Symbol of an Era." *The Modesto Bee*, September 17, 2008. https://www.modbee.com/latest-news/article3114135.html.

Jordan, Chris. "Albatross." 2017. Film 1:37:43. https://www.albatrossthefilm.com/watch-albatross.

Leshko, Isa. *Allowed to Grow Old: Portraits of Elderly Animals from Farm Sanctuaries*. University of Chicago Press. 2019.

Meijer, Eva. *When Animals Speak: Toward an Interspecies Democracy*. New York University Press, 2019.

Meister, Sarah. "Piecing Together Dorothea Lange's Migrant Mother." *Museum of Modern Art Magazine,* February 6, 2020. https://www.moma.org/magazine/articles/233.

Norris Webb, Rebecca. *My Dakota*, Radius Books, 2012.

Nussbaum, Martha. "Compassion: Human and Animal" In *Species Matters: Humane Advocacy and Cultural Theory*, edited by Marianne DeKoven and Michael Lundblad. Columbia University Press, 2012.

Ogden Museum of Southern Art. "Curated Conversations With Revelations Photographer, Lee Deigaard Part II: Featuring Photographs from the Unbidden Series.", June 3, 2020. https://ogdenmuseum.org/curated-conversation-with-revelations-photographer-lee-deigaard-part-ii/.

Oldham, Julia. "Laika's Lullaby." Artist's website, updated 2025. http://www.juliaoldham.com/index.php/portfolio/laikas-lullaby/.

Piraro, Dan. "25 Years of Scribbles." *BizarroBlog*, January 22, 2010. http://bizarrocomic.blogspot.com/2010/01/25-years-of-scribbles.html.

Scott, Traer. *Finding Home:Shelter Dogs & Their Stories.* Princeton Architectural Press, 2015.

Stockton, Richard. "The True Story of the Iconic 'Migrant Mother' Photograph." *ATI: All That's Interesting*, April 26, 2017, last modified March 19, 2018. allthatsinteresting.com/migrant-mother-photo.

Time Inc. "*Migrant Mother*, Dorothea Lange 1936." *100 Photographs: The Most Influential Images of All Time*. Time Inc., 2016. Accessed March 18, 2019. http://100photos.time.com/photos/dorothea-lange-migrant-mother.

Wispé, Lauren. "The Distinction Between Sympathy and Empathy: To Call Forth a Concept, a Word Is Needed." *Journal of Personality and Social Psychology* 50, no. 2 (1986).

CHAPTER 6

In Defense of "Sentimental Nonsense": Richard Adams's *The Plague Dogs* and Animal Compassion

Michael J. Gilmour

There is a long tradition of animal literature with a humane agenda, with novels as diverse as Anna Sewell's *Black Beauty* (1877) and David Duchovny's *Holy Cow* (2015) among them. A common complaint of such books is an overreliance on pathos.[1] We tend to be wary of stories

[1] Historical moment is a relevant consideration in the case of some writers on this point. In remarks about Jack London (1876–1916), Ernest Thompson Seton (1860–1946), and Charles G. D. Roberts (1860–1943), Marian Scholtmeijer notes that the Darwinian conception of nature as red in tooth and claw influenced some authors to "supplant Victorian neurosis over bloodshed among animals with assent to the ongoing battle in the wild" and to "undermine that anthropomorphism which allows wild animals to be assimilated by the tender sentiments of the reader." *Animal Victims in Modern Fiction: From Sanctity to Sacrifice* (Toronto: University of Toronto Press, 1993), 100. Such commitment to relentless realism is not necessarily easy for animal

M. J. Gilmour (✉)
Providence University College, MB, Canada
e-mail: michael.gilmour@prov.ca

C. Linzey and A. Linzey (eds.), *The Palgrave Companion to Humane Education*, The Palgrave Macmillan Animal Ethics Series,
https://doi.org/10.1007/978-3-032-19795-5_6

that play on our emotions, especially when there is an obvious agenda.[2] We also consider excessive emotion a cop-out, a literary and rhetorical shortcut, a way to bypass more demanding forms of story construction. Anyone can play on a reader's sympathies, but not everyone is able to develop plots and characters that simultaneously persuade.

Richard Adams's 1977 novel *The Plague Dogs* is part of this venerable, if oft-criticized, tradition of humane storytelling, but the author not only anticipates the complaints just described but embraces them, and his reason for doing so deserves notice as we consider the place of creative writing in humane education.[3] Adams uses self-parody to make his point. When Ronald Lockley enters the story, he grumbles about dewy-eyed

writers, or uniformly successful. Scholtmeijer suggests the writers mentioned "really want to sentimentalize over the wild animal victim, that they would like nothing better than to be allowed the luxury of pity. Sometimes they do allow themselves the luxury, but more often they take flight from what they would consider weak emotionality into forced neutrality or celebrations of heroism" (*Animal Victims* 119). On the turn away from sentiment and empathetic connection at the end of the Victorian period, at least in academic circles, see Suzanne Keen, *Empathy and the Novel* (Oxford: Oxford University Press, 2007), 55. Despite the disapproval of some educators, she observes, anthropomorphism "has proven difficult to stamp out" (68).

[2] Immanuel Kant emphasizes a rationalist approach to ethics in *Fundamental Principles of the Metaphysics of Morals* (1785) that distrusts sympathy and sentiment as a basis for ethical argument. For a summary and critique, see, e.g., Josephine Donovan, "Attention to Suffering: Sympathy as a Basis for Ethical Treatment of Animals," in *The Feminist Care Tradition in Animal Ethics: A Reader*, ed. Josephine Donovan and Carol J. Adams (New York: Columbia University Press, 2007), esp. 174–76; and Josephine Donovan, *The Aesthetics of Care: On the Literary Treatment of Animals* (New York: Bloomsbury, 2016), esp. chap. 1.

[3] It is reasonable to posit that animal advocacy is part of the author's agenda. Richard Adams was president of the Royal Society for the Prevention of Cruelty to Animals from 1980 to 1982, an organization mentioned in *The Plague Dogs* as supporting the cruel laboratory "on the grounds that the experiments and surgery would redound to the benefit of animals in general." Richard Adams, *The Plague Dogs* (1977, New York: Vintage, 2016), 9. This unexpected comment reflects controversial positions taken by the RSPCA at points in its history. Though soon after its founding in 1824, it denounced vivisection as "unchristian," by the 1870s, the organization lobbied for regulation of vivisection as opposed to an outright ban; see A. W. H. Bates, "A Spark Divine? Animal Souls and Animal Welfare in Nineteenth-Century Britain," in *The Routledge Handbook of Religion and Animal Ethics*, ed. Andrew Linzey and Clair Linzey (London: Routledge, 2019), 361, citing an SPCA Minute Book of 1832; and Diane L. Beers, *For the Prevention of Cruelty: The History and Legacy of Animal Rights Activism in the United States* (Athens: Swallow/Ohio University Press, 2006), 22–23. The RSPCA's lenient position proved an impetus for the rise of anti-vivisection societies; see, e.g., Darryl Jones, introduction to *The Island of Doctor Moreau*, by H. G. Wells, ed. Darryl Jones, Oxford World's Classics

animal stories, with obvious reference to Adams's better-known novel *Watership Down*:

> Ignorant sentimentality about animals and birds can be as bad as deliberate destruction.... Well-intentioned amateurs like that chap Richard Adams—fond of the country—reasonably good observer—knows next to nothing about rabbits—hopelessly sentimental—everyone starts thinking rabbits are marvellous when what they really need is keeping down if they're not to become an absolute pest to the farmer.[4]

Adams's self-deprecating humor effectively takes some punch out of the critics by weaving into the story the very objections they have, and in doing so he creates an opportunity to respond. Lockley's friend Peter Scott answers these charges about oversentimentality this way:

> I think that for ordinary, non-specialist people, a certain amount of anthropomorphism's probably useful in helping them to arrive at feeling and sympathy for animals—that's to say, readiness to put the good of a species, or even just the welfare of an individual creature, above their own advantage or profit. We can't all have scientific minds.[5]

Adams's defense is a simple one. Some kinds of animal writing are unable to address welfare questions, which is why anthropomorphic, sentimental, pathos-laden stories are not without purpose. By acknowledging the criticisms and explaining his reasons, Adams gives readers permission to engage the story on its own terms. He invites them to play along, as it were, and to entertain the moral arguments put forward. What Lisa Sainsbury writes of Adams's *Watership Down* applies to *The Plague Dogs* equally well. The novel "represents the natural world with ethical purpose and utilizes anthropomorphism to pull its reader into the land and consequently into the sphere of environmental ethics"[6]—and, I would add,

(Oxford: Oxford University Press, 2017), xxv–xxvi. In *The Plague Dogs*, scientists consider anti-vivisectionist protests "sentimental nonsense" (144).

[4] Adams, *Plague Dogs*, 485–86; cf. 487.

[5] Adams, *Plague Dogs*, 488.

[6] Lisa Sainsbury, *Ethics in British Children's Literature: Unexamined Life*, Perspectives on Children's Literature (London: Bloomsbury, 2013), 112. Sainsbury explores ways children's literature contributes to young readers' progress toward moral agency, which she describes as a kind of second birth: "the child wakes into a new phase of awareness that

to pull readers into moral questions about human treatment of animals specifically. I highlight here four ways this sympathy-generating anthropomorphic storytelling encourages readers to take the humane treatment of animals seriously.[7]

6.1 Realism Intrudes on Fantasy

The first strategy involves a kind of blurring of genre categories. *The Plague Dogs* is a flight of fancy with talking animals (i.e., talking among themselves), and as such we willingly suspend our disbelief as we enter the story, but at the same time, realism constantly intrudes. Adams interrupts readers frequently, never quite permitting full escape into fantasy. *The Plague Dogs* is a story about two test subjects who escape an English laboratory where they have been enduring horrific experiments. Rowf is a

transforms the individual into a being of a different (moral) order and it is a second birth that presses close upon the first, so that it occurs during childhood or adolescence" (2). She distinguishes cognitively restrictive moralizing from ethical discourse, which promotes philosophical inquiry: "I accept that while moralizing often seeks to tell children what to think, it can also contribute to a process whereby children are shown *how* to think" (6, italics original). Her book involves close reading of select texts that illustrate "*how* books encourage philosophical enquiry and seek to engage the reader in an exchange of ethical discourse" (9, italics original). Though she does not discuss *The Plague Dogs*, these observations describe Adams's novel well, even if it is not a children's book. He encourages serious consideration of the ethics of vivisection (and more) without simplifying the issues, and through characters debating and modeling alternative ways of viewing animals, he encourages moral reflection (shows "*how* to think").

[7] Scholtmeijer identifies an important concern about use of "sentimental anthropomorphism" for ethical purposes—namely, that it becomes another kind of violation: "When a mystic like Margery Kempe sees 'Christ being beaten' when she is in fact looking at a suffering horse, she is doing a disservice to the animal. With sentimental anthropomorphism, what suffers is not the animal *per se* but a phantom person standing in for the animal. The approach may well prevent abuse, and that is a fine thing, but animal autonomy has evaporated in the process" (*Animal Victims* 62–63). Arguably, Adams manages to avoid this. His rabbits in *Watership Down* and the dogs and fox in *The Plague Dogs* behave physically in ways consistent with the habits, strengths, weaknesses, and environments of their respective species. As he puts it in the introduction to *Watership Down*, his "anthropomorphic story" follows "the idea of Rudyard Kipling, in his two *Jungle Books*. That is to say, although my rabbits could think and talk, I never made them do anything physical that real rabbits could not do." Richard Adams, introduction to *Watership Down* (1972; New York: Scribner, 2005), xiii. Even if this insistence on realism lessens concern about the substitution of "a phantom person" for an animal in Adams's stories, the validity of Scholtmeijer's point about the limitations of this literary form for ethical argument remains.

black mongrel forced to tread water in a metal tank to test his endurance. When he succumbs and sinks hours later, they revive him, and after his recuperation, they put him back into the tank to repeat the experiment. And then again. And then again. Experiments on Snitter, a smaller fox terrier, focus on the brain. He has an enormous gash from the nape of his neck to his forehead. If Rowf suffers from severe posttraumatic stress as a result of repeated drownings, Snitter straddles the line between sanity and madness as a result of his injuries.

This novel differs from other stories about vivisection in one important respect. Unlike Wilkie Collins's *Heart and Science* (1883), Mark Twain's *A Dog's Tale* (1903), C. S. Lewis's *That Hideous Strength* (1945), or William Kotzwinkle's *Doctor Rat* (1976), to name a few, Adams explains in a preface that the research described has "actually been carried out on animals somewhere."[8] This is an unusual preamble for a work of fiction and certainly heightens the tension as we read because we remember that these cruel, often-meaningless tests occur on actual dogs, cats, birds, rats, and more. It also signals that this book is more than a spirited yarn. It is simultaneously a statement about the morality of animal research. Yes, the novel is fantasy, but real-world concerns are never far away.

Adams mingles fact and fancy in other ways too. As the novel draws to its satisfying conclusion, he again blurs the lines between fiction and nonfiction with the entrance into the story of Peter Scott, mentioned previously.[9] Sir Peter Scott is an actual person, a renowned ornithologist

[8] Adams, *Plague Dogs*, xviii. H. G. Wells reflects the vivisection debates of the late nineteenth century in his famous novel on the subject (see, e.g., Jones's introduction to *The Island of Doctor Moreau*, xxv–xxvii). Frequent reference to animal experiments occurring at Cornell, Harvard, Columbia, Creighton, Yale, Rochester, Johns Hopkins, and MIT lend a degree of verisimilitude to William Kotzwinkle's *Doctor Rat* (New York: Open Road, 1976), 21, 23, 24, 38, 42–43, 169, 174. For helpful analysis of fiction treating vivisection, see Hadas Marcus, "An Ecocritical Approach to Cruelty in the Laboratory," *Journal of Animal Ethics* 6, no. 2 (2016): 223–33. She focuses on Mark Twain's *A Dog's Tale* (1903) and *The Plague Dogs*, both the 1977 novel and the 1988 film adaptation.

[9] Adams, *Plague Dogs*, 481. The closing words of what we might call the novel's first ending suggest Rowf and Snitter drown in the ocean while trying to swim to the longed-for, mythical Isle of Dog: "Cold. Sinking. Bitter, choking dark" (480). The next page offers a colloquy between reader and author that pleads for an alternative conclusion to the story: "*Good author, please / Dredge those dark waters Stygian / And then, on some miraculous breeze, / Bring lost dogs home to vanished man!*" (481, italics original). The author acquiesces and offers a different conclusion involving the reconciliation of "bird, beast and man" by means of a savior, the wise and generous Peter Scott (481).

and conservationist who arrives at the crucial, climactic scene when the canine heroes Rowf and Snitter are about to drown in the ocean. Scott is the story's *deus ex machina*, conveniently on a yacht at the right place and time and able to pull the dogs to safety when all seems lost.[10] It is a remarkable moment. Rowf, for the first time, meets a kind human guardian, and Snitter—because of Scott's intervention—reunites with his devoted caregiver, long thought dead by the desperate dog. Adams even cites excerpts from the British *Who's Who*, as if to reinforce this interruption of the fantasy by a real-world hero.[11] Why do this? Animals suffer terribly in this world—the note about the actually occurring animal experiments reminds us of that—but human kindness exists too. By allowing a real-life *deus ex machina* to break into the story, Adams suggests there is room to reimagine human–animal relations outside the confines of fiction.[12]

6.2 Aligning the Plight of Suffering Animals with the Plight of Suffering People

A second strategic use of sympathy-generating anthropomorphism is the alignment of animal and human suffering. Perhaps the most touching moment in the story involves a fleeting encounter between the terrified, emotionally unstable Snitter and a heartbroken, lonely Jew named

[10] Adams uses the feminine "Dea ex Machina" as the title for chapter 48 of *Watership Down*. The farm girl Lucy Cane rescues the rabbit Hazel from a cat and then, with the family doctor's help, takes him back to Watership Down (456, 459). This is the only scene in the novel where a human does anything other than threaten and harm rabbits. The doctor's name is Adams, suggesting that the author playfully wrote himself into the story here too. Just as Dr. Adams encourages the child Lucy's natural kindness to animals, Adams the writer hopes to do the same for his implied child readers. Perhaps there is a nod here to Adams's father as well. He too was a nature-loving family doctor.

[11] Adams, *Plague Dogs*, 482–83.

[12] Peter Scott is not alone on the yacht. With him is the naturalist and ornithologist Ronald Lockley (Adams, *Plague Dogs*, 484). Readers of *Watership Down* recognize the name because Adams frequently mentions Lockley's book *The Private Life of the Rabbit* (see, e.g., *Watership Down*, acknowledgments page and 22, 161). As he does in *The Plague Dogs*, Adams also reinforces geographically the connection between fiction and the real world in *Watership Down*, telling readers in an opening note, "Nuthanger Farm is a real place, like all the other places in the book" (vi).

Mr. Ephraim.[13] Ephraim is a businessman who arranges a hunt to track down the sheep-killing dogs Snitter and Rowf. Initially, it is an advertising gimmick for his business, a way to generate goodwill with customers, many of whom are poor farmers concerned about the loss of livestock.

The crucial moment occurs when Ephraim is alone in the wilderness. Solitude is painful for him. As the narrator explains, "the memories induced by solitude ... speak with the voices of hell."[14] Those memories are horrors rooted in the Holocaust. Ephraim thinks about his parents during this hunt for the dogs, "gone without strength before the pursuer; then of his Aunt Leah, vanished more than thirty years ago into the night and fog of desolate Europe, slain by God alone knew what sword in the wilderness." In addition, his traumatized brother testified against Dr. Dering, "the self-styled experimental research expert of Auschwitz."[15] Adams ensures that readers do not miss the parallels between Ephraim and the dogs. Ephraim, we read, has the mind of a "frightened dog," and when he says aloud to himself that it is "a bad world for the helpless," he unwittingly echoes Rowf's oft-repeated phrase, "it's a bad world for animals."[16]

Immediately after the rehearsal of personal and family traumas brought on by his wilderness solitude, Mr. Ephraim sees Snitter approaching. He raises his gun to shoot, but as he slips off the safety latch, he has a change of heart. He notices the scars left by the scientists at the laboratory:

> More remarkable and arresting than all else [in Snitter's appearance]—and at this Mr. Ephraim stared, at first incredulously and then with growing horror and pity—was a deep, hairless cleft, barely healed, pink as the inside of a rabbit's ear and showing the white marks of stitches running clear across the skull from nape to forehead—a terrible gash, giving the dog an

[13] There is mention of his aversion to pork earlier in the story, which foreshadows the import of his religious heritage for the scene describing his encounter with Snitter (see *Plague Dogs* 163).

[14] Adams, *Plague Dogs*, 180.

[15] Adams, *Plague Dogs*, 180.

[16] Adams, *Plague Dogs*, 180, 27. Ephraim's bittersweet memories of hunted and lost loved ones also resemble the experience of the fox. When Snitter asks his fox friend what happened to his mother, he answers with a single word—"Hoonds" (169; i.e., hounds killed her during a fox hunt).

> unreal appearance, like some macabre creature from a Kafka fantasy or a painting by Hieronymus Bosch.[17]

Mr. Ephraim's binoculars blur with tears, his intent to kill gone, and he begins the slow process of coaxing the terrified dog to himself. Remarkably, mystically, deep suffering calls to deep suffering. The torture that Snitter endured, the ongoing pain, and the near madness left in its wake suggest to Mr. Ephraim a kindred spirit. He himself is an outsider, a lonely Jew among Gentiles who know nothing of his family's tragedy. But in this dog, he sees a fellow being who understands only too well what it is to be alone and one who understands the kind of medical experiments endured by Ephraim's older brother. This dog also understands what it means to be hunted, as Mr. Ephraim's parents and his aunt Leah were hunted during the war, and furthermore understands what it is to be alone in the wilderness—an important term in Jewish religious thinking—haunted by memories of death and loss, as Mr. Ephraim is at that very moment.

We also glimpse this spiritual connection between the two from Snitter's perspective. He hears the man's invitation to come. It is attractive because he longs for human company and protection, but he also associates the sound of the man's voice, a human voice, with "Death." The man is one of the many human hunters searching for him at that very moment, and yet Snitter approaches cautiously. As he does, the ringing in his head gradually takes the form of a "lament"—another important term in Judaism; it takes the form of a song given by the wind, which Snitter obediently carries toward the man:

> *From Warsaw and from Babylon*
> *The ghosts will not release the lives.*
> *A weary burden falls upon*
> *The groping remnant that survives.*
> *So this distracted beast contrives*
> *His hopeless search as best he can.*
> *Beyond the notebooks and the knives*
> *A lost dog seeks a vanished man.*[18]

The song returns to Snitter later when he first glimpses the sea from a distance. The song on the wind has its origin in suffering (vivisection,

[17] Adams, *Plague Dogs*, 181.

[18] Adams, *Plague Dogs*, 183, italics original.

scientists cutting him, the resulting madness), but now promises healing and wholeness: "*Come, lost dog; seek your vanished man.*"[19] Mr. Ephraim's voice is the voice of Death. To seek the vanished man, to seek Death itself, is to search for ultimate relief from suffering.

A dog cruelly tortured approaches a man cruelly tortured, and for a moment they find solace in one another. For a brief moment, readers see that violence to one species is violence to all. The encounter is short-lived, however. Seconds after Snitter gets into the car, hopeful for peace at last, he sees one of the other hunters approaching. In the startled mayhem that follows, Mr. Ephraim dies when his gun goes off accidentally.

6.3 Canine Theology: Longing for a Better World

A third use of sentimental anthropomorphism is the inclusion of canine spirituality. On one occasion, Rowf relates a story learned from his mother when a pup, a version of the biblical creation stories.[20] The creator in this story is a great dog in the sky, made of stars. This deity creates animals and sets them in their respective environments—birds in trees, rats and mice underground, cats in houses, fish in water, and so on.[21] The star dog entrusts the care of his animals to human beings, giving them dominion over the earth: "I want you to remember all the time that if I've made you the most powerful animal it's so that you can look after the others.... You're in charge of the world. You must try to act with dignity, like me."[22] Humans even name the animals in this version of the story.[23] The star dog gives humans permission to "kill what animals you need—not too many—for food and clothing and so on," but the caveat about care and imitating the creator's dignity puts restraints on that permission.[24]

[19] Adams, *Plague Dogs*, 256, italics original.

[20] Adams, *Plague Dogs*, 152–58. Adams does the same in *Watership Down*. For the rabbits' creation story, see esp. 26–29. For analysis of Adams's use of religious imagery in that novel, see Michael J. Gilmour, *Reading the Margins: Encounters with the Bible in Literature* (Minneapolis: Fortress, 2024), chap. 7.

[21] Adams, *Plague Dogs*, 152.

[22] Adams, *Plague Dogs*, 153. Cf. Gen. 1:26.

[23] Adams, *Plague Dogs*, 153–54. Cf. Gen. 2:20.

[24] Adams, *Plague Dogs*, 153.

It goes badly. When the creator visits, he does not find the human right away, echoing God's search for Adam and Eve in the Garden of Eden (Gen. 3:9). Instead, the star dog finds a frightened young badger who explains how men tore up the family's den and took his parents away. The star dog finds the mother and father badger in the company of humans, placed in a ring and made to fight dogs. The star dog ends the violence, and a human explains what they are up to: "You said we could make use of the animals, so we were just having a bit of sport. After all, animals are given us for our amusement, aren't they?"[25] There are other crimes. The star dog finds a river with its fish destroyed, and men "sticking iron, pointed things into a wretched bull and making it rip the stomachs out of a lot of poor old broken-down horses" while they watch, laughing.[26]

The star dog finally has enough of the humans' abuse of power. "Because thou hast done this thing," he says, in what is an unambiguous allusion to Genesis 3:14,

> thou art cursed above every beast of the field. *They* will continue to live their lives as before, without reflection or regret, and I will speak to them in their hearts, in hearing and in scent and instinct and in the bright light of their perception of the moment. But from you I shall turn away for ever [*sic*], and you will spend the rest of your days wondering what is right and looking for the truth that I shall conceal from you and infuse instead into the lion's leap and the assurance of the rose.[27]

This is a poignant retelling of the Genesis stories because in Christian tradition, the serpent in the Garden of Eden cursed by God in Genesis 3:14 is the devil, the deceiver who tempts Adam and Eve. Humans are the devilish villains in Rowf's tale. The latter story also indicates that the animals are innocent of moral failings, even if they suffer the reality of a creation distorted by sin.[28]

[25] Adams, *Plague Dogs*, 155.

[26] Adams, *Plague Dogs*, 156.

[27] Adams, *Plague Dogs*, 157.

[28] As Rowf says throughout, "it's a bad world for animals" (27; cf. 61, 158, 207, 282, 313, 388, 481). Cf. *Watership Down*, in which the rabbit Captain Holly observes that other animals eat animals because of need, but "Men will never rest till they've spoiled the earth and destroyed the animals" (151).

Religion insists there is more than what we see, something beyond the physical senses, something other than the experiences of this life. Adams's playful introduction of an animal theology creates an imaginative space for readers to entertain the idea of something startling—namely, that the realities of the human–animal relationships we know (animals hunted, animals subjected to experiments, animals despised, animals disregarded) need not be the only possibility.

6.4 A Conversion Story: From Indifference to Compassion

Fourth and finally, I turn to Adams's use of a dramatic conversion story to urge engagement with ethical questions. Mr. Stephen Powell is a scientist at the Lawson Park facility and is responsible for most of the experiments described in the novel, including the repeated drownings of Rowf. As mentioned earlier, the dog is left struggling to stay afloat for more than two hours, and after he sinks to the bottom of the tank, exhausted, Powell pulls him out and resuscitates him so they can repeat the experiment. Though we wince at descriptions of the work, almost immediately in the opening pages of the book, we suspect Powell has reservations about it too. When the mongrel sinks to the bottom of the tank, he presses his supervisor for permission to pull out the drowning dog, and there is "a shade of anxiety creeping into his voice" as he does so.[29]

Powell evolves as the story unfolds, and he increasingly finds his complicity in a vast array of animal tortures troubling. When told about a new experiment involving dogs in refrigeration units with a built-in escalator and movement deterrents, we see further cracks in Powell's stoic demeanor. The idea is to replicate freezing Arctic conditions, forcing dogs to travel thirty to sixty miles to get to their food. This brutal test disturbs Powell, who asks for another to take the assignment, and Dr. Boycott finds his protégée's reluctance troubling: "To his horror, Dr. Boycott saw—or thought he saw—tears standing in Mr. Powell's eyes."[30] This is a crucial moment. Boycott knows that emotion has no place in this kind of work, and he begins to doubt Powell's competence.

[29] Adams, *Plague Dogs*, 4.

[30] Adams, *Plague Dogs*, 380.

In the laboratory there is one animal in particular who wins Powell's sympathy. The scientists put a monkey in a cylinder for what is called a "social deprivation experiment."[31] They leave the animal alone, with no light and limited room to move, for an extended period. Boycott asks about the monkey and how long "it" has been left in these conditions: "'Forty and a half days,' replied Mr. Powell. 'I believe it's going to die. I wish—I wish to God—.'"[32] Here again there is a flicker of emotion, an acknowledgment of animal suffering, and an implied question about the value of such brutal research methods. Boycott notices and shortly afterward recommends that Powell be fired because "he has allowed himself to express inappropriately emotional feelings about a proposed experimental project."[33] When Powell discovers this, he releases the monkey from his cylinder in an act of defiance and smuggles him out of the lab.[34]

There are other ways Adams maps out Powell's progress on the issue of animal compassion. It is revealing to discover that he reads Doctor Dolittle stories to his daughter.[35] Hugh Lofting's children's books are about a character who interacts kindly with animals, helping them when able. Adams here suggests that reading animal literature has the potential to transform the indifferent, and indeed, he knows this from personal experience. In his 1990 memoir *The Day Gone By*, Adams credits literature for his lifelong interest in animal welfare. When discussing the books his parents read to him during his boyhood, he mentions the Doctor Dolittle stories with particular fondness and credits them with awakening ethical concern. Hugh Lofting "wrote with warmth and humor, and again, the characters are likeable and well-drawn. In the best of the books the narrative grip is powerful. Above all, the author obviously felt real compassion for animals. If I am up to the neck in the animal rights movement today, Dr. Dolittle must answer for it."[36] He develops this point again later: "there is nothing amiss with the Doctor's passionate concern about the abuse of animals. He turned me against circuses, fur

[31] Adams, *Plague Dogs*, 6.

[32] Adams, *Plague Dogs*, 428.

[33] Adams, *Plague Dogs*, 432.

[34] Adams, *Plague Dogs*, 442–45.

[35] Adams, *Plague Dogs*, 354, 496–97.

[36] Richard Adams, *The Day Gone By: An Autobiography* (1990; London: Penguin, 1991), 22.

coats and other such evil things—for life."[37] In *The Plague Dogs*, the most unlikely of characters, the very scientist who torments Rowf and Snitter, undergoes just such a moral awakening, owing in part to oft-maligned, anthropomorphic, sentimental stories not unlike *Watership Down* and *The Plague Dogs*.

Adams juxtaposes the fantasy world of the talking, feeling, thinking, and at times even theologically and philosophically reflective dogs and a fox with at least four other forms of discourse about animals. Rowf and Snitter tell their own story, but others tell it too. There is the language of scientific research that focuses on the animals' bodies, with their biological makeup contributing to medical studies and more. There is also the language of British politicians and military leaders for whom the escape of potentially dangerous, plague-ridden dogs has implications for policy, public opinion, poll numbers, and election outcomes. For newspaper editors and reporters, the mystery surrounding the runaway dogs and the secretive, possibly nefarious activities at a little-known laboratory is a media sensation. With some selective reporting, rhetorical flourishes, and appropriate spin, they generate a wider readership, selling more newspapers. And then there is the story told by farmers in the region. The escaped dogs kill sheep and are a real, if sometimes exaggerated, threat to farmers' bottom line, which views the plight of the dogs through a merely economic lens. And this brings us back to Adams's remark cited earlier. These other forms of storytelling—the biological/scientific, the political, the sensational, the economic—are ineffective when it comes to addressing welfare concerns. By anthropomorphizing the dogs and a fox in a Hugh Lofting–Doctor Dolittle fashion, by generating sympathy for the physical and emotional distress of other sentient beings, the author confronts readers with a moral dilemma. As Rowf says, it is a bad world for animals. Humane fiction holds up a radical alternative, positing it need not be.

Bibliography

Adams, Richard. 2005. *Watership down. 1972.* New York: Scribner.
Adams, Richard. 2007. Attention to suffering: Sympathy as a basis for ethical treatment of animals. In *The feminist care tradition in animal ethics: A reader*,

[37] Adams, *Day Gone By*, 106.

ed. Josephine Donovan and Carol J. Adams, 174–197. New York: Columbia University Press.
Adams, Richard. 2016a. *The plague dogs. 1977*. New York: Vintage.
Adams, Richard. 2016b. *The aesthetics of care: On the literary treatment of animals*. New York: Bloomsbury.
Adams, Richard. 1991. *The day gone by: An autobiography. 1990*. London: Penguin.
Bates, A. W. H. 2019. A spark divine? Animal souls and animal welfare in nineteenth-century Britain. In *The Routledge handbook of religion and animal ethics*, ed. Andrew Linzey and Clair Linzey, 361–370. London: Routledge.
Beers, Diane L. 2006. *For the prevention of cruelty: The history and legacy of animal rights activism in the United States*. Athens: Swallow/Ohio University Press.
Jones, Darryl. 2017. Introduction to *the Island of doctor Moreau*, by H. G. Wells, ix–xxxiii. Edited by Darryl Jones. Oxford World's Classics. Oxford: Oxford University Press.
Keen, Suzanne. 2007. *Empathy and the novel*. Oxford: Oxford University Press.
Kotzwinkle, William. 1976. *Doctor rat*. New York: Open Road.
Marcus, Hadas. 2016. An ecocritical approach to cruelty in the laboratory. *Journal of Animal Ethics* 6 (2): 223–233.
Sainsbury, Lisa. 2013. *Ethics in British children's literature: Unexamined life*. Perspectives on Children's Literature. London: Bloomsbury.
Scholtmeijer, Marian. 1993. *Animal victims in modern fiction: From sanctity to sacrifice*. Toronto: University of Toronto Press.

CHAPTER 7

Webs of Empathy: Animal Subjectivities and the Teaching of Charlotte's Web

John Drew

Animal-themed books abound in literary education, and many of them are enduring classics. However, in the teaching of these works, the animal concerns at the surface of the text are subdued in favor of a supposedly more important human subtext that lies beneath the narrative surface. In this chapter I concentrate on E. B. White's *Charlotte's Web* and unpack the barriers to and possibilities for engendering interspecies empathy, even among young children.

Conventional readings and teachings of *Charlotte's Web* paradoxically displace the experiences of animals despite it ostensibly being an animal-centered book. This contributes to the devaluing of animals in educational discourse, as the animals' experiences and the related ethical questions are erased in plain sight. Yet I posit that *Charlotte's Web* can be differently conceptualized and enlisted by children's literary scholars and educators alike. The book can be understood not only as illuminating

J. Drew (✉)
King's University College at Western University, ON, Canada
e-mail: jwdrew@uwo.ca

C. Linzey and A. Linzey (eds.), *The Palgrave Companion to Humane Education*, The Palgrave Macmillan Animal Ethics Series,
https://doi.org/10.1007/978-3-032-19795-5_7

animal subjectivities and relations, but also as a work that challenges the carnophallogocentric culture[1] within the story's world—a disruptive, denaturalizing dynamic that could extend into the broader world of the reader. This crucial yet strikingly underacknowledged and virtually untheorized dimension is what renders White's classic such urgent classroom reading in the context of the climate emergency.

Humanism's anthropocentric grip extends far and wide over education, and generations of teachers—and scholars—have rationalized their interpretations within these narrow ontological confines. As such, scholarly readings have framed the book as a rumination on the ephemerality of life in all its stages.[2] More recent scholarship has theorized the book as a meditation on human subject formation and developmental progress, as exemplified by the perceived emotional growth of the book's central human character, Fern.[3] These dominant humanist readings block White's subversive counter-anthropocentric vision that subjectifies an animal designated for slaughter while critiquing the ravenous agricultural economy that seeks to "murder" and ultimately consume him. While this is the surface story of *Charlotte's Web*, it has been obfuscated by the imposition of humanist symbolic interpretive frames that encourage us, as Susan McHugh has argued, to circumvent the animal.[4] Under these imposed humanist rubrics, *Charlotte's Web* is deprived of its more-than-human vitality and, more importantly, its pedagogical potential to challenge the pernicious instrumentalist logics underpinning human

[1] Jacques Derrida, "'Eating Well,' or the Calculation of the Subject: An Interview with Jacques Derrida," trans. Peter Connor and Avital Ronell, in *Who Comes after the Subject?*, ed. Eduardo Cadava, Peter Connor, and Jean-Luc Nancy (New York: Routledge, 1991); A. Ratelle, "Ethics and Edibility in Charlotte's Web," *The Lion and the Unicorn* 38, no. 3 (2014): 327–41. Derrida (like other scholars, including ecofeminists) argued that meat consumption, patriarchy, and capitalism are culturally entwined and mutually reinforcing.

[2] Norton D. Kinghorn, "The Real Miracle of Charlotte's Web," *Children's Literature Association Quarterly* 11, no. 1 (1986): 4–9; John Griffith, *Charlotte's Web: A Pig's Salvation* (New York: Twayne Publishers, 1993); Anne Devereaux Jordan, "Appreciating *Charlotte's Web*," *Teaching and Learning Literature with Children and Young Adults* 7, no. 2 (1997): 32–42.

[3] Karen Coats, "Lacan with Runt Pigs," *Children's Literature* 27, no. 2 (1999): 105–28; C. W. Yu, "Reading Children and Animal–Human Relations in *Charlotte's Web* and *The One and Only Ivan*," *Papers: Explorations into Children's Literature* 24, no. 1 (2016): 132–48.

[4] Susan McHugh, *Animal Stories: Narrating across Species Lines* (Minneapolis: University of Minnesota Press, 2011).

exceptionalism and carnophallogocentrism in the increasingly precarious context of the Anthropocene.

Accordingly, I propose possibilities for re-animalizing *Charlotte's Web*, to guide readers and educators beyond the anthropocentric and invite them to grapple with animals' lives and deaths as they develop intellectually and ethically. With an ever-increasing awareness of the cruelties inherent in animal consumption, educators are increasingly compelled to confront or at least consider the ethics of animal slaughter as a social justice issue. In the context of the climate emergency, Greta Thunberg implores educators to understand that such issues are magnified existentially and that the only relevant pedagogies are the ones that crystallize the truth: that animal agriculture is pushing our species, and many others, onto a path to extinction. White himself understood this, and as we shall see, his enduring friendship with Rachel Carson, with whom he shared a deep and abiding concern for people's destructive influence over the natural environment, is testament to his commitment to a more-than-human ecological ethic. *Charlotte's Web* thus becomes an indispensable text for educators committed to climate and animal pedagogies, one that challenges the inherent violence of human exceptionalism while proposing new possibilities for all life.

7.1 Pigs in Context

Before we delve into specific textual analysis, it is important to contextualize this discussion—and pigs—conceptually, culturally, and socioeconomically. Pigs have long been incorrectly associated with filth and squalor, a result of their subjugation at the hands of human farmers, and correspondingly have accrued a pejorative cultural reputation as the dirtiest and lowliest of all farmed animals. According to Harriet Ritvo, pigs were long subordinated in the symbolic hierarchy of domesticated farmed animals.[5] With horses and dogs representing the upper echelon of animal domesticity (a prestige marked by their separation from the meat economy in many countries of the Global North), and with cattle and

[5] Harriet Ritvo, "Learning from Animals: Natural History for Children in the Eighteenth and Nineteenth Centuries," *Children's Literature* 13, no. 1 (1985): 72–93.

sheep embodying an exploitable animal proletariat, pigs have been "routinely castigated as stupid, filthy, and sordid"[6] and came to represent the unsavory underclass of "livestock" whose value was exclusively confined to that of slaughter animal. Ritvo posits that the historical devaluation of pigs relative to other farmed animals is rooted in their distinctive obliviousness to human authority, one often perceived as an active defiance of their human overlords. Given their misbegotten status, pigs were ideally situated as revolutionaries committed to inverting this animal hierarchy in *Animal Farm*. Along these lines, pigs also were perceived as a fundamentally amoral species, one willing to eat their young and disregard their mothers. Ritvo challenges such stereotypes by pointing to persistent anecdotal accounts of the fierce loyalty exhibited by pigs coming to the rescue of "companions in distress … to the utmost of their power."[7] Pigs' subordinated status in the agricultural economy would then seem to be rooted in their failure, or refusal, to flatter the sensibilities and temperaments of their human controllers.

In the context of the contemporary agri-industrial economy, these negative perceptions reduce pigs to even less enviable circumstances. In much of the Global North, they are largely kept out of sight inside factory farms in gestation crates and other forms of intensive confinement within which they can stand up and lie down, but not turn around or move for much of their short lives. When they are temporarily "visible," they are packed densely into poorly ventilated truck trailers as they are moved to slaughterhouses. Yet there is growing interest in pigs and their advanced cognitive capacities. Biologist Françoise Wemelsfelder has elaborated on the distinctly expressive sentience of pigs, which often exceeds the cognitive capacities of dogs—a fact, she notes, that long has been recognized, although reluctantly, by many farmers.[8] Indeed, it may be this anecdotal recognition of pigs' natural intelligence that informs their wider cultural representation in children's stories and entertainment. Regardless, their symbolic presence is refracted perversely from their real-world existence, where the cruelty of their living (and slaughter) conditions is intensified

[6] Ritvo, "Learning from Animals," 85.

[7] Harriet Ritvo, *The Animal Estate: The English and Other Creatures in Victorian England* (Cambridge: Harvard University Press, 1987), 37.

[8] Françoise Wemelsfelder, "A Science of Friendly Pigs: Carving Out a Conceptual Space for Addressing Animals as Sentient Beings," in *Crossing Boundaries: Investigating Human–Animal Relationships*, ed. Lynda Birke and Jo Hockenhull (Leiden: Brill, 2012), 32.

by their cognitive awareness. Of course, such contradictory dynamics are not reserved for pigs. A variety of other animals used in agriculture figure in the cultural imaginary provided for children and youth, notably horses. But pigs are undoubtedly prevalent. From "The Three Little Pigs" to *Babe*, from Porky Pig to Miss Piggy, and as we shall examine, from Wilbur to Esther the Wonder Pig, our ambivalence toward pigs is revealed in many popular texts and cultural artifacts.

So why does a carnistic culture[9] choose to symbolically reanimate, as it were, the vanquished victims of the "animal industrial complex"?[10] Nicole Shukin uses a symbolic economic analysis to contrast the physical rendering of animals with their symbolic rendering—namely, the cute, anthropomorphized, simulated animals decorating popular culture. She argues that these two phenomena are mutually reinforcing as our culture's disneyfied avatars offer a spectral alibi that works to obfuscate the system's physical rendering of animals in the economy.[11] In other words, we decorate our surface culture with endless spectral animal "carcasses" devoid of animal souls, infused instead with synthetic human voices, precisely as real animal bodies are used to serve our economic ends.

I suggest that hegemonic literary education prepares us to instrumentalize (most) animals by teaching young people that the animals have value only insofar as they can reflect anthropocentric concerns back to us. This exploitative and exclusionary logic is embedded in Western educational practices, contributing to an educational silencing of animals.[12] This means that animals may be present—physically in the biology lab or representationally as characters in the numerous stories, films, and books decorating students' classrooms—but that they are invariably reduced to the status of manipulable objects and drained of any authentic subjectivity.

Helena Pedersen argues that ideological messaging embedded in official and hidden curricula works actively to codify boundaries between

[9] Melanie Joy, *Why We Love Dogs, Eat Pigs, and Wear Cows: An Introduction to Carnism* (Newbury Port: Conari Press, 2020).

[10] Barbara Noske, *Humans and Other Animals: Beyond the Boundaries of Anthropology* (London: Pluto Press, 1989), 22.

[11] Nicole Shukin, *Animal Capital: Rendering Life in Biopolitical Times* (Minneapolis: University of Minnesota Press, 2009).

[12] Helena Pedersen, "Is 'the Posthuman' Educable? On the Convergence of Educational Philosophy, Animal Studies, and Posthumanist Theory," *Discourse: Studies in the Cultural Politics of Education* 31, no. 2 (2010): 237–50.

human and animal interests while also serving to "sustain and reinforce the incorporation of animals into capitalist-specific modes of production and consumption."[13] Given these dynamics, it remains peculiar that so many of our educational texts would depict—and often anthropomorphize—animals destined for slaughter and human consumption. Some scholars[14] argue that an anthropomorphized treatment of animals has contributed to the erasure of authentic animal experiences. However, it also could be argued, as Sam Cadman has offered, that these anthropomorphic treatments offer animals something they are desperately missing: a voice through which to communicate their experience, albeit an artificial and imperfect one.[15]

From my experience teaching such texts to both elementary and secondary students, the practice of reading animals as symbolic abstractions is not an intuitive process. Such abstractions require the imposition of hermeneutic frames that must be pedagogically instilled in students and modeled by teachers. The cultural and ideological factors informing how we "read" animals in literary education involve what I call the anthropo-allegorical frame.[16] Here young people are guided to read animals as symbolic avatars and animal stories as allegory for human experience—the allegorical and symbolic frames are presented as the *sophisticated* way to address animals in literature. Animals are thereby subordinated, even displaced, from their own stories, allowing scholars and educators to conveniently sidestep troubling animal experiences and related ethical questions.[17] Susan McHugh refers to this effective silencing of animals

[13] Pedersen, "Is 'the Posthuman' Educable?," 242.

[14] Jane. Desmond, "Performing 'Nature': Shamu at Sea World," in *Cruising the Performative: Interventions into the Representation of Ethnicity, Nationality and Sexuality*, ed. S. E. Case, P. Brett, and S. L. Foster, 217–36 (Bloomington: Indiana University Press, 1995); E. Crist, *Images of Animals: Anthropomorphism and Animal Mind* (Philadelphia: Temple University Press, 2000).

[15] Sam Cadman, "Reflections on Anthropocentrism, Anthropomorphism, and Impossible Fiction: Towards a Typological Spectrum of Fictional Animals," *Animal Studies Journal* 5, no. 2 (2016): 161–82.

[16] John Drew, "Re-animalizing *Animal Farm*: Challenging the "Anthropo-allegorical" in Literary and Pedagogical Discourse and Practice." *Humanimalia*. 13, no. 1 (2022): 163–201.

[17] Erica Fudge, "At the Heart of the Home: An Animal Reading of Mikhail Bulgakov's *The Heart of a Dog*," *Humanimalia* 1, no. 1 (2009); O. Oerlemans, *Poetry and Animals* (New York: Columbia University Press, 2018).

in plain representational sight "as the disappearing animal trick ... now you see the animal in the text, now you don't."[18] Indeed, this "disappearing animal trick" is so deeply entrenched in literary interpretation and education that it has mostly gone unnoticed. Animals are thoroughly instrumentalized in Western carnophallogocentric culture such that even their symbolic incarnations cannot exist beyond their exploitable value.

Such anthropocentric attitudes are reflected in conventional pedagogical approaches to *Charlotte's Web*. Teachers sometimes express trepidation regarding the text's themes, understanding that the novel may be introducing many of their students to questions of mortality and the transitory nature of life. Yet within the context of Western human exceptionalism, the pedagogical prospect of grappling with the complex subject of death when transposed onto animal stand-ins would seem to be significantly less risky than confronting ethical issues pertaining to farmed animals and food systems. A scene from an episode of the acclaimed HBO series *Big Little Lies* satirizes such a conflict, but one highlighting the environmental entanglements of industrial agriculture more than animal subjectivities. In this scene, a second-grade teacher at an affluent California school is using *Charlotte's Web* as a pedagogical tool for teaching about climate change. One student's anxiety prompts her to pass out, ultimately summoning the wrath of the child's mother, who proceeds to attack the school's climate change pedagogical program. It is later revealed that the fainting child's anxiety attack was brought on as much by her parents' extensive personal and financial problems as it was by incipient apocalyptic fears. But interestingly, the teacher's question to the children "Why does Charlotte not want Wilbur to be eaten?"—elicits a canned response: "sus-tain-a-bil-i-ty." The children then recite the amount of water consumption used to produce a single pound of sausage. Even here, the value of Wilbur's life and his perspectives are not addressed.

Notably, the scene became fodder in the American culture wars and sparked a range of reactionary op-eds and blog pieces castigating the "woke" perversion of the novel's values of friendship and personal sacrifice in favor of "leftist educational propaganda." However, I suggest that such reactionary perspectives offer a deeper truth, just not in the way these reactionary critics would assume. *Charlotte's Web*'s greatest

[18] McHugh, *Animal Stories*, 42.

power rests in its capacity to inspire "empathetic engagement"[19] with Wilbur and his tragic and unjustly assigned position within the agricultural economy while promoting the power of friendship and personal sacrifice to confront the logic and morality of this system.

7.2 Where Is Wilbur?

E. B. White's *Charlotte's Web* has, in equal measure, delighted and perturbed generations of children since its publication in 1952. Indeed, scholars have argued that the book offers children a means of confronting the realities of temporality, change, and, most compellingly, death.[20] Not surprisingly, the vast majority of literary scholarship devoted to *Charlotte's Web* emphasizes its pedagogical role in preparing children for a decidedly anthropocentric perspective on death. It would seem strange that a book depicting the horrors of impending animal slaughter from an animal's perspective could be so easily distorted through the filter of humanism. But before the relatively recent incursions of animal studies and posthumanism into literary scholarship, an ethico-ontological hermeneutic that Robert McKay calls "compulsory humanity"[21] dominated our interpretive frameworks. Borrowing from Judith Butler and her notion of compulsory heterosexuality, McKay convincingly argues that animals have been radically excluded from literary theory and pedagogy, even in the examination of works in which they occupy considerable presence.

Indeed, by building on McKay's analysis, I propose the term *anthropo-allegorical* to capture the specific ways that animals are normally constructed in literary education. Symbolism, metaphor, and allegory become devices enlisted to erase animal presence. For texts such as *Charlotte's Web*, the normalized use of an anthropo-allegorical lens allows scholars and teachers to sidestep the sticky animal issues in favor of a

[19] Lori Gruen, "Attending to Nature: Empathetic Engagement with the More than Human World," *Ethics & the Environment* 14, no. 2 (2009): 23–38.

[20] John Griffith, "Charlotte's Web: A Lonely Fantasy of Love," *Children's Literature* 8, no. 2 (1980): 111–17; Kinghorn, "The Real Miracle"; Anne Devereaux Jordan, "Appreciating *Charlotte's Web*," *Teaching and Learning Literature with Children and Young Adults* 7, no. 2 (1997): 32–42.

[21] Robert McKay, "'Identifying with the Animals': Language, Subjectivity, and the Animal Politics of Margaret Atwood's *Surfacing*," in *Figuring Animals: Essays on Animal Images in Art, Literature, Philosophy, and Popular Culture*, ed. M. S. Pollock and C. Rainwater. (New York: Palgrave Macmillan, 2005), 209.

cleaner, anthropocentric interpretation. The animals are not animals; they are allegories. Consequently, to enlist Susan McHugh, Wilbur's experience in *Charlotte's Web* thus reflects the disappearing pig trick: now you see him, now you don't. Crucially, his primal plea ("I don't *want* to die") to avoid the farmer's blade and dinner table is also evaded.

In his book *Creaturely Forms in Contemporary Literature: Narrating the War against Animals*, Dominic O'Key argues that such interpretive evasions and displacements serve a wider cultural and political agenda to normalize an economic system dependent on the instrumentalized destruction of animal lives.[22] Drawing on Derrida's premise (and Dinesh Wadiwal's[23] more extensive theorization) that humans are engaged in an all-out war against animals, O'Key argues that literary humanism serves surreptitiously as a sort of vanguard supporting this anthropocentric campaign of species-based destruction. Accordingly, "compulsory humanity" becomes the precondition determining the boundaries of acceptable literary pursuit, as well as subsequent analysis and instruction, regardless of authorial intention. O'Key's analysis reveals how the scholarly guardrails of literary humanism deflect, or otherwise disregard, the counter-anthropocentric interventions of Mahasweta Devi, J. M. Coetzee, and W. G. Sebald. Important to O'Key's analysis is that such anomalous animalized literary interventions can be read as attempts to narrate the war against animals. It is in this light that we should (re)read and (re)teach *Charlotte's Web* by engaging directly with its overt proto-carnophallogocentric critique of the meat economy.

In 1948, E. B. White wrote an essay based on his personal experiences called "Death of a Pig" that would ultimately inspire *Charlotte's Web*.[24] The essay recounts White's midlife goal to retreat from the city to become a "gentleman farmer" and the personal conflict he faces in his role as animal slaughterer. When his pig becomes ill with erysipelas, White vainly endeavors to nurse the pig back to health but is struck by the emotional bond he develops with an animal he is supposed to kill anyway. In a subsequent essay titled "Pigs and Spiders," White interrogates the farmer–animal relationship more incisively, arguing that the farm poses a

[22] Dominic O'Key, *Creaturely Forms in Contemporary Literature: Narrating the War against Animals* (London: Bloomsbury Publishing, 2022).

[23] Dinesh Wadiwel, *The War Against Animals* (Leiden: Brill, 2015).

[24] E. B. White, "Death of a Pig," *Atlantic Monthly* 181, no. 1 (January 1948): 28–33.

"peculiar problem" for animal lovers because the "fate of most livestock is that they are murdered by their benefactors. The creatures may live serenely but they end violently and the odour of doom hangs about them always."[25] It is noteworthy that animal slaughter is similarly referred to as murder in *Charlotte's Web* when the old sheep first reveals to Wilbur the fate that awaits him, stating, "Almost all young pigs get murdered by the farmer as soon as the real cold weather sets in. There's a regular conspiracy to kill you at Christmastime."[26] One would be hard-pressed to find a stronger declaration that "meat is murder" than this proclamation in one of the most widely read children's books in history. According to Amy Ratelle, White's personal struggle to "redress the wrong he feels he is doing to a trusting fellow creature provides a direct impetus for creating Wilbur."[27] In doing so, White offers himself "a pig that he gets to save by granting him a unique subjectivity, thus removing Wilbur from what Derrida calls the 'carnophallogocentric' paradigm of Western meat consumption."[28]

Fern, the story's ostensible protagonist, stands in for White and like-minded animal sympathizers, challenging the carnophallogocentric logic. In this respect, it is significant that Fern is a girl. Her older brother, Avery, is presented as less intelligent than Fern and as deeply invested in reproducing carnophallogocentrism by persistently engaging in the hunting and capturing of animals. He is also notably less empathetic. When Fern intervenes in her father's intended slaughter of the runt pig who will become "Wilbur," she empathizes viscerally by putting herself in the pig's position, declaring the act to be "unfair.... The pig couldn't help being born small, could it? If I had been born very small at birth, would you have killed me?"[29] In doing so, she is disrupting the accepted hierarchical order that ranks humans above animals, men above women, and adults over children. She could just as easily be pointing out that Wilbur had no choice in being born a pig—or herself being born a human,

[25] E. B. White, "Pigs and Spiders," *McClurg's Book News*, January 1953, 49.

[26] E. B. White, *Charlotte's Web* (New York: Harper, 1980), 49.

[27] Amy Ratelle, "Ethics and Edibility in Charlotte's Web," *The Lion and the Unicorn* 38, no. 3 (2014), 332–33

[28] Ratelle, "Ethics and Edibility."

[29] White, *Charlotte's Web*, 3.

for that matter. Fern's intervention contests the legitimacy of normalized inequities in the system by positing the potential interchangeability of the objects subordinated within its hierarchy (i.e., children, girls, and animals). Her father promptly rejects this conflation, stating, "Certainly not.... This is different. A little girl is one thing, a little runty pig is another." Fern responds, "I see no difference.... This is the most terrible case of injustice I've ever heard of."[30] By invoking the concept of justice, Fern seems to be elevating the stakes beyond the immediate and interpersonal realm to the larger ethico-social domain. Ultimately, it is this appeal that prompts her father to yield the pig's fate to her hands. Under Fern's care, the pig is subjectivized as Wilbur. He is offered respite from the objectified status of meat animal and cared for as a fellow sentient creature. This is powerful pedagogical material for engaging students' animal empathy and for considering the ethics of Fern's position versus that of her father.

Still, once Wilbur grows too big for Fern to care for, her father reasserts the dominant, carnophallogocentric paradigm by suggesting she sell Wilbur to her uncle's farm. Fern, innocent of Wilbur's intended fate, reluctantly agrees and sells him for six dollars but promises to visit him regularly. Thus, Wilbur is thrust back into the meat economy. Gradually, Fern's attachment to Wilbur wanes. Wilbur soon learns that he cannot always count on Fern for emotional support, and though she saved him once, she will be powerless to protect him from Farmer Zuckerman's blade when the time comes. He is reduced to what Agamben calls the "bare life" status of killable resource,[31] and the horror of this realization registers viscerally.

Many literary scholars have argued that Wilbur's presence serves merely as a metaphor for Fern's developmental progress.[32] Yu has gone so far as to argue that *Charlotte's Web* conforms to the wider tradition of "talking animals and their relations with human child characters to express the idea of children-as-animals."[33] He argues that Fern is "forced to save

[30] White, *Charlotte's Web*, 3.

[31] Alex Murray, *Giorgio Agamben* (New York: Routledge, 2010), 11.

[32] Karen Coats, *Looking Glasses and Neverlands: Lacan, Desire, and Subjectivity in Children's Literature* (University of Iowa Press, 2004); Yu, "Reading Children and Animal–Human Relations."

[33] Yu, "Reading Children and Animal–Human Relations," 145.

her animal by speaking in Charlotte's voice."[34] However, this argument does not account for Fern's growing ambivalence toward Wilbur, which culminates in her apparent disinterest in Wilbur's triumph of survival by the novel's conclusion. When all of the other human characters are at last engaged with Wilbur and his blue-ribbon success, Fern is off riding the Ferris wheel with Henry.

7.3 Re-Animalizing Wilbur

Given what we know of White's experiences with animals from "Death of a Pig" and "Pigs and Spiders," we might consider Fern's ultimate withdrawal from Wilbur to be in fact a broader indictment of the social pressures inflicted on children to wean themselves off their animal attachments in order to become fully integrated into the adult order. Indeed, a telling exchange occurs in the middle of the book, when Fern's mother visits the family doctor (Dr. Dorian) to relay her concerns about her daughter's animal attachment. Mrs. Arable informs the doctor that Fern is talking to animals and asks him if he believes animals can talk. Dr. Dorian responds, "I never heard one say anything.... But that proves nothing. It is quite possible that an animal has spoken civilly to me and that I didn't catch the remark because I wasn't paying attention. Children pay better attention than grownups. Perhaps if people talked less, animals would talk more."[35] Dr. Dorian's perspective reflects a nonconformist attitude toward the place of animals and their broader treatment by humans, one preceding the growing interest in other species' voices and agency in the field of animal studies. Although subtle, the doctor's statement offers a fairly incisive critique of humanity failing in its duty to recognize the sentience of our fellow creatures. There is the sense of the Lacanian *real* pervading the doctor's assessment of the animal realm, one that is ironically silenced by the human imposition of the symbolic order of language. Fittingly, the language barrier separating human and animal discourse is closed only on the human end and is only temporarily breached by Fern.

As noted, Fern's intervention to save Wilbur is only partially successful. Once he has grown bigger, he is sold to Zuckerman's farm, where

[34] Yu, "Reading Children and Animal–Human Relations," 142.

[35] White, *Charlotte's Web*, 110.

Wilbur's life will be devalued. It is finally left to a member of the arachnid species, Charlotte, to save Wilbur from slaughter. Upon Wilbur's arrival at Zuckerman's farm, he is immediately confronted with the desolation of the space and the loneliness he feels without Fern's companionship. He is initially shunned by the other animals, specifically by the sheep, and Wilbur is oblivious to the fact that he has been returned to the status of consumable resource. Charlotte sees the wonder, sadness, and capacity for joy in Wilbur's heart and becomes the friend he desperately needs. Initially, Wilbur is repulsed by Charlotte's predatory feeding habits, being that she must trap, kill, and eat flies and other insects for her own sustenance. Charlotte explains that trapping and killing insects is the way spiders have fed themselves since the first spider got the idea. It is a strangely social interpretation of her species' biologically essential feeding habits. When Charlotte later reveals that she offers the flies "an anesthetic so they won't feel pain"—a "little service I throw in," she says[36]—this suggests an imperfect but well-intentioned "empathetic engagement" with her prey, one that can be learned.

This attempt to mitigate the pain of others then informs Charlotte's attempt to intervene on Wilbur's behalf to spare him from becoming a consumable food animal. To this end, Charlotte employs a savvy, "web-based" public relations campaign to save Wilbur's life. Written on a literal web, Charlotte's message must be pithy. The resulting declarations, beginning with "some pig," resemble the condensed, if somewhat impoverished, linguistic economy pervading current social media platforms. Charlotte weaves her web to proclaim the distinctiveness of Wilbur.

Griffith has argued that the simplicity of Charlotte's messaging reveals the superficiality of Wilbur's value and, by extension, the value of other farmed animals within White's narrative. Why not declare "friend not food" or "meat is murder" to better communicate Charlotte's plight?[37] I suggest that the superficial messaging reflects more incisively humans' demonstrated deficiency toward self-refection and interrogation, a motif that abounds throughout the novel. Charlotte seems to implicitly understand that people respond more positively to shameless self-promotion than they do to demands for personal and social examination (as current

[36] White, *Charlotte's Web*, 48.

[37] Griffith, *Charlotte's Web: A Pig's Salvation*, 48.

political and social media trends attest). Thus, Charlotte employs a proto–social media PR savvy to reconfigure Wilbur into a someone rather than a something. This is a powerful assertion of animal subjectivity, one that requires careful pedagogical emphasis.

Reading *Charlotte's Web* anthropocentrically and anthropo-allegorically not only drains the story of its essential richness but also, I would argue, offers a misrepresentative and even disingenuous interpretation of the book to young readers. The anthropo-allegorical reading that positions Wilbur as an animal avatar for children themselves requires the imposition of Wilbur's fear of slaughter onto human children, who may or may not fear death themselves. The explicit reference to "murder" problematizes this mapping immensely. Few children could relate to such a primal fear, but many children register concern for the deaths of animals. The fact that the story's protagonist is a farmed animal who wants to live presents a visceral dimension that defies anthropo-allegorical interpretation and represents a reality beyond the pages.

Yet the hegemonic imperatives of "compulsory humanity" have compelled generations of critics to interpret *Charlotte's Web* through this very lens. Yu and other critics have developed elaborate hermeneutical frames to map both Wilbur and Charlotte onto Fern, in efforts to explain away the dominant animal voices in the text. Such readings provide a twisted irony given that this is a novel devoted largely to the assertion of animals' voices and subjectivity in the context of farming and food production. Indeed, Dr. Dorian's suggestion that "if people talked less, animals would talk more" could be equally applied to the humanist imposition of the anthropo-allegorical lens onto this narrative. Reading and teaching the text beyond the anthropo-allegorical allows the animals to speak for themselves and children to engage directly with the animals' experiences and material conditions. This animalized approach offers more volatile, and less ideologically containable, possibilities but is one that promotes a deeper interspecies connectivity and more authentic engagement with the representations and ideas presented in the text.

7.4 Wilbur and Friends (Not Food)

A more recent example of a real pig (and her very popular story) resonates with these issues and dynamics in compelling ways. Steve Jenkins and Derek Walter adopted a pig named Esther, and in a promotional strategy not dissimilar from the one deployed by Charlotte, helped her to become

Esther the Wonder Pig. Steve and Derek, like Charlotte, provided Esther with the platform to become a someone rather than a something, materially and virtually. In this way, Esther's social media presence has offered a strong contemporary connection to *Charlotte's Web*. Sadly, Esther passed in 2023 but teachers can still use the informative and entertaining web videos to engage student interest further and extend the discussion of the issues raised in *Charlotte's Web* into the present context. In doing so, teachers can help generate discussion of more current and relevant issues related to farming and specifically factory farming. Derek and Steve discuss their reasons for adopting Esther and for creating a farm sanctuary for other rescued farmed animals in videos where they speak in sensitive, accessible, and "non-preachy" language about how their devotion to animals has changed their lives for the better. Such videos demonstrate the power and potential of empathetic engagement and demonstrate that there are real-world possibilities for human–animal relations that extend beyond consumption.

Derek and Steve's championing of Esther has not come without criticism, however. Some have accused them of reproducing sexual-politics-of-meat themes and ideas because of various videos that feminized Esther, in at times rather garish fashion. Steve and Derek have attributed some of these early mishaps to their own emerging animal rights perspectives and previous lack of education on these issues. They also have been criticized for diving too deeply into the celebrity-obsessed waters of social media, perpetuating neoliberal myths of individual exceptionalism that lead to charismatic animals being seen by the public as more deserving of life than others. Although these may be valid critiques of the Esther phenomenon, it must be noted that intentions and goals matter, as do outcomes. There are hundreds of anecdotal testimonials from people who reevaluated their consumption of pigs, or "pork" (and often meat in general), because of their exposure to Esther. Like Charlotte, Derek and Steve found themselves in an unexpected position—with the opportunity to become rescuers of a pig who otherwise would have been doomed to destruction. Like Charlotte, Steve and Derek worked tirelessly—and desperately—to raise consciousness of the plight of pigs and other farmed animals; and like her, they were a little shameless at times to raise the necessary consciousness.

Charlotte's public relations triumph is not without problematic implications either. The fact that Wilbur is positioned competitively against another pig for first prize (and hence the opportunity for survival) places the reader in a position to determine which life has more value. Ratelle argues that despite White's best efforts to confront the logic and morality of carnism, he reproduces entrenched binaries of worthy and unworthy lives by positioning Wilbur as the model show animal worthy of the blue ribbon, thereby "reinforcing the carnophallogocentrism of which he was so critical both in this novel and his other writings."[38] Ratelle argues that Wilbur is presented as "radiant" while Uncle (the opposing pig) is described as coarse, brutish, and ultimately less sympathetic, thereby emphasizing Wilbur's case for individual exceptionalism rather than a broader interrogation of the ethics of animal slaughter.

This argument, however, would seem to underestimate the cynicism and broader skepticism of humans that belies White's narrative. It is not that Wilbur is actually more "radiant" than Uncle—he is described as being quite ordinary, indeed runtish, throughout the novel. The perception of Wilbur's superior appearance is a mere construction successfully perpetuated by Charlotte's "spin." She is positioned as a master campaigner, figuratively spinning Wilbur's narrative through her literal web spinning, and in so doing she successfully manipulates public opinion such that people perceive a special, even "supernatural" aura around Wilbur's presence. Charlotte exploits the credulity of her human audience (or marks) to save Wilbur while also undermining the anthropocentric standards that determine the beauty, value, and right to life of nonhuman animal species. Indeed, when Mrs. Arable asks Dr. Dorian if he understands how there could be writing in a spider's web, he responds, "I don't understand it. But for that matter I don't understand how a spider learned to spin a web in the first place. When the words appeared, everyone said they were a miracle. But nobody pointed out that the web itself is a miracle."[39] Here, Dr. Dorian reveals the parochial limits of the anthropocentrism that conditions people to see the more-than-human world through the prism of humanist instrumentalism.

White also undermines the zero-sum logic that the competition places on the selection process—there can be only one winner; the rest will perish. Indeed, Uncle wins the prize, but Charlotte's final campaign succeeds in forcing the judges to subvert the rules of their own competition by creating a special prize (the blue ribbon) that will finally spare

[38] Ratelle, "Ethics and Edibility," 338.

[39] White, *Charlotte's Web*, 108–9.

Wilbur's life as well. Wilbur's victory is indeed an exception, but one that also subverts the supposedly objective standards guiding the competition. As the runt of the litter, Wilbur offers no future value for competitions, whereas Uncle (with his larger size) embodies the ideal pig. Wilbur can be safely removed from his precarious place in the carnophallogocentric economy only by an intervention that would seem to approach the divine, when in fact it is an act of "interspecies solidarity"[40] that subverts the intended goals of the competition. Even at the point that Wilbur is awarded the blue ribbon, he unceremoniously collapses into a heap that resembles the lifeless corpse he almost became. The judge declares, "We can't give a prize to a *dead* pig.... It's never been done."[41] The judge's statement enacts an unintended tautology (a dead pig cannot win a prize because his exclusion assures his death, and his being dead would negate his eligibility), bringing to the surface the repressed carnophallogocentric ethic guiding the fair celebrations. When Templeton revives Wilbur by biting his tail, the threat of Wilbur's disqualification is farcically resolved, further undermining the legitimacy of the proceedings. Charlotte therefore succeeds in altering the carnophallogocentric logic, if only for this agricultural pageant. I suggest that by interrogating these dynamics more closely, it becomes possible to expand students' own webs of empathy, even without the magical interventions of an arachnid spin doctor, because people's carnistic attachments are similarly constructed.

It is noteworthy that Charlotte succeeds in presenting Wilbur as a "special pig" even though there is nothing discernibly special about him to the human characters who cannot, or will not, hear him speak. White thus gives his young readers access to Wilbur's thoughts and fears, first through the proxy of Fern. When Fern's interest and attention shift beyond animals to the boy named Henry, the reader can forge a direct connection to the animals, one no longer mediated by a human protagonist. As the Dr. Dorian suggests, our world and experience become considerably narrowed and less rich when we do not listen to or empathize with animals. The novel's ending invites children to find communion with animals and points to another possibility, one that now sadly seems lost to Fern, as it is to so many other children who are socialized away from

[40] Kendra Coulter, *Animals, Work, and the Promise of Interspecies Solidarity* (New York: Palgrave Macmillan, 2016), 3.

[41] White, *Charlotte's Web*, 159.

animal interests. Similarly, through Esther, Steve and Derek reveal that empathetic engagement with animals can be restored, even after it has been lost. They have demonstrated that all of us can listen to and care for animals, even those who may otherwise be destined for the abattoir, on their way to the dinner plate. Linking these narratives in a classroom environment allows children to see possibilities that extend beyond the imaginary and into the actual world, where they and grown-ups can take a stand too.

While it is vital that educators help to empower and activate children ethically and ecologically, it is essential that they take the lead by ensuring critical curricular space and opportunities to foster a counter-anthropocentric consciousness. We must heed Greta Thunberg's Fridays for the Future call and its direct admonition of education—that it is obliviously preparing children for a world that may not exist in their lifetimes. In this light, Dr. Dorian's intimation that we are diminished as a species by our myopic disregard for the multispecies beauty and richness around us stands as a powerful lesson for children, one that contrasts starkly with the unflattering examples of animal agricultural subjugation depicted throughout the novel. This message aligns with a more elegiac statement made by White as referenced by Rachel Carson in the epigraph to her seminal work documenting the looming ecological catastrophe, *Silent Spring*:

> I am pessimistic about the human race because it is too ingenious for its own good. Our approach to nature is to beat it into submission. We would stand a better chance of survival if we accommodated ourselves to this planet and viewed it appreciatively instead of sceptically and dictatorially.[42]

Here, White's lament of human supremacy and its destructive influence on the environs—and the animals within—also serves to narrate the war against nature, including animals. He is speaking to the consequences of the Anthropocene and the climate emergency before we came to know them in these terms. In this light, *Charlotte's Web* offers a vision of entwined animal worlds, domesticated and non-domesticated, at the periphery of human experience, one that encourages children (and teachers) to challenge issues of human autonomy over animals' bodies and lives in ways that position children to extend those questions into the

[42] Rachel Carson, *Silent Spring* (Greenwich: Penguin Books, 1962).

wider natural world. In so doing, we may begin to establish a necessarily counter-anthropocentric climate curriculum that allows us to appreciate the more-than-human realm as miracles on their terms rather than mere objects or projections of human mastery.

Bibliography

Cadman, Sam. 2016. Reflections on anthropocentrism, anthropomorphism, and impossible fiction: Towards a typological Spectrum of fictional animals. *Animal Studies Journal* 5 (2): 161–182.

Carson, Rachel. 1962. *Silent Spring*. Greenwich: Penguin Books.

Coats, Karen. 1999. Lacan with runt pigs. *Children's Literature* 27 (2): 105–128.

Coats, Karen. 2004. *Looking glasses and neverlands: Lacan, desire, and subjectivity in children's literature*. University of Iowa Press.

Coulter, Kendra. 2016. *Animals, work, and the promise of interspecies solidarity*. New York: Palgrave Macmillan.

Crist, Eileen. 2000. *Images of animals: Anthropomorphism and animal mind*. Philadelphia: Temple University Press.

Derrida, Ja. "'Eating Well,' or the Calculation of the Subject: An Interview with Jacques Derrida." Translated by Peter Connor and Avital Ronell. In *Who Comes after the Subject?*, edited by Eduardo Cadava, Peter Connor, and Jean-Luc Nancy, 96–119. New York: Routledge, 1991.

Desmond, Jane. 1995. Performing 'nature': Shamu at sea world. In *Cruising the performative: Interventions into the representation of ethnicity, nationality and sexuality*, ed. S. E. Case, P. Brett, and S. L. Foster, 217–236. Bloomington: Indiana University Press.

Drew, John. 2022. Re-animalizing *animal farm*: Challenging the "Anthropo-allegorical" in literary and pedagogical discourse and practice. *Humanimalia*. 13 (1): 163–201.

Fudge, Erica. 2009. At the heart of the home: An animal Reading of Mikhail Bulgakov's *the heart of a dog*. *Humanimalia* 1, no. 1.

Griffith, John. 1980. Charlotte's web: A lonely fantasy of love. *Children's Literature* 8 (2): 111–117.

Griffith, John. 1993. *Charlotte's web: A pig's salvation*. New York: Twayne Publishers.

Gruen, Lori. 2009. Attending to nature: Empathetic engagement with the more than human world. *Ethics & the Environment* 14 (2): 23–38.

Jordan, Anne Devereaux. 1997. Appreciating *Charlotte's web*. *Teaching and Learning Literature With Children and Young Adults* 7 (2): 32–42.

Joy, Melanie. 2020. *Why we love dogs, eat pigs, and Wear cows: An introduction to carnism*. Newbury Port: Conari Press.

Kinghorn, Norton, and D. 1986. The real miracle of Charlotte's web. *Children's Literature Association Quarterly* 11 (1): 4–9.
McHugh, Susan. 2011. *Animal stories: Narrating across species lines*. Minneapolis: University of Minnesota Press.
McKay, Robert. 2005. 'Identifying with the animals': Language, subjectivity, and the animal politics of Margaret Atwood's *surfacing*. In *Figuring animals: Essays on animal images in art, literature, philosophy, and popular culture*, ed. M. S. Pollock and C. Rainwater, 207–227. New York: Palgrave Macmillan.
Murray, Alex. 2010. *Giorgio Agamben*. New York: Routledge.
Noske, Barbara. 1989. *Humans and other animals: Beyond the boundaries of anthropology*. London: Pluto Press.
O'Key, Dominic. 2022. *Creaturely forms in contemporary literature: Narrating the war against animals*. London: Bloomsbury Publishing.
Oerlemans, Onna. 2018. *Poetry and animals*. New York: Columbia University Press.
Pedersen, Helena. 2010. Is 'the posthuman' educable? On the convergence of educational philosophy, animal studies, and Posthumanist theory. *Discourse: Studies in the Cultural Politics of Education* 31 (2): 237–250.
Ratelle, Amy. 2014. Ethics and edibility in Charlotte's web. *The Lion and the Unicorn* 38 (3): 327–341.
Ritvo, Harriet. 1985. Learning from animals: Natural history for children in the eighteenth and nineteenth centuries. *Children's Literature* 13 (1): 72–93.
Ritvo, Harriet. 1987. *The animal estate: The English and other creatures in Victorian England*. Cambridge: Harvard University Press.
Shukin, Nicole. 2009. *Animal capital: Rendering life in biopolitical times*. Minneapolis: University of Minnesota Press.
Wadiwel, Dinesh. 2015. *The war against animals*. Leiden: Brill.
Wemelsfelder, Françoise. "A science of friendly pigs: Carving out a conceptual space for addressing animals as sentient beings." In *Crossing Boundaries: Investigating Human–Animal Relationships*, edited by Lynda Birke and Jo Hockenhull, 223–250. Leiden: Brill, 2012.
White, E. B. 1948. Death of a pig. *Atlantic Monthly* 181 (1): 28–33.
White, E. B. 1953. Pigs and Spiders. *McClurg's Book News*.
White, E. B. 1980. *Charlotte's web*. New York: Harper.
Yu, C. W. "Reading Children and Animal–Human Relations in *Charlotte's Web* and *The One and Only Ivan*." Papers: Explorations into Children's Literature 24, no. 1 (2016): 132–148.

CHAPTER 8

A New Binary: Creator and Created

Matthew Webber

Christian education at the highest levels of study remains extensively anthropocentric, and this human-centeredness bears the potential to neglect some of the most important lessons we can learn about God and our—that is, humans'—place in God's creation. Anthropocentrism also leads people to neglect and engage in violence toward nonhuman animals and even fellow humans, all while claiming divine mandate. While this does not mean that all mention of nonhuman animals has been eliminated from the conversation, the emphasis on humankind has relegated the topic of nonhuman animals to the fringes, or perhaps worse, it has led to any discussion of care for creation being branded as "extremism." Albert Schweitzer, in somewhat archaic terms, once wrote, "Just as the housewife who has scrubbed the parlour takes care that the door is shut so that the dog may not get in and spoil the work she has done by the marks of his paws, so do European thinkers watch carefully that no animals run

M. Webber (✉)
Iliff School of Theology, University of Denver, Denver, CO, USA
e-mail: matthew.webber@du.edu

C. Linzey and A. Linzey (eds.), *The Palgrave Companion to Humane Education*, The Palgrave Macmillan Animal Ethics Series,
https://doi.org/10.1007/978-3-032-19795-5_8

about in the field of their ethics."[1] Similar statements surely apply today, extending beyond European thinkers alone and into a vast array of fields of study. In this chapter I propose that a change is needed within Christian higher education, through which the false binary that has been erected separating human from nonhuman animals can be exposed and replaced with the originally intended binary that is found in scripture—Creator and created.

The debate may not be entirely settled regarding whether something can create itself; for the purposes of this discussion, I will work under the assumption that nothing is self-created. Quite simply, this means that there are two categories of existence: that which creates and that which is created. Furthermore, I concur with Ryan McLaughlin, who writes in his book *Preservation and Protest*, "Traditionally, the term 'creation' refers to all that is not God."[2] While this should go without saying in Christian higher education—that is, Christian colleges and universities as well as seminaries—it is worth reemphasizing. Few have made this statement clearer than nineteenth-century German philosopher Ludwig Feuerbach or Swiss Reformed theologian Karl Barth. Both go to extremes to establish that God is drastically different from what and whom God has created—including humans. The two thinkers' arguments begin in opposing points of origin, but the result is the same. Creation occupies one realm of existence, while the Creator occupies the other.

Through Barth and Feuerbach, humans are confronted with the fact that we are more like the creation that surrounds us than we would like to admit. Humans have established a binary of human beings and everything else in the universe. Be it nonhuman animal life, plant life, or any other life-form that may exist in the universe, humans occupy one category, whereas the rest of the universe occupies another. What is too often forgotten is that humans are not set apart, over and against creation, but are part of it. One need not take a literal six-day interpretation of creation in order to acknowledge the truth of this statement, yet one need not ignore the biblical accounts of creation entirely either. One need simply acknowledge that what exists in the universe now was created by something. For the Christian, God is that which is uncreated. Everything else

[1] Albert Schweitzer, *The Philosophy of Civilization* (Amherst, NY: Prometheus Books, 1987), 297.

[2] Ryan Patrick McLaughlin, *Preservation and Protest: Theological Foundations for an Eco-Eschatological Ethics* (Minneapolis: Fortress Press, 2014), 27.

is created. Humans do not make up a third category. Furthermore, as the opening story of creation unfolds, the Hebrew poet describing the events appears to at least acknowledge some similarities between human and nonhuman animals. Humans and nonhuman animals—excluding birds and fish—were created on the same day.

Before delving too deep, it is necessary to acknowledge that the human–nonhuman binary arises from the rendering of Genesis 1:28 wherein humans are granted dominion over nonhuman animals by God. The term "רָדָה" or "dominion" lies at the root of this binary. Far too often, there is a tendency to associate "dominion" with "domination." Similar translations of "רָדָה" elsewhere in scripture depict instances in which individuals are subject to the reign of others (Lev. 25:43, 46, 53; 27:17; Isa. 14:2, 6; 41:2) while other passages describe dominion as something akin to land ownership (1 Kings 4:24 and Ps. 72:8). How one translates and interprets this verb has great potential to influence one's regard for nonhuman animals, yet this single verb is but a part of a greater edifice we must examine if we hope to find a remedy. What remains consistent regardless of how one interprets "רָדָה" is that humanity is subject to God's divine command and must not act contrary to God's will when exercising dominion over creation.

It is, however, worth inquiring about the aim of creation and whether or not the intention of the creative act was accomplished solely for the sake of humans.[3] Admittedly, the language of theology is human. In the second volume of his *Church Dogmatics*, Karl Barth said, "The Word of God is concerned with God and man."[4] Similarly, as he begins his volume on the doctrine of creation, Barth says, "The purpose and therefore the meaning of creation is to make possible the history of God's covenant with man."[5] One must not, however, understand this to mean that

[3] An additional question emerging from the discussion of the aim of creation might be the aim of salvation and whether or not humans alone are the focus of God's atoning work through the life, death, and resurrection of Jesus. Were humans alone saved through this act of atonement, or was all of creation atoned for as well? However, this is a topic for another time and publication.

[4] Karl Barth, *Church Dogmatics*, ed. G. W. Bromiley and Thomas F. Torrance (London: T & T Clark, 2010), vol. 2, pt. 2, p. 6.

[5] Barth, *Church Dogmatics*, vol. 3, pt. 1, p. 42.

God's focus is solely anthropocentric. Theology is not anthropology.[6] For someone like Immanuel Kant, the ultimate purpose behind God's creation of the world lies not in human happiness—especially that which comes at the expense of other created beings[7]—but in individuals striving to be worthy of happiness through the fulfillment of their moral duties.[8] I acknowledge that Kant is no theologian, though few thinkers—if any—have had a similar impact in the West regarding metaphysics or ethics. In fact, references to Kant can be found throughout Barth's magnum opus. Additionally, the relationship between humans and nonhuman animals within the Kantian school of thought is being reexamined and may well continue to inform how we relate to our fellow created beings.[9] Setting aside what might sound like "work's righteousness," Kant's suggestion that human happiness is not the purpose of creation appropriately turns the focus away from humanity. The center around which all creation revolves is not humanity itself, but the will of God, which may very well include the duty of caring for one another, including humans caring for—opposed to exploiting—nonhuman animals.[10]

That humans require human language in order to understand anything, theology included, should simply suffice as a given. The notion that human language is the only means by which God speaks or that humans are the only recipients of God's attention is a misunderstanding, to say the least. One should not think that human language is the sole transmitter of God's message. What should happen if human language were to disappear? According to scripture, "I tell you, if these were silent, the stones would shout out" (Luke 19:40). Moreover, according to Karl Barth's oft-quoted passage, "God may speak to us through Russian

[6] Andrew Linzey, *Creatures of the Same God: Explorations in Animal Theology* (Brooklyn: Lantern Books, 2009), 10.

[7] Immanuel Kant, *Grounding for the Metaphysics of Morals*, trans. James W. Ellington (Indianapolis: Hackett Publishing, 1993), 36 (Akademie edition IV, 429).

[8] Immanuel Kant, *Critique of Practical Reason*, trans. Werner S. Pluhar (Indianapolis: Hackett Publishing, 2002), 165 (Akademie edition V, 130).

[9] See, for instance, Christine M. Korsgaard, *Fellow Creatures: Our Obligations to the Other Animals* (Oxford: Oxford University Press, 2018) and J. John Callanan and Lucy Allais (eds.), *Kant & Animals* (Oxford: Oxford University Press, 2020).

[10] Kant himself does not state this; however, I hope to demonstrate that such care for nonhuman animals is indeed one of our duties when we shift from a human–nonhuman binary to one of Creator–created.

Communism, a flute concerto, a blossoming shrub, or a dead dog."[11] How much more might God speak to us through a living dog? Or a cat? Or an ape? Or a slug? We know the story of God's Word spoken through Balaam's donkey (Num. 22:21–39) and the message, heard loud and clear, that we not mock baldheaded prophets, that was spoken through two she-bears (2 Kings 2:23–24). What might we find God revealing to us if we focused less on ourselves and more on our fellow created beings?[12]

Despite scripture stating time and time again that humans are part of the created element and are not divine, humans have demonstrated a tendency to elevate humankind above the rest of the world, coming dangerously close to regarding humanity as godlike. The extreme of this is idolatry, wherein humans occupy the center of a religion that seeks to meet humankind's every desire, giving little or no care to what it might cost to do so. This is exemplified in the critique of the faith offered by Ludwig Feuerbach, who states that "nature, the world, has no value, no interest for Christians. The Christian thinks only of himself and the salvation of his soul."[13] Nothing else matters to the Christian aside from their personal salvation. Sadly, this may mean that God—the one from whom and through whom salvation is given—is pushed to the margins of the individual's life. The individual, rather than God, is at the heart of each action. What exacerbates this problem is when individuals believe that their central role in the world is ordained by God, which potentially leads one to believe that God's desire is for the individual human to have whatever he or she wants. Moreover, one might believe that God blesses the means by which human desires are satisfied regardless of the costs. Feuerbach sums up this misplaced focus in the phrase "homo homini Deus est," or "the human is God to the human."[14]

Andrew Linzey offers that inclusion of nonhuman animals might "save Christians from the idolatry of self-worship."[15] Humbling as it may be, admitting that humans are not the pinnacle of existence allows humanity

[11] Barth, *Church Dogmatics*, vol. 1, pt. 1, p. 55.

[12] This perhaps exemplifies the complexities of Karl Barth's dialectical work as it unfolds across thousands of pages, offering more conversation than definitive answers.

[13] Ludwig Feuerbach, *The Essence of Christianity*, trans. George Eliot (New York: Harper & Brothers, 1957), 287.

[14] Feuerbach, *Essence of Christianity*, 159.

[15] Linzey, *Creatures of the Same God*, 15.

to step down from the pedestal upon which we have situated ourselves. Rather than worshiping humanity as the highest of all creation, we instead worship the Creator. As Hannah Reichel points out, the former not only is dangerous to interhuman relationships but also is detrimental to the human–divine relationship.[16] The danger of believing that humans occupy a position of supremacy in the universe is precisely why Barth recommended that his students read Feuerbach.[17] By simply admitting that we are part of creation and not, as I said, some third category hovering between Creator and everything nonhuman in the universe, we save ourselves from idolatry.

Criticism of such a binary system need not be regarded as anti-human. Critique of the belief that humans are so uniquely different from any and all nonhuman life in the universe may actually serve to help us better understand who and what humans are and what we require. Understanding that we and the lives that surround us are more alike and that we are not foreign to this world will perhaps lead to greater insight into our past, shed light on our present situation, and inspire how humans live in the world in the future. We can learn from our fellow beings in the journey of life, gleaning information about how others adapt to changes in their environment, so that we too might adapt. Seeing that humans are not vastly different from the rest of the universe might serve to benefit humanity, which is far from anti-human.

Moreover, eliminating a human–nonhuman binary is not anti-Christian; rather, it poses a more accurate understanding of the Judeo-Christian teaching. The distinction between Creator and created is of far greater significance than any perceived differences between human and nonhuman. Individuals may continue to espouse human uniqueness where it is supported, such as the ability to write poetry or play the violin. Yet we will never be as unique as that which created the universe. Thus, as unique as humans may believe ourselves to be, our uniqueness is not a reason to abuse or exterminate any of our fellow creation. Were uniqueness a reason to violate and destroy those who fail to demonstrate that they are similarly unique, humanity's logic would allow for beings of higher consciousness to exercise the same power over us, should they ever

[16] I am especially grateful to Dr. Reichel for her permission to cite her paper "The Other Question between Projection, Colonial Imagination, and Liberation," given at the 2018 Karl Barth Conference.

[17] Feuerbach, *Essence of Christianity*, viii.

arrive. No, humans have more in common with nonhuman animals than we are often willing to admit or notice, even when it is glaringly obvious. This is best demonstrated in the religious sacrifices of the Hebrew people.

Animals were regarded as similar enough to humans to bear the punishment for human transgressions, yet different enough to be relegated to the category of "other." Under Jewish sacrificial practices, nonhuman animals would suffice as worthy substitutes in order that humans might atone for their sins against God and their fellow humans. Countless nonhuman animals were slaughtered so that humans could make up for their failure to uphold their part in covenantal relationships. Interestingly, nonhuman animals were held responsible for some acts perpetrated against humans, such as an ox goring a human, yet few laws were in place prohibiting human mistreatment of nonhuman animals.[18] Thus, similarities are acknowledged only when they benefit the human agents. When similarities might benefit nonhuman animals, the differences become the focus. Were the differences as significant to the Christian believer as the human–nonhuman binary would suggest, Christians would be at the front lines arguing that any results of tests on nonhuman animals that are extrapolated and applied to humans are obscene.[19] It is worth asking, then, whether humans are of greater concern to God than anyone or anything else.[20]

From the very beginning of scripture to the prophetically described end, we see that God is concerned with both the human and the nonhuman elements of creation. God establishes a covenant with all living creatures (Gen. 9:9–11) while the eschatological imagery is that of a land in which humans dwell peacefully with nonhumans, where predators and prey sleep side by side and eat together (Isa. 11:6–9). One of the seldom-referenced passages noting God's compassion for nonhuman animals is

[18] See Exod. 21:28–29 regarding the penalty for an ox goring a person resulting in death. See also Exod. 20:10 and Deut. 5:14, requiring that nonhuman animals rest on the Sabbath.

[19] Note that this is *not* an argument in favor of scientific, medical, or cosmetic testing on nonhuman animals. The biological or anatomical similarities between human and nonhuman animals are one aspect of the argument against—not in favor of—such tests.

[20] One question might be that if all creation fell through human sinfulness, did Jesus atone for humans alone? Or was all of creation atoned for in Christ? It would seem unfair if nonhuman animals bore punishment for human sin and never received the grace humans claim. Despite human desires to be first in God's sight, perhaps we are not God's sole care in the universe.

the closing line of the book of Jonah. Justifying the divine mercy given to Nineveh, God tells the angry prophet Jonah God's reason: "And should I not have concern for the great city of Nineveh, in which there are more than a hundred and twenty thousand people who cannot tell their right hand from their left—and also many animals?" That the contrite and repentant people are worthy of forgiveness is to be expected, but that the nonhuman animals are also a reason for God's mercy is worthy of our attention. It is not simply the humans—ignorantly fumbling their way through life—who concern God, but also the nonhuman animals.

In addition to ignoring portions of scripture that might inform our lives, an anthropocentric Christianity allows those who profess to be followers of Jesus to participate in acts of violence against those they regard as "other." Despite claims that God created all life, humans remain uniquely blessed, and any association with nonhuman animals beyond that of dominance is an affront to God's plan. In 2010, a Vacation Bible School curriculum put out by Answers in Genesis stated that there are not any monkeys in the human family tree.[21] Granted, Answers in Genesis may not be a definitive authority nor represent the majority of Christians in the world, yet it remains worth noting that designating humans as separate from nonhuman animals and acknowledging any similarities between the two may potentially be deemed heretical.

By "othering" animals, humans juxtapose themselves over and against nonhuman animals, claiming the former to be superior and making the latter inferior, controllable, usable, a means to human ends, and finally, disposable. Sadly, such designation of superior over inferior "other" does not stop with the human–nonhuman binary. Extending from the othering of animals comes the animalizing of the other. When nonhuman animals are understood to be other, everyone who falls within the category of "other" can be easily animalized as well, including women, people of color, or people of a different nationality or religion; even within these categories, when someone is regarded as "other," they too are regarded as inferior, usable, and too often, disposable.[22] Worse yet is when such individuals have been "othered" to the point where they are renamed with

[21] Ken Ham, "Egypt File: Decoding the Mystery of Life," *Answers in Genesis*, September 18, 2009, accessed June 24, 2019, http://answersingenesis.org/blogs/ken-ham/2009/09/18/egypt-file-decoding-the-mystery-of-life/

[22] Carol J. Adams, *The Sexual Politics of Meat: A Feminist-Vegetarian Critical Theory* (New York: Bloomsbury Publishing, 2016), 20–25.

slurs or derogatory language.[23] Strangely, given that this behavior stems from human superiority, it has been described in animalistic terms, stemming back to the proverbial "homo homini lupus est," or "the human is wolf to the human," evoking images of a different binary—that of predator and prey.[24]

Too often, the differences that lead to the categorization of the "other" as inferior are subjective and established to benefit the ones making the designation. In many cases, humans are inferior to nonhuman animals. In such areas as speed, strength, sight, hearing, camouflage, and lifespan, humans are an inferior species. Thus, humans created a measurement in which we are deemed superior and other, nonhuman species are found wanting—rationality. And when nonhuman animals demonstrate the ability to reason or problem-solve, we again draw the lines of distinction, with humans inside the realm of superiority while the "other" remains inferior. Should we ever encounter a species who is superior in the ability to reason, surely we will yield to their superiority and allow them to treat us the way we have treated nonhuman animals. Or perhaps we will opt instead to designate a new means by which humans are deemed superior.

One particularly problematic pericope is when Jesus himself animalizes the other in calling a Canaanite—that is, a non-Jewish woman—a "dog" (Matt. 15:21–28). How might one interpret Jesus calling a Canaanite woman a dog, as opposed to the children of Israel whom he came to feed? The woman does not go away hungry, as her request is fulfilled, but how one wrestles with this Gospel account is certainly worth greater exploration if we are to emphasize a binary of Creator and created over a human–nonhuman one.

One might be forgiven for "animalizing the other" when even Jesus falls prey to similar tendencies. Yet it may be that in addition to reevaluating one teaching moment found in the Gospels, we also need to reevaluate what it means to be "created in the image of God." Perhaps instead of looking to humans as the image bearers, we might benefit from looking at Jesus, the image of God par excellence. Jesus, recognized as

[23] Adams, *Sexual Politics of Meat*, 24–43. Adams points out that we do this to nonhuman animals when we relabel their bodies as "beef," "pork," "steak," etc., opposed to identifying the bodies and body parts.

[24] Associating certain nonhuman animals with violence and the subsequent violent retribution must be acknowledged here, but remains a topic for another paper.

morally superior to humans, opts not to reign over humanity by exercising dominion with force, but rather to serve and sacrifice for the sake of those over whom he has dominion. Those who claim to be created in the image of God might better demonstrate this by exercising dominion in a similar fashion.

Again I turn to Karl Barth, who, while describing humankind as more noble than nonhuman animals, nonetheless calls for nonhuman animals to be understood as humanity's "fellow-animals" and "constant companion[s]," dwelling within creation.[25] Barth goes so far as to say that nonhuman animals will join alongside the humans, participating with humans in the covenant between humanity and God.[26] Furthermore, Barth states that humans' dominion over nonhuman animals must be confined within humanity's covenant with God, which means that what is done to nonhuman animals must never be done by human authority. Killing a nonhuman animal and citing human authority is no less than murder according to Barth.[27] To falsely claim that one's authority comes from God would be committing a similarly grave sin.

Barth does not shy away from designating an "other" in the created world, however. The wholly other, according to Barth, is not those who differ from humanity or within humanity, but those who differ from Jesus. Thus, while humanity is wholly other, Jesus is the *holy* other.[28] If we focus on the fact that we are unlike God, we are better able to acknowledge our need for God and therefore the importance of Jesus's atoning acts on our behalf. Again, humans do not occupy some third category of existence, floating, as it were, between that which is created and the Creator. Humans must reside in one category or the other. If Barth is correct—and I believe he is—humankind falls into the category of "created" along with the vast majority of that which exists.

By no means do I wish to belittle humanity; however, I will admit that, as humans, we may need to think less highly of ourselves. Barth goes to great lengths describing the nature of humanity as that which not only differs from God but is an enemy of God, opposing God's will for

[25] Barth, *Church Dogmatics*, vol. 3, pt. 1, pp. 177–78.

[26] Barth, *Church Dogmatics*, vol. 3, pt. 1, p. 178.

[27] Barth, *Church Dogmatics*, vol. 3, pt. 3, 355.

[28] Reichel, "The Other Question between Projection, Colonial Imagination, and Liberation."

humankind and for creation.[29] Barth then goes on to describe the care God has for that which God created, humbling God's self and dwelling in the midst of creation. The juxtaposition of Creator and created is bridged not by human beings, but by God alone. Thus, humans do not and cannot occupy a third category somewhere between Creator and creation without likening ourselves to Jesus. The echoes of Feuerbach's criticism, in calling the arrogance of perceived human superiority over the rest of creation nothing less than idolatry, ring loud and clear.

If we are uncomfortable bearing the label of idolaters, a theologically sound solution is readily available. Shifting from a binary in which humans are set apart from and above all other created beings, we should employ a binary of Creator and created. Again, this is not meant to belittle humans. Instead, it brings to our attention that humans are not the end-all of God's creation, nor are humans the only creatures about whom God cares.

Is Barth alone in his willingness to speak of our fellow nonhuman animals as worthy of God's attention? While more noted for his willingness to agree with the Thomistic understanding that nonhuman animals were created in order to benefit humans, Martin Luther has strangely been deemed a dog lover. Whether the attribution is accurate or apocryphal, countless instances of Luther being quoted as saying, "Be thou comforted, little dog, thou too in Resurrection shall have a little golden tail" have persisted since the 1800s. What can we make of this? Perhaps simply the idea that there is more to some of our past theologians than we may have first thought? Or perhaps the message is that we have been too focused on the humans and not enough on the nonhuman aspects of theology. What might we learn from reexamining the works of other theologians through a similar lens?

In short, what this proposed new binary suggests is not that we rewrite the history of the Christian Church. Neither does this intimate that theologians who posit more anthropocentric arguments should be ignored or that their positions should be reimaged into something that resembles our own. Rather, the shifting from a human–nonhuman binary to a Creator–created one allows individuals not only to reread scripture through a different lens than before but also to reread and understand

[29] Barth, *Church Dogmatics*, vol. 2, pt. 2, pp. 211–83.

existing theology through a similarly new lens. Much like reading scripture through the eyes of a Gustavo Gutierrez or a Mary Daly, reading the Bible and theologians with nonhuman animals in mind may expand our understanding of God's Kingdom. Feuerbach offers insight to the believer by questioning the embodiment of the deity one worships, potentially inspiring us to read significant thinkers differently than before. This means that Luther need not necessarily be the dog lover his reader is, but that this reader might glean new insights once Luther is reexamined. The same can be said of Thomas Aquinas, who, while remaining anthropocentric, still speaks volumes. It would be just as much a mistake to ignore these individuals as it would be to attempt to create them in our own image. Thus, we stand on the shoulders of those who have gone before us, grateful for their work yet not conceding that they spoke the final word on any given matter.

Recognizing a new binary would have practical implications not only for those who teach in upper levels of Christian higher education but also for how Christians care for one's fellow created beings. Care for one's fellow created beings would open the door to how Christians interpret scripture and challenge long-held concepts of what it means to exercise "dominion" over other beings. Similarly, one might begin to expand the circles of inclusion and recognize that God's Kingdom might include far more humans than many might initially believe and that the Kingdom might also extend beyond humans alone. Furthermore, through acknowledgment of God's love for nonhuman animals, care for one's fellow humans might be improved. In her recent book *Our Symphony with Animals*, Aysha Akhtar points out that many who are on the fringes of society put the needs of their beloved nonhuman companions before their own needs. Thus, caring for the people experiencing homelessness who risk their lives sleeping in the cold because shelters do not allow dogs inside, or for a battered spouse who returns to her abuser, fearing for the safety of her beloved dog or cat, might encourage new ways to care for these individuals, whether human or animal.[30] Christians who profess to care for these individuals must recognize the needs of the animals that the people they desire to help place before their own.

Christian universities and seminaries are lagging behind secular universities in recognizing the ethical responsibilities we have toward nonhuman

[30] Aysha Akhtar, *Our Symphony with Animals: On Health, Empathy, and Our Shared Destinies* (New York: Pegasus Books, 2019), 22–34.

animals. The anthropocentrism of Christianity has done great harm to our relationship with nonhumans and our understanding of the breadth of God's love. By focusing less on a human–nonhuman binary and more on a binary that is true to scripture—the Creator and the created—we are able to participate in the discussion. This is not to be taken as yielding to secular influence, but as righting a wrong that has been allowed to spin out of control. In the end I pray not only that we look back upon theological studies and see the footprints of our fellow created beings, but also that we welcome their presence in our lives and in our study. May these paw, hoof, and claw prints not only remind us of our status as created beings, but also teach us more about who we are as humans in the world and in relationship with our Creator, preventing the atrocities that accompany human arrogance.

Bibliography

Adams, Carol J. 2016. *The sexual politics of meat: A feminist-vegetarian critical theory*. New York: Bloomsbury Publishing.

Akhtar, Aysha. 2019. *Our symphony with animals: On health, empathy, and our shared destinies*. New York: Pegasus Books.

Barth, Karl. 2010. In *Church dogmatics*, ed. G. W. Bromiley and Thomas F. Torrance, vol. 4 vols. London: T & T Clark.

Callanan, John J., and Lucy Allais, eds. 2020. *Kant & Animals*. Oxford: Oxford University Press.

Feuerbach, Ludwig. 1957. *The essence of Christianity*. Translated by George Eliot. New York: Harper & Brothers.

Ham, Ken. "Egypt File: Decoding the Mystery of Life." *Answers in* Genesis, September 18, 2009. Accessed June 24, 2019. http://answersingenesis.org/blogs/ken-ham/2009/09/18/egypt-file-decoding-the-mystery-of-life/.

Kant, Immanuel. 1993. *Grounding for the metaphysics of morals*. Translated by James W. Ellington. Indianapolis: Hackett Publishing.

Kant, Immanuel. 2002. *Critique of practical reason*. Translated by Werner S. Pluhar. Indianapolis: Hackett Publishing.

Korsgaard, Christine M., and Fellow Creatures. 2018. *Our obligations to the other animals*. Oxford: Oxford University Press.

Linzey, Andrew. 2009. *Creatures of the same god: Explorations in animal theology*. Brooklyn: Lantern Books.

McLaughlin, Ryan Patrick. 2014. *Preservation and protest: Theological foundations for an eco-eschatological ethics*. Minneapolis: Fortress Press.

Reichel, Hannah. 2018. *The other question between projection, colonial imagination, and liberation*. Lecture given at the Annual Karl Barth Conference.
Schweitzer, Albert. 1987. *The philosophy of civilization*. Amherst, NY: Prometheus Books.

CHAPTER 9

Moral Education and Animal Rights

Kai Horsthemke

The topic of other-than-human animals is a surprisingly novel theme within philosophy of education. This chapter focuses both on the moral and pedagogical implications of psycho-physical continuity between humans and other animals and on the promise of animal-centered (as opposed to human-centered) education. Does anti-racist and anti-sexist education logically entail anti-speciesist education? Is animal rights education logically linked to human rights education? Various approaches have been suggested for including the ethical treatment and moral status of animals as an urgent concern within pedagogy and within teaching and learning generally:

- environmental education and education for sustainability, ecophilia, and biophilia;
- humane education and theriophilia (that is, love of/for animals);
- philosophical posthumanism;
- critical pedagogy and ecopedagogy;

K. Horsthemke (✉)
University of the Witwatersrand, Johannesburg, South Africa
e-mail: Kai.Horsthemke@gmail.com

C. Linzey and A. Linzey (eds.), *The Palgrave Companion to Humane Education*, The Palgrave Macmillan Animal Ethics Series,
https://doi.org/10.1007/978-3-032-19795-5_9

- critical animal studies and animal standpoint theory; and
- vegan education.

Each of these has obvious strengths and also considerable flaws. As a viable alternative I propose an account of animal rights education that takes seriously matters of both social justice and "moral feeling," cultivation of (appropriate) moral sentiments. Given most children's natural interest in and feeling for animals, this may seem easier than is commonly assumed. However, it does require effort, commitment, and consistency on the part of caregivers and educators, of parents and teachers alike.

9.1 Animals, Education, and Philosophy

For centuries, the only teaching and learning about other-than-human animals that took place in educational institutions happened in the medical and biological sciences and in comparative psychology, with focal areas such as animal anatomy/physiology, behavior, and cognition. Since the early 1970s, animal ethics has been an influential area of scholarly concern and research among professional philosophers.[1] It entered university philosophy courses as an increasingly popular study option, while humane education was introduced in numerous schools. It is all the more puzzling, then, that for the longest time the subject of nonhuman animals was largely ignored within philosophy of education. It is only in recent years that educational philosophers and theorists of moral education have begun to pay much overdue heed to the pedagogical implications of the psycho-physical continuity between humans and other animals.[2]

The case for the inclusion of other animals in the sphere of direct or immediate moral concern is as follows.[3] Differences between "us" and

[1] See, for example, Stanley Godlovitch, Roslind Godlovitch, and John Harris, eds., *Animals, Men, and Morals* (London: Gollancz, 1971); Peter Singer, *Animal Liberation* (New York: Random House, 1975); Tom Regan, *The Case for Animal Rights* (London: Routledge & Kegan Paul, 1983).

[2] Kai Horsthemke, *Animal Rights Education* (Cham, Switzerland: Palgrave Macmillan, 2018); Horsthemke, "Non-Human Animals," in *International Handbook of Philosophy of Education*, ed. Paul Smeyers (Dordrecht: Springer, 2018), 1401–14.

[3] See Kai Horsthemke, *The Moral Status and Rights of Animals* (Johannesburg, South Africa: Porcupine Press, 2010).

"them" are not radical. They are gradational, differences in degree, and in many morally relevant respects, humans and nonhumans are relevantly similar. Animals are conscious subjects of a life that can be better or worse for them, they experience pleasure and pain, and they normally seek to avoid not only sources of pain and nociceptive stimuli but also death. They may lack full-fledged moral agency and accountability, but like human moral agents, they are moved by altruism, affection, tenderness, prudence, deference, dominance, and aggression. Moreover, there are many human beings who are not full-fledged moral agents and who cannot reasonably be held (fully) morally responsible. Any criterion advanced to justify the moral demotion or exclusion of animals also could be used to justify downgrading or excluding those humans who fall short with regard to a relevant capacity, be it moral agency, rational autonomy, or language use—any criterion, that is, except the criterion of "being human." But to use this as a moral demarcation is as arbitrary as citing "being a man" or "being white." Given evolutionary biological and psychological kinship between the human and nonhuman species, speciesism is as morally questionable as sexism and racism. We thus obtain a two-pronged case for animals' inclusion in the sphere of direct/immediate moral concern, if not equality/equal moral status, consisting of the "argument from marginal cases"[4] and the "argument from speciesism."[5]

Learning about animals, then, has been and continues to be an integral feature of education. Humans are the primary beneficiaries of studying animals, in terms of learning about their anatomy and behavior. But pedagogical and educational encounters between humans and animals also have the potential of benefiting animals, over and above enjoyment of appropriate veterinary care and attention. Human beings can learn to interact and engage with other animals in morally defensible ways, to appreciate their abilities and respect their needs and interests, to coexist with them in a caring, noninvasive fashion. I now examine a variety of approaches that have been suggested for including the ethical treatment and moral status of animals as a pressing concern within pedagogy and within teaching and learning generally.

[4] Jan Narveson, "Animal Rights," *Canadian Journal of Philosophy* 7 (1977): 161–78.

[5] Richard. D. Ryder, "Experiments on Animals," in *Animals, Men, and Morals*, ed. Godlovitch, Godlovitch, and Harris (London: Gollancz, 1971), 41–82. See also Horsthemke, *Animal Rights Education*; and Horsthemke, *Moral Status*.

9.2 Environmental and Sustainability Education, Ecophilia, and Biophilia

For several decades, experts have cautioned against overexploitation of natural resources. In recent years, with the accumulation of facts about climate change, including global warming in particular, the quest for solutions has acquired a new and profound sense of urgency in education, as elsewhere. What is at stake here is nothing less than the survival of our planet in its present state, as a place that is inhabitable, and therefore also the survival of all species that inhabit it. At the very least, it is a matter of the quality and conditions of human and other-than-human lives, present and future. If this is correct, then one of the priorities of academic research—scientific, philosophical, educational, and other—may be how to arrest and possibly reverse the present decline.

Environmental education and sustainability education[6] have been at the forefront of pedagogical initiatives and have enjoyed a fair amount of publicity and popularity. The idea of Earth, as our only home, being under unprecedented threat has led to curricula, modules, and courses geared toward environmental awareness, ecological mindfulness, and the fostering of a sustainable lifestyle. It will come as no surprise that the popularity of these initiatives is due to their generally anthropocentric orientation. It is human beings who are at the center of the environment and the ecosphere and who stand to benefit from sustainable development and use of "natural resources," including nonhuman animals. Such a human-centered orientation is likely to be deeply problematic and ultimately unacceptable for anyone concerned for animals and the natural environment in and for themselves—that is, independently of human interests that are, more often than not, driven by the profit motive.

An additional problem is that environmental and sustainability education are more focused on collective/communal interests than on human and nonhuman individuals. Ecophilia and biophilia are characterized, similarly, by communalism and holism in society and education. Both are essentially to involve nature-friendliness or love of nature. The difference is that biophilia refers exclusively to organic, living nature, while ecophilia also covers non-organic, non-living nature that forms part of

[6] See, for example, Moacir Gadotti, "Education for Sustainability: A Critical Contribution to the Decade of Education for Sustainable Development," *Green Theory & Praxis: The Journal of Ecopedagogy* 4, no. 1 (2008): 15–64.

the ecosphere. The communalist, holistic approach inherent in both views places greater emphasis on the community, on the whole, and on harmonious relations than on individual natural entities such as animals. These orientations will hold little appeal for those who maintain that the way to worry about rare or endangered species is to think of their members as individuals whose lives and well-being are threatened. Nor will it persuade those who worry about the normative force of feelings such as love.

9.3 Humane Education and Theriophilia

Like environmental education, humane education has a fairly long and well-documented history. Extending back into the nineteenth century, its driving force—then and now—has been the conviction that (most) children's natural interest, attraction, and feelings toward animals constitute a fertile foundation for the transmission of knowledge and information about species-appropriate keeping and care of animals and promotion of empathy, sympathy, and enthusiasm for animal protection. The central idea at work here is that teaching children to treat animals with kindness and respect will encourage them to treat humanity in the same way, once they grow older and stronger. By the same token, cruelty toward animals in early life will lead to cruelty toward human beings in later life. If there is evidence to show that children who maltreat nonhuman animals more often than not grow up to be abusive adults (even serial killers), then perhaps children who grow up caring for animals (i.e., in a climate of compassionate fellowship with them) will, more often than not, develop into caring, compassionate adults. Clearly, empirical evidence is considerably more difficult (if not impossible) to supply for this idea than for the case of abusive children who later become serial killers—for which there now exists an impressive body of evidence.[7] But perhaps the very least one can say is that exposing children to the needs and interests of nonhumans is very likely not only to benefit the animals in question but also to promote in children an attitude of empathy, compassion, responsibility, and respect—which is bound to make them better people.[8]

[7] See, for example, Andrew Linzey, ed., *The Link between Animal Abuse and Human Violence* (Brighton: Sussex Academic Press, 2009).

[8] See Horsthemke, "Non-Human Animals," 1413.

An initial problem with the standard defense of humane education is its ultimate focus on human benefits—that is, its latent anthropocentrism: treating animals with kindness and respect is likely to ensure that children will treat human beings in the same way, especially once they develop full-fledged moral agency. In a sense, then, kind and respectful treatment of animals is seen as instrumental for the (present and future) kind and respectful treatment of one's fellow humans. There are additional concerns, especially with the emphasis on kindness and theriophilia, or love for animals. The problem with such appeals is that *how one feels* about what one does is logically distinct from the moral assessment of *what one does*. Not only is it possible to kill with kindness and to smother with empathy and compassion, but love for animals also is quite compatible with rearing and killing them for food, with using them in biomedical and other research, and with fishing and hunting.[9] Thus, appeals to kindness and injunctions against cruelty constitute a basis neither for morality nor for moral education. Most importantly, embodying (as they do) reference to agents' mental states, motives, or intentions, they fail to account for our positive and negative duties. At most, such appeals and injunctions characterize a virtue ethic's identification of "rightness" and "wrongness," respectively, without these necessarily translating into action. This is not to cast doubt on the ethical significance of feelings. It is questionable, however (and this is where I disagree with virtue ethicists and care ethicists), that empathic and sympathetic imagination, caring, and so on can provide a sound moral *basis*—that is, a guarantee or consistent prescription—for right *action*. As I will argue later, empathy and compassion might be the heart of a comprehensive "humane" movement, but rights are its backbone.

9.4 Philosophical Posthumanism

At the heart of philosophical posthumanism is the endeavor to venture beyond humanism, an examination of the philosophical and educational implications of expanding the sphere of moral concern, and the acknowledgment of subjectivities beyond the human species. In other words,

[9] For stark examples, see Kathy Rudy, *Loving Animals: Toward a New Animal Advocacy* (Minneapolis: University of Minnesota Press, 2011), 75, 77, 157, 170–71, 194–95; Roger Scruton, *Animal Rights and Wrongs*, 3rd ed. (London: Metro/Continuum, 2000), 100, 115, 153, 197; Gadotti, "Education for Sustainability," 40, 41.

philosophical posthumanism and its affiliates (such as transhumanism, anti-humanism, and ahumanism) target the anthropocentrism that has been so pervasive in, for example, philosophy, ethics, and education. This has led theorists such as Patricia MacCormack[10] to postulate a principle of radical non-interference or non-intervention regarding other animals, in order to end the cycle of oppression and exploitation.

However, the problem with a radical "hands-off" approach to animal pedagogies is that children are unlikely to learn anything about animals. While this arguably would curb abusive practices in teaching and learning, it also would be highly likely to erode any basis for concern, empathy, and compassion in future generations. MacCormack may be prepared to bite this particular bullet. After all, she rejects any pedagogical desire to know "the animal" in favor of the unmaking of "man."[11] An additional problem is the implicit denial that pedagogical and educational "interventions" are often mutually beneficial. They benefit not only humans (in learning about animals) but often also other-than-human individuals who are presently alive. Perhaps the most plausible understanding of *post*humanism involves imagining a world in the absence of its most disruptive, aggressive, and destructive species: "the world does not need humans, and is likely to thrive ignorant of human existence."[12] This tallies with Steven Best's contentions that whereas earthworms, dung beetles, butterflies, and bees are important to the integrity and diversity of nature, "human beings could be removed from earth ecosystems with positive effect,"[13] and that *Homo sapiens* is "the one species the earth could well do without."[14] But is the world also likely to thrive ignorant of human education? As long as *Homo sapiens* is around, education arguably will

[10] Patricia MacCormack, "Ethics, Animality and Ahuman Theory" (presentation at the University of California–Santa Barbara, November 5, 2013); MacCormack, "Gracious Pedagogy," *Journal of Curriculum and Pedagogy* 10, no. 1 (2013): 13–17. See also Karin Gunnarsson Dinker and Helena Pedersen, "Critical Animal Pedagogies: Re-Learning Our Relations with Animal Others," in *The Palgrave International Handbook of Alternative Education*, ed. Helen E. Lees and Nel Noddings (Basingstoke: Palgrave Macmillan, 2016), 415–30.

[11] MacCormack, "Gracious Pedagogy", 13.

[12] Helena Pedersen and Barbara Pini, "Educational Epistemologies and Methods in a More-than-Human World," *Educational Philosophy and Theory* 49, no. 11 (2017): 1053.

[13] Steven Best, *The Politics of Total Liberation: Revolution for the twenty-first Century* (New York: Palgrave Macmillan, 2014), 119.

[14] Best, *Politics*, 166.

continue to be a significant component of the struggle against disruption, aggression, and destruction.

9.5 Critical Pedagogy and Ecopedagogy

In their quest for holistic, ecological, and animal-centered models of pedagogy, philosophers and educational theorists have built on Paulo Freire's critical pedagogy and have defended a range of approaches, such as revolutionary critical pedagogy,[15] critical animal pedagogy,[16] and ecopedagogy.[17] Revolutionary critical pedagogy is critical of the view that education and legislation are the only defensible tools in the struggle against oppression and for advancing liberatory goals. Revolutionary critical pedagogy, as defended by Best, McLaren and Nocella, involves dialogue with radicals, militants, and dissidents and is inextricably connected with inspiration to act.

Like philosophical posthumanism, critical animal pedagogy involves taking "an epistemological and pedagogical step aside" and "unthinking," as it were, "our parasitic selves"—that is, "unthinking the human."[18] It constitutes an alternative approach to education "where students at all levels across the curriculum are invited to explore *both* a critical analysis *and* a radically transformative approach to animals and affect in education."[19] Karin Gunnarsson Dinker and Helena Pedersen note that MacCormack's "'hands-off' approach to animal pedagogies"[20] appears to

[15] Steven Best, Peter McLaren, and Anthony J. Nocella II, "Revolutionary Peacemaking: Using a Critical Pedagogy Approach for Peacemaking with 'Terrorists,'" *Journal for Critical Education Policy Studies* 5, no. 2 (2007): 494–528.

[16] Dinker and Pedersen, "Critical Animal Pedagogies."

[17] See Richard Kahn, "From Education for Sustainable Development to Ecopedagogy: Sustaining Capitalism or Sustaining Life?," *Green Theory & Praxis: The Journal of Ecopedagogy* 4, no. 1 (2008): 1–14; Kahn, "Towards an Animal Standpoint: Vegan Education and the Epistemology of Ignorance," in *Epistemologies of Ignorance in Education*, ed. Erik Jalewski and Nathalia Jaramillo (Charlotte, NC: Information Age Publishing, 2011), 53–70.

[18] Dinker and Pedersen, "Critical Animal Pedagogies," 416–17.

[19] Dinker and Pedersen, "Critical Animal Pedagogies," 418.

[20] Dinker and Pedersen, "Critical Animal Pedagogies," 416.

"*a priori* rule out the possibility of encounters with animals or intersubjectivity in human–animal relationships."[21] It is puzzling, therefore, that they too should argue "for the *absence* … of animals in education" and that "unthinking the human … asks us to leave alone, to exercise non-intervention."[22] As a purely liberationist project, this makes sense, as it does in the Freirean understanding of critical pedagogy leading to the educator's or researcher's own liberation,[23] but it is unclear what then remains of education, of learning with, from, and for animals.

Ecopedagogy is a critical response to environmental and sustainability education,[24] targeting both its romanticized and idealized domain of tendentious and biased representations of "wild nature" and its proximity to (essentially capitalist) political and corporate interests. Unfortunately, Kahn's references to "diverse cultures' different epistemological relationships to nature," "forms of environmental knowledge that accord only with Western science and citizenship values," "cultural ways of knowing," "traditional ecological knowledge," and "multiculturally-relevant knowledge"[25] take for granted what ought to be established by way of careful argument. Thus, the ideas of diverse epistemologies and/or epistemological diversity, of indigenous, local, or traditional knowledge, and of culture-specific ways of knowing are simply *assumed* to make sense. A central problem is the lack of clarity about the meaning or understanding of "epistemology" and "knowledge." A distinction is commonly made between "skills" and "knowledge"—which suggests, in the absence of any definition, that at least part of the understanding problematized here concerns factual (propositional or theoretical) knowledge. Insofar as "knowledge" in this sense includes reference to "truth," this invites the perception of the latter also being "indigenous," "traditional," or "culturally determined." A problem that would need to be addressed is that of relativism (about both knowledge and truth) and the implications of epistemological relativism. A further question concerns the basis,

[21] Dinker and Pedersen, "Critical Animal Pedagogies," 417.

[22] Dinker and Pedersen, "Critical Animal Pedagogies," 426.

[23] See Anthony J. Nocella II, "Unmasking the Animal Liberation Front Using Critical Pedagogy: Seeing the ALF for Who They Really Are," *Animal Liberation Philosophy and Policy Journal* 5, no. 1 (2007): 55–64.

[24] See Kahn, "From Education"; Gadotti, "Education for Sustainability."

[25] Kahn, "From Education," 10.

if there is one, for distinguishing between knowledge and superstition within indigenous or traditional belief systems. Kahn thinks that a "critical ecoliteracy as deployed by ecopedagogy would ultimately attempt to mobilize diverse peoples to engage with culturally appropriate forms of ecological politics and to engage in movement building on these issues through critical dialogue and constructive alliances."[26] "In this way," he adds, "people and groups can then recognize *their own ecopedagogy* as a form of ethical epiphany that serves to individuate the state of planetary ecology … within a given historical time period."[27] Would the notion of peoples' and groups' "own ecopedagogy" not relativize ecopedagogy beyond meaningfulness and usefulness? Surely, it is not just a matter of *understanding* "foundational cosmologies that work ideologically in ways that are less sustainable to life," "technologies that are less appropriate to the support of biological diversity and social flourishing across history," and "traditions and institutions that are less democratic and integral to the daily life experiences of the people and places such knowledge is meant to support,"[28] but a matter also of criticizing and judging them and bringing education to bear on their respective improvement or replacement.

9.6 Critical Animal Studies and Animal Standpoint Theory

Critical animal studies[29] emerged in higher education as a result of mainstream animal studies increasingly being perceived as having sold out, as having been domesticated, colonized by organizations and individuals without any real or practical commitment to animal rights and liberation or to a vegan and generally anti-speciesist lifestyle. However, the novel domain of critical animal studies is not without problems. As Best himself acknowledges, it is—like mainstream animal studies—under threat of co-option, appropriation, and domestication by academics and "university

[26] Kahn, "From Education," 11.

[27] Kahn, "From Education," 11 (emphasis added).

[28] Kahn, "From Education," 10.

[29] Steven Best, "The Rise (and Fall) of Critical Animal Studies," *Liberazioni*, n.d., 1–35, accessed March 22, 2018, http://www.liberazioni.org/articoli/BestS-TheRise%28and%20Fall%29ofCriticalAnimalStudies.pdf; Helena Pedersen, *Animals in Schools: Processes and Strategies in Human–Animal Education* (West Lafayette, IN: Purdue University Press, 2010).

bureaucrats" and of absorption "into personal agendas and conservative paradigms and programs."[30] Best's blanket dismissal, in this regard, of individualism[31] is both puzzling and unfortunate for someone who presumably cares for animals not as groups or species but as individuals.[32] Animals are "beings who live and suffer now."[33] If "speciesism strips animals of all intrinsic value to reduce them to instrumental value, to mere tools and objects whose cosmic purpose is to satisfy human purposes,"[34] then why lambaste individualism—which acknowledges and prizes this intrinsic value? The fact that animals, too, are conscious individuals, subjects of a life that matters to them and that can be better or worse *for them*—individuals who can be benefited and harmed, individuals with needs and interests—best accounts for the initial intuitions behind "the animal standpoint."

Animal standpoint theory stands alongside other standpoint theories, in their commitment to represent the perspectives and viewpoints of those who historically have been marginalized, oppressed, and exploited. The Cartesian skeptic's and behaviorist's objection that animals do not have a standpoint is swiftly dispensed with. An impressive, steadily expanding array of empirical evidence is available for establishing that animals have rich conscious and conative lives, experiences, and abilities that are often on a par with those of human beings—and sometimes indeed transcend these. The morally significant aspects of animals' standpoints are not inaccessible. They concern the need and ability to live in peace, without being subjected to physical and psychological discomfort, stress, distress, and trauma and without their lives being prematurely terminated. These needs and preferences are essential features of animals' standpoints that are readily determinable and that can be taught to children and adult learners, can be engaged with, understood, and valued.

There are other concerns, however, such as Best's endorsement of Nietzschean perspectivalism, the view that perception, cognition and

[30] Best, "The Rise (and Fall)," 29.

[31] Best, "The Rise (and Fall)," 32–33n5.

[32] See also Best, "The Rise (and Fall)," 7, 11, 27; Best, McLaren, and Nocella, "Revolutionary Peacemaking," 515.

[33] Steven Best, "The Rise of Critical Animal Studies: Putting Theory into Action and Animal Liberation into Higher Education," *Journal for Critical Animal Studies* 7, no. 1 (2009): 22.

[34] Best, "The Rise (and Fall)," 7.

corresponding judgements always emanate from particular perspectives, which amounts to little more than epistemological and moral relativism. His rejection of "absolutist truths" and "universal values"[35] is problematic for several reasons. For one thing, the dismissal of objective, absolutist knowledge and "truths" is contradicted by Best's own knowledge and truth claims. If they are also perspectival and subjective, with no objective and general explanatory purchase, it would be difficult to understand why he bothers advancing them in the first place. Moreover, the abjuration of universal values is glaringly at odds with Best's anti-speciesism, revolutionary environmentalism, and "politics of total liberation"—all of which express universal values. To put it another way, it is certainly remarkable for someone who rejects absolutes—notably, moral absolutes—to evoke a "politics of *total* liberation."[36] It may be correct that our inquiry and knowledge acquisition are necessarily limited by certain assumptions and proclivities, but this does not mean that they cannot also be based on good reasons and outflank, or be better justified than, competing arguments.

A further problem is the hermeneutical and conceptual overloading of "the animal standpoint." When Best refers to animals' "righteous anger and profane diatribes" against us (humans),[37] these emotions inform his own anti-speciesist standpoint, quite understandably. Yet the relevant "animal standpoint" is at best one of incomprehension and bewilderment, of helplessness and perplexity in the face of nearly insurmountable odds. And when Best speaks of the "standpoint of the earth,"[38] there is a potential weakening of the relevant concepts here. Does (or could) the earth have a standpoint? It is difficult, if not impossible, to make sense of these kinds of claims in the absence of, or without reference to, conscious individuals who are the subjects of a life that is experienced by them. That there is *something that it is like to be* a particular animal, that there is a particular vantage point or standpoint from which life and the world are perceived and experienced—a perspective that is communicated in a

[35] Best, *Politics*, 52.

[36] Best, *Politics* (emphasis added).

[37] Best, *Politics*, 11.

[38] Best, *Politics*, 16, 104, 165.

variety of ways and informed by more or less complex needs, beliefs, preferences, desires, and emotions—also constitutes a suitable range of topics for education.

9.7 Vegan Education

Unlike Best, an outspoken non-pacifist who endorses "direct action" beyond "open rescue,"[39] Gary Francione favors a pacifist stance, with veganism as the baseline in the fight against animal abuse and exploitation[40]: we cannot consistently oppose the clubbing of seals, the testing of cosmetics and household products on animals, or fox hunting if we are not prepared at the same time to change our diet, to forgo the pleasure of ingesting and imbibing animal products. Any such campaign is arbitrary and inconsistent in the absence of an accompanying vegan lifestyle. Both Best and Francione advocate education. Yet whereas Francione is confident that education (coupled with active veganism) will bring about substantial changes, Best favors a critical pedagogy approach coupled with tactics far beyond mere education. Vegan education is not limited to opposing the use of animals for food but targets all abuse by humans of nonhumans, which it seeks to remedy by nonviolent, pacifist education.

Surely, unqualified "direct action" is as difficult to justify as absolute pacifism. The former tends to condone not only life-threatening violence (which is inconsistent with an essentially life-preserving stance) but also preemptive and excessive violence. The latter, which would reject even open rescue and euthanasia of non-viable individuals, fails to accommodate individuals who are suffering and who are threatened with death *now*, *presently*, and who would be benefited through direct action–type interventions and processes. Equally surely, a defensible position must lie somewhere between qualified direct action and qualified pacifism. When it comes to education, Francione (although he may lack a coherent theory of education and a more sophisticated grasp of educational practice) is

[39] See Best, *Politics*; Best, McLaren, and Nocella, "Revolutionary Peacemaking."

[40] Gary Francione, "On Violence," *Animal Rights: The Abolitionist Approach* (blog), February 14, 2010, accessed March 29, 2018, http://www.abolitionistapproach.com/on-violence/; Francione, "The Problem with Single-Issue Campaigns and Why Veganism Must Be the Baseline," *Animal Rights: The Abolitionist Approach* (blog), March 15, 2011, accessed May 31, 2018, http://www.abolitionistapproach.com/problem-with-single-issue-campaigns-and-why-veganism-must-be-the-baseline/.

arguably in a stronger position than direct-action apologists. It is both easier and more consistent to educate for nonviolence and peaceful negotiation, to develop mindfulness and dispositions to care and nurture, than it is to educate for a circumspect, measured, nuanced use of violence—for how do we teach that violence is undesirable but occasionally necessary to bring about desirable ends? However, whether Francione's favored pedagogical approach is *transformatively effective*, notwithstanding the comparative ease with which it can be accomplished, is questionable. Further criticisms concern what is perceived to be vegan education's elitist and prescriptive stance that is out of touch with the plight of the poor in the Global South or those in remote arctic regions (who are highly unlikely to survive, let alone flourish, on a vegan diet), as well as its logical and practical separation from other radical and liberatory social movements.

9.8 Morality, Rights, and the Promise of Animal Rights Education

Each of the approaches examined thus far has undeniable strengths as well as considerable weaknesses. As a viable alternative I now propose an account of animal rights education.[41] Animal rights education is clearly contingent on animals having (moral) rights—or, in principle, being ascribable such rights. The promise of animal rights education, in turn, depends on the possibility of animal rights education. In a fairly trivial sense, talk of animal rights conveys the idea that animals matter morally, that humans have certain responsibilities in this regard, that animals should not be made to suffer gratuitously, and so on. In a more controversial and philosophically challenging sense, "animal rights" leads into deep deontological territory. If animal rights education is philosophically and educationally meaningful, then it arguably is a powerful tool pedagogically and also for effecting lasting pro-animal changes.

Rights can be taken to exist not only in law; they are correctly seen also as binding moral precepts that do not depend on legal institution for their validity. An interest model of rights (as opposed to a choice conception) advocates protection of all those who have interests and a welfare and guarantees the pursuit of nonthreatening interests, by means

[41] See Horsthemke, *Animal Rights Education*.

of (equal) rights. At the level of basic moral rights, all right-holders (human and nonhuman) have the same rights—for example, subsistence rights, liberty rights, and welfare rights. Non-basic moral rights are not necessarily shared by moral agents and moral recipients—indeed, not even by all agents. Thus, the right to vote is not a right that applies to very young and relevantly incapacitated humans or to other-than-human animals, and the right to pregnancy termination on demand is not a right that applies to men. Although rights confer prohibitions and restrictions, with regard to agency, they are not absolute. It is permissible to override them in situations where right-holders cannot reasonably be called "non-threatening" or "innocent." On the other hand, the obligation to provide assistance and the duties of beneficence obtain only if such assistance and beneficence do not themselves involve violation of rights.[42]

Not only is rights language meta-ethically and normatively defensible and compatible with other moral considerations such as sympathy, empathy, care, and appeals to kindness, but nonhuman animals can also plausibly be seen to be among the sorts of beings to whom rights can be meaningfully ascribed. One of the goals of human rights education is awareness of one's own rights and development of the ability to take an active stand for one's rights as well as for the rights of others. Animal rights education is concerned especially with the latter ability—insofar as animals are seen not as agents but as recipients of pedagogical and ethical agency. It involves an approach to education with rights as the backbone of concern and agency for other-than-human animals and compassion as its heart.

Bibliography

Best, Steven. 2009. The rise of critical animal studies: Putting theory into action and animal liberation into higher education. *Journal for Critical Animal Studies* 7 (1): 9–52.

Best, Steven. 2014. *The politics of Total liberation: Revolution for the 21st century.* New York: Palgrave Macmillan.

Best, Steven. n.d. The rise of critical animal studies. *Liberazioni*1–35. Accessed March 22, 2018. http://www.liberazioni.org/articoli/BestS-TheRise%28and%20Fall%29ofCriticalAnimalStudies.pdf.

[42] See Horsthemke, *Moral Status*, part 2.

Best, Steven, Peter McLaren, and Anthony J. Nocella II. "Revolutionary Peacemaking: Using a Critical Pedagogy Approach for Peacemaking with 'Terrorists.'" *Journal for Critical Education* Policy Studies 5, no. 2 (2007): 494–528.

Dinker, Karin Gunnarsson, and Helena Pedersen. 2016. Critical animal pedagogies: Re-learning our relations with animal others. In *The Palgrave international handbook of alternative education*, ed. Helen E. Lees and Nel Noddings, 415–430. Basingstoke: Palgrave Macmillan.

Francione, Gary. "On Violence." *Animal Rights: The Abolitionist Approach* (blog), February 14, 2010. Accessed March 29, 2018. http://www.abolitionistapproach.com/on-violence/.

Francione, Gary. "The Problem with Single-Issue Campaigns and Why Veganism Must Be the Baseline." *Animal Rights: The Abolitionist Approach* (blog), March 15, 2011. Accessed May 31, 2018. http://www.abolitionistapproach.com/problem-with-single-issue-campaigns-and-why-veganism-must-be-the-baseline/.

Gadotti, Moacir. 2008. Education for sustainability: A critical contribution to the decade of education for sustainable development. *Green Theory & Praxis: The Journal of Ecopedagogy* 4 (1): 15–64.

Godlovitch, Stanley, Roslind Godlovitch, and John. Harris, eds. Animals, men, and morals. London: Gollancz, 1971.

Horsthemke, Kai. 2010. *The moral status and rights of animals*. Johannesburg, South Africa: Porcupine Press.

Horsthemke, Kai. 2015. *Animals and African ethics*. Basingstoke: Palgrave Macmillan.

Horsthemke, Kai. 2018a. *Animal rights education*. Cham, Switzerland: Palgrave Macmillan.

Horsthemke, Kai. 2018b. Non-Human Animals. In *International handbook of philosophy of education*, ed. Paul Smeyers, 1401–1414. Dordrecht: Springer.

Kahn, Richard. 2008. From education for sustainable development to Ecopedagogy: Sustaining capitalism or sustaining life? *Green Theory & Praxis: The Journal of Ecopedagogy* 4 (1): 1–14.

Kahn, Richard. 2011. Towards an animal standpoint: Vegan education and the epistemology of ignorance. In *Epistemologies of ignorance in education*, ed. Erik Jalewski and Nathalia Jaramillo, 53–70. Charlotte, NC: Information Age Publishing.

Linzey, Andrew, ed. 2009. *The link between animal abuse and human violence*. Brighton: Sussex Academic Press.

MacCormack, Patricia. 2013a. Ethics, animality and Ahuman theory. *Presentation at the University of California–Santa Barbara, November* 5.

MacCormack, Patricia. 2013b. Gracious pedagogy. *Journal of Curriculum and Pedagogy* 10 (1): 13–17.

Narveson, Jan. 1977. Animal rights. *Canadian Journal of Philosophy* 7:161–178.
Nocella, Anthony J., II. 2007. Unmasking the animal liberation front using critical pedagogy: Seeing the ALF for who they really are. *Animal Liberation Philosophy and Policy Journal* 5 (1): 55–64.
Pedersen, Helena. *Animals in Schools: Processes and Strategies in Human–Animal Education*. West Lafayette, IN: Purdue University Press, 2010.
Pedersen, Helena, and Barbara Pini. 2017. Educational epistemologies and methods in a more-than-human world. *Educational Philosophy and Theory* 49 (11): 1051–1054.
Regan, Tom. 1983. *The case for animal rights*. London: Routledge & Kegan Paul.
Rudy, Kathy. 2011. *Loving animals: Toward a new animal advocacy*. Minneapolis: University of Minnesota Press.
Ryder, Richard D. 1971. Experiments on animals. In *Animals, men, and morals*, ed. Stanley Godlovitch, Roslind Godlovitch, and John Harris, 41–82. London: Gollancz.
Scruton, Roger. 2000. *Animal rights and wrongs*. 3rd ed. London: Metro/Continuum.
Singer, Peter. 1975. *Animal liberation*. New York: Random House.

PART II

Making Progress

CHAPTER 10

Concerning Critical Humane Education

Kay Peggs and Barry Smart

10.1 Introduction: Anthropocentrism—Pedagogy, Practice, and Critique

We live in the epoch of the humans—that is, the Anthropocene, the current geological epoch that is defined by human-induced ecological crisis.[1] Humans have tended to think of themselves—and to operate—as though they are "radically separated from the world around them and can therefore regard it as a realm that provides goods and services for human

[1] Ariane Conty, "The Politics of Nature: New Materialist Responses to the Anthropocene," *Theory, Culture & Society* 35, nos. 7–8 (2018): 73–96.

K. Peggs (✉)
School of Law, Social and Behavioural Sciences, Kingston University, London, UK
e-mail: k.r.peggs@kingston.ac.uk

B. Smart
Faculty of Humanities and Social Sciences, University of Portsmouth, Portsmouth, UK
e-mail: Barry.Smart@port.ac.uk

C. Linzey and A. Linzey (eds.), *The Palgrave Companion to Humane Education*, The Palgrave Macmillan Animal Ethics Series,
https://doi.org/10.1007/978-3-032-19795-5_10

benefit."[2] Human social, cultural, economic, political, and scientific institutions are centered on the pursuit of human interests and well-being and the exploitation and use of other species and the environment as resources. But the Anthropocene is the symptom; the cause is the anthropocentric worldview.[3] Anthropocentrism is hegemonic. It privileges the human species above all others, and it is the foremost paradigm of human thought in Western culture.[4] Its consequences are violence against and oppression of nonhumans[5] and the catastrophe of climate change and global heating, as well as other significant transformations of the planet that have led to the mass extinction of species.[6] One possible urgent address of this state of affairs lies in education, which could effect changes in attitudes and behavior to help us live well with other species and help alleviate climate catastrophe.[7] But education as an institution and science continues to propagate anthropocentric values and, in turn, associated assumptions, as we have argued:

> The pedagogical site all too frequently legitimizes the oppression of nonhuman animals in society. It is one of the significant sites where the ethically subordinate being-as-resource or property status of the nonhuman animal is constituted, displayed, exploited, and scientifically endorsed.[8]

[2] Clive Hamilton, *Requiem for a Species: Why We Resist the Truth About Climate Change* (London: Earthscan, 2010), 54.

[3] Francesca Ferrando, "The Party of the Anthropocene: Post-Humanism, Environmentalism and the Post-anthropocentric Paradigm Shift," *Relations* 4, no. 2 (2016): 159–73.

[4] Neil Badmington, "Theorizing Posthumanism," *Cultural Critique* 53, no. 1 (2003): 10–27.

[5] Núria Almiron, "Beyond Anthropocentrism: Critical Animal Studies and the Political Economy of Communication," *The Political Economy of Communication* 4, no. 2 (2016): 54–72.

[6] Gerardo Ceballos, Paul. R. Ehrlich, and Rodolfo Dirzo, "Biological Annihilation via the Ongoing Sixth Mass Extinction Signalled by Vertebrate Population Losses and Declines," *PNAS* 114, no. 30 (2017): 6089–96, https://www.pnas.org/content/pnas/114/30/E6089.full.pdf.

[7] Cecilia Boakye, "Climate Change Education," *SAGE Open* 5, no. 4 (December 2015).

[8] Kay Peggs and Barry Smart, "Nonhuman Animal Suffering: Critical Pedagogy and Practical Animal Ethics," *Society and Animals* 25, no. 2 (2017): 183.

The pedagogical site has a default orientation that occludes the status of nonhuman animals as beings with lives and purposes of their own.[9] In this chapter we ask the question "Does a 'humane'[10] approach to education address this, or does it leave the instrumental relationship toward nonhuman animals largely undisturbed?"[11]

The idea of humane education assumes an approach that is kinder to all beings, both human and nonhuman. This chapter offers a critical viewpoint on the notion of humane education by proposing an approach that challenges the instrumental orientation of humans to nonhuman animals and eliminates animal suffering that is human-caused. To proceed we need to consider, and consider critically, what has been—and is—meant by being humane, by reflecting on the etymology of "humane," in respect of its genealogy and representation as an institutional and educational practice, and by exploring the links between the idea of "humane" and nonhuman animal "welfare." We move on to consider the purpose of education, giving examples from higher education, and the intention of critical pedagogy, which seeks to encourage students to become independent learners who challenge social norms and strive to transform and emancipate. We conclude that a critical humane education—effectively, in our terms, a "post-humane" education—should counter anthropocentric bias in favor of cultivating awareness of the often injurious effects of human–nonhuman animal entanglements, thereby offering a way forward for a positive transformation of our relationships with nonhuman animals.

10.2 Being Humane

By calling into question the idea that humane education necessarily benefits nonhuman animals, in this section we explore conceptualizations of "being humane." Humane education has a number of manifestations, including a focus on the humanities and a "resistance to scientific

[9] Helena Pedersen, "Schools, Speciesism, and Hidden Curricula: The Role of Critical Pedagogy for Humane Education Futures," *Journal of Futures Studies* 8, no. 4 (2004): 1–14.

[10] In line with our critical approach to the notion of "humane," we have enclosed the term in quotation marks on this first use in the chapter. For subsequent uses, quotation marks should be assumed. We have frequently omitted them for the sake of presentation and ease of reading.

[11] See Peggs and Smart, "Nonhuman Animal Suffering," 188.

method."[12] For Roger Scruton it involves the pursuit of understanding "ourselves and others," but where "others" is anthropocentrically limited to human animals, and the understanding it promotes "is the origin of *our* moral sense."[13] In his considerations on the question of phenomenology, the German philosopher Martin Heidegger observed that "there is no such thing as *the one* phenomenology,"[14] and it is evident that there is no such thing as *the one* humane education. Perhaps more usually, humane education is associated with "teaching compassion and caring for all living beings,"[15] whether they are human or nonhuman.

Despite this, the similarity between "humane" and "human" suggests an inherent anthropocentrism in the meaning of the former. Although not concerned with anthropocentrism, Raymond Williams considered this similarity in his examination of the history of the word "humane." In *Keywords*, Williams notes,

> It is necessary first to understand the distinction between *human* and *humane*, which only became settled in its modern form from eC18. Before this *humane* was the normal spelling for the main range of meanings which can be summarized as the characteristic or distinct elements of ... the *human* species ... Early uses of *humane* referred to *human nature*, *human language*, *human reason* and so on, but there was also from eC16 a use of *humane* to mean kind, gentle, courteous, sympathetic. After eC18 the

[12] For example, see Roger Scruton, "Humane Education," *The American Scholar* 49, no. 4 (1980): 497.

[13] Scruton, "Humane Education," 498 (emphasis added). A potential alternative to the limitation of moral sense to human animals is proposed by Barbara Davy, who suggests drawing on and developing Emmanuel Levinas's notion of the primacy of an "ethical responsibility towards the other" (Emmanuel Levinas and Richard Kearney, "Dialogue with Emmanuel Levinas," in *Face to Face with Levinas*, ed. Richard A. Cohen [New York: State University of New York, 1986], 29), beyond the proximity of face-to-face relations between humans to encompass nonhuman animals. What is necessary, Davy contends, for "interspecies ethics is not a transcendence of animality or nature, but a transcendence of anthropocentrism." Barbara Davy, "An Other Face of Ethics in Levinas," *Ethics & the Environment* 12, no. 1 (2007): 46.

[14] Martin Heidegger, *The Basic Problems of Phenomenology*, trans., introd., and lexicon by Albert Hofstadter (Bloomington: Indiana University Press, 1982), 328.

[15] Rose Arbour, Tania Signal, and Nicola Taylor, "Teaching Kindness: The Promise of Humane Education," *Society & Animals* 17, no. 2 (2009): 136.

> old spelling was specialized to the now distinct word *humane*, in this latter range of senses, while *human* became standard for the most general uses.[16]

This focus on kindness, gentleness, and empathy can be found in institutional manifestations of the use of the word—for example, as it is employed by philanthropic organizations such as the Royal Humane Society in the United Kingdom, which has retained it focus (since its inception in 1774 as the Society for the Recovery of Persons Apparently Drowned) on "the saving of human life and for the restoration of life by resuscitation."[17] What comes across here is the close articulation of *humane* with *human*, exemplifying the continuing persistence of the conflation identified by Williams in the early genealogical roots of the term. In contrast, the Humane Society of the US (HSUS), founded in 1954, has, since its foundation, focused on nonhumans, with the ambition of eliminating "all animal cruelty throughout the world."[18] This global ambition led, in 1991, to HSUS founding its international division, Humane Society International, which seeks to "protect all animals" across the globe.[19] The emphasis on nonhuman animals shifts the humane gaze toward nonhuman species who are often occluded and whose position as "brute" and purely "for human use" has a long—if rather more nuanced than is often appreciated—history.[20] The confluence of a concern for humans and nonhumans was voiced decades before the launch of HSUS by Albert Schweitzer. In 1919 at the University of Uppsala, Schweitzer

[16] Raymond Williams, *Keywords: A Vocabulary of Culture and Society*, rev. ed. (New York: Oxford University Press, 1983), 148 (boldface emphasis in original, changed to italics here).

[17] Royal Humane Society, home page, https://www.royalhumanesociety.org.uk. Accessed August 2, 2019. The Royal Humane Society was founded in London in 1774 by two doctors, William Hawes (1736–1808) and Thomas Cogan (1736–1818). Hawes and Cogan were concerned about the number of people mistakenly assumed to be dead and subsequently buried alive, and they sought to promote and support the use of resuscitation techniques. See Steven B. Harris, "The Society for the Recovery of Persons Apparently Dead," *Alcor Life Extension Foundation*, 1990, https://alcor.org/Library/html/PersonsApparentlyDead.htm.

[18] Humane Society of the United States, "Our Mission," Accessed July 30,2019. https://www.humanesociety.org/our-mission.

[19] Humane Society International, "Protect All Animals," Accessed July 31, 2019, https://www.hsi.org/.

[20] Rod Preece, "Thoughts Out of Season on the History of Animal Ethics," *Society & Animals* 15, no. 4 (2007): 365–78.

gave his first detailed public statement on his drive for a "Reverence for Life"[21] In 1969 this would become the title for a collection of Schweitzer's writings in which he stressed the need for "love for all creatures, reverence for all being, compassion with all life, however dissimilar to our own,"[22] and insisted, "That is the beginning and foundation of morality,"[23] which is much more inclusive than Scruton's subsequent anthropocentric notion of morality's underpinning, as described previously.[24] It was Schweitzer's philosophy that opened the way for animal welfare associations such as HSUS.[25] Additionally, the Albert Schweitzer Foundation was established in 1981 in his name and supported, among other projects, the development of humane education.[26] However, this connection between "animal welfare" and being humane points to the problematical connotations associated with being humane when the idea is applied to nonhuman animals.

Animal welfare is the dominant paradigm used for the consideration of human moral obligations to nonhuman animals.[27] The welfare approach accepts that (many) nonhuman animals suffer but suggests that they suffer less so than humans, and consequently, the anthropocentric hierarchy prioritizes human interests over those of nonhuman animal subjects.[28] This leads to the view that "unnecessary suffering" should not be inflicted on nonhuman animals, with the disclaimer that "animals do not have an interest in continuing to live and their interests in not suffering have lesser value than do competing human interests."[29] In acceding that humans can inflict suffering that is "necessary," the animal welfare

[21] George Marshall and David Polling, *Albert Schweitzer: A Biography* (London and New York: John Hopkins Press). 180.

[22] Albert Schweitzer, *Reverence for Life: Sermons 1900–1919*, trans. Reginald. H. Fuller (New York: Irvington, 1969), 116.

[23] Schweitzer, *Reverence for Life*, 116.

[24] Scruton, "Humane Education."

[25] Ann Cottrell Free, *Animals, Nature and Albert Schweitzer* (Washington, DC: Flying Fox Press, 1982), 55.

[26] Muskan Khurana "The Wilderness Live " Albert Schweitzer Animal Welfare Fund (rjyrosesaksena.wixsite.com) 2021. Accessed July 30, 2021.

[27] Gary Francione, "Animal Welfare and the Moral Value of Nonhuman Animals," *Law, Culture and the Humanities* 6, no. 1 (2010): 24–36.

[28] Peggs and Smart, "Nonhuman Animal Suffering."

[29] Francione, "Animal Welfare," 27.

approach accepts that humans can "use" nonhuman animals for human benefit[30] and, most importantly in terms of animal welfare, that there can be a "'humane use' of animals by humans."[31] As Sue Donaldson and Will Kymlicka observe, this means that although the animal welfare approach accepts that "animal welfare matters," it "subordinates animal welfare to the interests of human beings."[32] This is indicated in educational institutions such as universities. For example, in the UK university setting, the subordination of nonhuman animal welfare to human interests is evident in the ways in which "Animal Technologists" (ATs) participate in what is called "humane animal experimentation" by implementing two of the three key guiding principles of the "3Rs."[33] Beth Greenhough and Emma Roe summarize the work of ATs as follows:

> They are involved in the breeding and preparation of animals for use in research and are sometimes called upon to carry out licensed experimental procedures such as tail injections or gavaging (administration of drug or similar directly into the stomach through a tube inserted down the throat); they receive animals back into their care following operations and/or their participation in experiments and they cull animals who are no longer needed.[34]

Greenhough and Roe report that one of the senior ATs summarizes the technician's role as follows:

> Everything we do is governed by the Home Office ... all the experimental work being done is under a licence. And that licence has limits ... what we get trained for, is to look for these signs and if [we] see it's exceeding the

[30] Kay Peggs, *Animals and Sociology* (Basingstoke: Palgrave Macmillan, 2012); Peggs and Smart, "Nonhuman Animal Suffering."

[31] Sue Donaldson and Will Kymlicka, *Zoopolis: A Political Theory of Animal Rights* (Oxford: Oxford University Press, 2011), 3.

[32] Donaldson and Kymlicka, *Zoopolis*, 3.

[33] See Beth Greenhough and Emma Roe, "Exploring the Role of Animal Technologists in Implementing the 3Rs: An Ethnographic Investigation of the UK University Sector," *Science, Technology, & Human Values* 43, no. 4 (2018): 700.

[34] Greenhough and Roe, "Exploring the Role," 695.

licence, then report it quickly so that animal doesn't suffer. So that's an animal technician's job, to police the researchers, the scientists.[35]

Millions of nonhuman animals are used in biomedical experiments every year,[36] and William Russell and Rex Burch's 3Rs approach[37] is endorsed by biomedical scientists as a way of trying to "minimiz(e) the welfare costs to animals used in research."[38] The 3Rs seek to (1) replace (i.e., find alternatives to the use of nonhuman animals), (2) reduce (the numbers of nonhuman animals used), and (3) refine (i.e., keep nonhuman animal suffering to a minimum where it is "unavoidable"). Replacement is outside the expertise of ATs; their focus is on reduction and refinement.[39] ATs implement a constrained view of "humane" nonhuman animal experimentation. ATs concentrate on trying to lower the number of nonhuman animals involved and on trying to keep their suffering to a minimum. However, experimentation continues to take place.

As Gary Francione observes, while the animal welfare approach recognizes the importance of the welfare of nonhuman animals and thus might offer improvements to the living conditions of incarcerated nonhuman animals, it gives no consideration to the intrinsic value of their lives and fails to address the property status of nonhuman animals.[40] In consequence, and returning to Donaldson and Kymlicka's rejection of the welfare approach for its assumption that there can be a "humane use" of nonhuman animals,[41] we seek to offer a critique of the idea that humane education necessarily counters anthropocentric values. In doing so, we

[35] Mark, senior AT, interviewed on September 5, 2013, in Greenhough and Roe, "Exploring the Role," 706.

[36] Katy Taylor, Nicky Gordon, Gill Langley, and Wendy Higgins, "Estimates for Worldwide Laboratory Animal Use in 2005," *ATLA* 36 (2008): 327–42 http://www.buav.org/_lib/userfiles/files/Science_Reports/Estimates_For_Worldwide_Laboratory.pdf.

[37] William Russell and Rex Burch, *The Principles of Humane Experimental Technique* (London: Methuen, 1959).

[38] Colin Blakemore, Judy MacArthur Clarke, Timo Nevalainen, Mike Oberdorfer and Arthur Sussman "Implementing the 3Rs in Neuroscience Research: A Reasoned Approach," *Neuron* 75, no. 6 (2012): 948–50.

[39] Greenhough and Roe, "Exploring the Role," 700.

[40] Gary Francione, *Animals, Property, and the Law* (Philadelphia: Temple University Press, 1995).

[41] Donaldson and Kymlicka, *Zoopolis*.

consider an alternative to the default notion of humane, one that counters anthropocentric values.

10.3 Humane Education

Humane education has various applications and meanings. It also has a long history. In this section we do not seek to offer an overview of the history or of the variations in meanings and applications—that would be impossible in the space available. Rather, what we hope to do is draw out a few elements to explicate some of the values and principles of humane education as we see them. Of course, these are not exhaustive.

The first value of humane education centers on human kindness to other beings, including nonhuman animals. In 1889 George T. Angell founded the American Humane Education Society (AHES). Angell summarized the aims of AHES as follows:

> to carry unsectarian humane education gratuitously *outside the State of Massachusetts, throughout the country and the continent*, and by the employment of suitable agents to establish *Bands of Mercy* and *Humane Societies* wherever they are most needed.[42]

The Bands of Mercy movement has its roots in the United Kingdom. Dating from 1875, the Bands of Mercy were "organizations educating children in kindness to animals."[43] In 1882, George Angell and the Rev. Thomas Timmins introduced the organizations into America. Interest grew. By July 1897, Angell notes, the combined Massachusetts Society for the Prevention of Cruelty to Animals, AHES, and Bands of Mercy had printed eighty thousand copies of their pamphlet *Our Dumb Animals* and sent twenty thousand packages of humane education materials to the National Educational Association at Milwaukee.[44]

[42] George T. Angell, quoted in National Museum of Animals & Society, "American Humane Education Society," *Be Kind: A Visual History of Humane Education 1880–1945*, curated by Keri Cronin (National Museum of Animals & Society, 2013), https://bekindexhibit.org/exhibition/american-humane-education-society/ (emphasis in original).

[43] Brian Harrison, "Animals and the State in Nineteenth-Century England," *The English Historical Review* 88, no. 349 (1973): 801.

[44] George T. Angell, *Our Dumb Animals* 30, no. 3 (August 1897): 1. Our Dumb Animals 1897–10: Vol 30 Iss 5 (archive.org).

In the late nineteenth century, not only was American humane education a central objective of social movements concerned with nonhuman animal protection, but also its inclusion in organized public education was considered to be conducive to reductions in criminal activity, social unrest, and other problems.[45] Concern about this connection remains today, as more recent empirical evidence indicates the coexistence of the abuse of nonhuman animals and human interpersonal violence.[46] Clifton P. Flynn argues that "a childhood socialization that includes exposure to interpersonal violence is also associated with childhood animal cruelty."[47] In reflecting on this, we draw on Norbert Elias, who suggested that kindness to nonhuman animals is, in Keith Tester's paraphrase, "a product of humans' social life, of civilisation and human embarrassment at our own animality."[48] We return to Elias's suggestion of human discomfort with our own animality later, but in the meantime we center on his notion of the "civilizing process."[49] Humans have to be taught about and learn how to be social individuals. The "civilizing process" is the process by which the individual learns to be social. Elias notes that "no human beings come into the world civilized," and "'growing up'... is nothing other than the individual civilizing process to which each young person, as a result of the social civilizing process over many centuries, is automatically subjected from earliest childhood, to a greater or lesser degree and with greater or lesser success."[50] This resonates with Angell's concerns, since for Angell and many successive humane educators, humane education aims at a greater degree of focus on care and compassion for others in order to achieve greater success in the control of human violence against humans. In this way, a commitment to human rights and social justice is a valued aspect of humane education. For this reason, Angell and Francis

[45] National Museum of Animals & Society, "American Humane Education Society."

[46] For example, see Clifton P. Flynn, "Examining the Links between Animal Abuse and Human Violence," *Crime, Law and Social Change* 55, no. 5 (2011): 453–68.

[47] Flynn, "Examining the Links," 454–55.

[48] Keith Tester, *Animals in Society: The Humanity of Animal Rights* (London: Routledge, 1991), 60.

[49] Norbert Elias, *The Civilizing Process: Sociogenetic and Psychogenetic Investigations*, rev. ed., ed. Eric Dunning, Johan Goudsblom, and Stephen Mennell (Oxford: Blackwell, 2000; first published 1939).

[50] Elias, *The Civilizing Process*, xi.

Rowley held the view that "animal advocacy and civil rights activism were inseparable."[51]

Because humane education is viewed as an essential element of how individuals learn how to be social, learn compassion, and learn to care, the rolling out of humane education across educational establishments is an essential principle for humane educators. For this to be achieved, humane education must be taken seriously, be promoted, and be subsidized. This has been an essential element of humane education in the United States of America since the early twentieth century, which saw a national campaign to make it compulsory within the school curriculum.[52] However, notwithstanding the reported growth in the number of the branches of the Bands of Mercy, the significance and reach of humane education diminished from around the middle of the twentieth century. The campaign failed to institutionalize humane education within the curriculum, as Bernard Unti and Bill DeRosa note: "few of the initiatives launched by humane organizations gained the lasting attention of teacher-training institutions, and humane education certainly did not become a regular element of teacher preparation."[53] It did not help that a $100,000 donation to Columbia University in 1907, specifically designed to promote and develop humane education, was not employed for that purpose. The president of the university instead used the donation for a faculty post in social legislation. As Unti and DeRosa explain, "the funds disappeared into Columbia's general accounts and, with the exception of several historical studies, no progress toward the goal of the donor was realized."[54]

By the 1930s and the onset of economic depression, other social and economic priorities and initiatives rendered the more ambitious conception of humane education relatively marginal, and what endured was "the simple lesson of kindness to pets, carried into the schools by SPCA staff members and volunteers."[55] In this regard, newer organizations such as

[51] Janet M. Davis, *The Gospel of Kindness: Animal Welfare and the Making of Modern America* (Oxford: Oxford University Press, 2016), 100.

[52] Bernard Unti and Bill DeRosa, "Humane Education: Past, Present and Future," in *The State of the Animals II*, ed. Deborah J. Salem and Andrew. N. Rowan (Washington, DC: Humane Society Press, 2003), 27–48.

[53] Unti and DeRosa, "Humane Education," 31.

[54] Unti and DeRosa, "Humane Education," 31.

[55] Unti and DeRosa, "Humane Education," 32.

the Humane Education Coalition have provided grant funding and other support, including educational programs "for agencies that teach about human rights, animal protection, and environmental ethics."[56] The coalition has a "no fee" policy, which means that fees are not charged for the programs offered.[57] But such organizations can only do so much. Despite the drive to institutionalize humane education, the National Anti-Vivisection Society of America reports that as of 2019, "only seventeen states [had] laws requiring schools to adopt policies regarding humane education and of these, only nine require[d] education on the humane treatment of animals."[58]

The deficiency in institutionalizing humane education occurred in the context of what might be described as its growing antithesis—namely, the increasing institutionalization of nonhuman animal exploitation. Wide-ranging transformations in the character of modern everyday life, including the growth in population, towns, and cities, increasing specialization, and the growing distance between consumer-citizens and the sites and conditions under which consumer goods and services were being produced, occluded many of the forms of cruelty and abuse to which nonhuman animals were being subjected. In addition, forms of self-censorship and limitations imposed by educational institutions in regard to "newer... socially sanctioned" uses and practices to which nonhuman animals were being subjected meant, as Uni and DeRosa note,[59] that new institutionalized forms of exploitation, cruelty, and abuse—for example, in scientific research, in the industrialized agricultural mass production of food, and in other consumer goods, fashion, and fur industries—were not reflected on within humane education programs. Furthermore, the growth and development of science education, influenced by and predicated on the Cartesian assumption of the difference between human and nonhuman animals, served to standardize the science curricula in high

[56] Humane Education Coalition, "What Is Humane Education?," Accessed August 2, 2019, https://www.hecoalition.org/what-is-humane-education.html.

[57] Humane Education Coalition, "Who We Are," Accessed August 2, 2019, https://www.hecoalition.org/about.html.

[58] National Anti-Vivisection Society, "Laws That Recognize the Importance of Humane Education," August 2, 2019, https://www.navs.org/what-we-do/keep-you-informed/legal-arena/education/humane-education/#.XURGrvJKiM8.

[59] Unti and DeRosa, "Humane Education," 32.

schools and universities and "left little room for the empathy-building emphasis of nature-study and humane education approaches."[60]

In consequence, humane education in America is now largely limited to matters related to companion animals and animal care and control. This means that the first principle of human kindness to nonhuman animals is now reserved for nonhuman animals who are seen as closest to us, "domesticated" nonhuman animals from whom we gain something and who retain their property status. Indeed, David Nibert's amending of the term domestication to "domesecration" makes clearer the violent practice of domestication in its colonization of the lives and bodies of nonhuman animals.[61] Humane education often fails to challenge this instrumental orientation of humans to nonhuman animals.

Jacques Derrida reflected on the instrumental positioning of humans to nonhuman animals. Specifically, he considered critically "the way animals are treated: in industrial production, in slaughter, in consumption, in experimentation"[62] and the way they are subjected to "medico-industrial exploitation, overwhelming interventions... [in their] milieus and reproduction, genetic transplants, [and] cloning."[63] In *The Future of the Profession or the Unconditional University*, Derrida argues that "the modern university *should* be without condition"; it should be an institution that embodies the "*unconditional* freedom to question and to assert."[64] Humane education might strive to be part of this "university without condition" (by means of its concern for the civilizing of humans into human kindness to human and nonhuman animals, commitment to civil rights activism, and wish to roll out humane education across educational establishments), but the question of the property status of nonhuman animals and the related issues raised by Derrida are omitted from humane

[60] Unti and DeRosa, "Humane Education," 33.

[61] David Nibert, *Animal Oppression and Human Violence: Domesecration, Capitalism, and Global Conflict* (New York: Columbia University Press, 2013).

[62] Jacques Derrida and Elisabeth Roudinesco, *For What Tomorrow: A Dialogue*, trans. Jeff Fort (Stanford, CA: Stanford University Press, 2004), 64.

[63] Jacques Derrida, *The Animal That Therefore I Am* (New York: Fordham University Press, 2008), 80.

[64] Jacques Derrida, "The Future of the Profession or the Unconditional University (Thanks to the 'Humanities,' What Could Take Place Tomorrow)," trans. Peggy Kamuf, in *Deconstructing Derrida*, ed. Peter Pericles Trifonas and Michael A. Peters (New York: Palgrave Macmillan, 2005), 11 (emphasis in original).

education programs because schools are unwilling to accept issues into the curriculum "that may be considered age-inappropriate, inflammatory, or inimical to a community's values, traditions, or economic base."[65] In addition, as we have discussed, within universities nonhuman animals continue to be exploited in scientific teaching, research, and laboratory experimentation. An increasingly frequent response to the criticism directed toward industrialized agricultural processes of production of nonhuman animals for human consumption and the appropriation and "design" of nonhuman animals for use as tissue resources in scientific experimentation is to appeal to the introduction of what are presented as compensatory "humane" or "welfare" measures.[66] As Derrida observed, the tendency is for the corporations and institutions involved to do all they can to "dissimulate this cruelty or to hide it from themselves; in order to organise on a global scale the forgetting or misunderstanding of this violence."[67]

Humane education often fails to challenge the property status of nonhuman animals. We attempt to give consideration to a *critical* form of humane education that would counter anthropocentric bias in favor of cultivating awareness of the often injurious effects of human–nonhuman animal entanglements, thereby offering a way forward for a positive transformation of our relationships with nonhuman animals.

10.4 "Normal" Education and Critical Pedagogy

The anthropocentric conceptualization of, and use of, nonhuman animals remain largely unchallenged in what we refer to here as "normal education." Although the idea of "normal education" might arouse thoughts about long-established "normal schools" that offer teacher training to graduates,[68] we employ the term in a more critical way. In this regard, we draw on Thomas Kuhn's use of "normal" in relation to science.[69]

[65] Unti and DeRosa, "Humane Education," 34.

[66] Peggs and Smart, "Nonhuman Animal Suffering."

[67] Derrida, *The Animal*, 26.

[68] H. A. Brown, "The Normal-School Curriculum," *The Elementary School Journal* 20, no. 4 (1919): 276–84.

[69] Thomas Kuhn, *The Structure of Scientific Revolutions: Fiftieth Anniversary Edition* (Chicago: University of Chicago Press, 2012; first published 1962).

Kuhn explains that when he refers to "normal science," he means science as taking place within an established paradigm, within parameters, assumptions, and understandings that are not generally questioned or challenged. Normal science is "predicated on the assumption that the scientific community knows what the world is like."[70] In a similar vein, we argue that "formal" or "normal" education is based on the assumption that teachers know what the world is like and that this knowledge should be transmitted to those who do not know. But drawing on Plato's allegory of the cave, Richard White contends that education is not just about training and "filling people with information"; it is also about encouraging "enlightenment, orientation towards the good and what really matters."[71] What really matters to Paul Gibbs is compassion. Gibbs explores the place of compassion in higher education, focusing on how the university "might become a compassionate place to be"[72] and on the wider effects of the compassionate university, which can stretch well beyond the institution itself.

In our work to date, we have been cautious and critical about the way the notion of "being humane" is invoked, preferring instead to try to distance our analyses from concepts and ideas that betray residual anthropocentric assumptions, insofar as they signify in some respects a sense of humans as central, as the primary species. In respect of education, particularly higher education and the institution of the university and its use of nonhuman animals in scientific experimentation, we have drawn on Henry Giroux's exemplification of critical pedagogy as capable of fostering the development of "a language for thinking critically about how culture deploys power and how pedagogy as a moral and political practice enables students to focus on the suffering of others,"[73] and we proceeded to propose a significant recasting of the aim of critical pedagogy—namely, to cultivate a non-speciesist education and promote

[70] Kuhn, *Structure of Scientific Revolutions*, 5.

[71] Richard White, "Compassion in Philosophy and Education," in *The Pedagogy of Compassion at the Heart of Higher Education*, ed. Paul Gibbs (Cham: Springer, 2017), 19.

[72] Paul Gibbs, "Higher Education: A Compassion Business or Edifying Experience," in *The Pedagogy of Compassion at the Heart of Higher Education*, ed. Paul Gibbs (Cham: Springer, 2017), 1.

[73] Henry Giroux, *On Critical Pedagogy* (New York: Continuum International, 2011), 5.

a non-speciesist way of being-in-the-world.[74] In short, we advocated "a radical shift away from the 'humane use' approach [to nonhuman animals] to a non-anthropocentric alternative."[75]

Critical pedagogy invites educators to be more observant and analytical of the ideologies embedded in the curriculum.[76] But theorists who advocate critical pedagogy often overlook nonhuman animals when considering its role and methods; nonhuman animals remain largely invisible within critical pedagogical practice. As Helena Pedersen shows,[77] theories of education are immersed in what Robert McKay refers to in a different context as "compulsory humanity." McKay explains that "it is compulsory that we 'become' human, and this very becoming is a function of our renunciation of the animal."[78] In our view this resonates with Elias's suggestion that the civilizing process, even in the form of kindness to nonhuman animals, is associated with human discomfort with our own animality.[79] Rather than disturbing or dismantling speciesist boundaries, formal education usually reinforces them.[80]

However, we might expect more of critical pedagogy because, as Giroux makes clear it offers a way of challenging deeply embedded assumptions and the prospect of overturning oppressive social practices[81] Giroux, like many theorists of critical pedagogy, does not extend his thinking to nonhuman animals.[82] But Abraham DeLeon insists, "Critical educators interested in building a society based upon principles of

[74] Peggs and Smart, "Nonhuman Animal Suffering," 184.

[75] Peggs and Smart, "Nonhuman Animal Suffering," 186.

[76] For example, see Stanley Aronowitz and Henry Giroux, "Radical Education and Transformative Intellectuals," *CTheory* 9, no. 3 (1985): 62.

[77] Helen Pedersen, "Is 'the Posthuman' Educable? On the Convergence of Educational Philosophy, Animal Studies, and Posthumanist Theory," *Discourse: Studies in the Cultural Politics of Education* 31, no. 2 (2010): 237–50.

[78] Robert McKay, "'Identifying with the Animals': Language, Subjectivity, and the Animal Politics of Margaret Atwood's *Surfacing*," in *Figuring Animals: Essays on Animal Images in Art, Literature, Philosophy and Popular Culture*, ed. Mary Sanders Pollock and Catherine Rainwater (New York: Palgrave Macmillan, 2005), 218.

[79] For example, see Norbert Elias, "Problems of Involvement and Detachment," *The British Journal of Sociology* 7, no. 3 (1956): 226–52.

[80] Pedersen, "Is 'the Posthuman' Educable?"

[81] Giroux, *On Critical Pedagogy*, 172–174.

[82] Peggs and Smart, "Nonhuman Animal Suffering," 184.

social justice are compelled to look outside the human milieu to explore how oppression is enacted on those who seem to occupy the lowest rungs of our collective social world."[83] We agree. As we have argued, critical pedagogy offers "the educational and cultural preconditions for nurturing critical, self-reflexive, and knowledgeable citizens who have the capacity to exercise moral judgements and act in an ethically responsible manner towards all forms of life."[84] This is because critical pedagogy seeks to confront common-sense assumptions by providing the opportunity for students to acquire "skills and knowledge to expand their capacity both to question deep-seated assumptions and myths that legitimate the most archaic and disempowering social practices that structure every aspect of society and to take responsibility for intervening."[85] Nonhuman animals are subjected to anthropocentric assumptions and suffering through pedagogic practices in primary and secondary schools, colleges, and universities, and they can be—indeed, should be—assigned a prominent place within a critical pedagogy, accorded moral and legal status, and not regarded as property or treated as resources to serve human interests.[86] Because critical pedagogy aims to address and counter the exclusion of oppressed subjects, it is appropriate, and indeed politically and ethically necessary, to include nonhuman animals, the multiplicity of species subjected to human-induced forms of domination, oppression, suffering, and genocide.[87]

Increasing academic and public consideration of human–nonhuman animal entanglements has stimulated an "animal turn" in a number of academic disciplines,[88] including in approaches to education. It is

[83] Abraham DeLeon, "What's That Nonhuman Doing on Your Lunch Tray: Disciplinary Spaces, School Cafeterias, and Possibilities for Resistance," in *School Food Politics: The Complex Ecology of Hunger and Feeding in Schools Around the World*, ed. Sarah A. Robert and Marcus B. Weaver-Hightower (New York: Peter Lang, 2011), 196.

[84] Peggs and Smart, "Nonhuman Animal Suffering," 184.

[85] Giroux, *On Critical Pedagogy*, 172.

[86] Steven Best, "Legally Blind: The Case for Granting Animals Legal Rights." Animal Rights Online. (2003) Legally Blind: The Case For Granting Animals Legal Rights - Animal Writes Newsletters From 15 June 2003 Issue From Animal Rights Online (all-creatures.org); Steven Best, *The Politics of Total Liberation: Revolution for the 21st Century* (New York: Palgrave-Macmillan, 2014).

[87] Pedersen, "Schools, Speciesism," 4.

[88] Reingard Spannring, "Animals in Environmental Education Research," *Environmental Education Research* 23, no. 1 (2017): 63–74.

to this development that we turn in order to consider how a critical humane education could counter anthropocentric bias in favor of cultivating awareness of the often injurious effects of human–nonhuman animal entanglements, thereby offering a way forward for a positive transformation of our relationships with nonhuman animals.

10.5 (Critical) Posthuman Education

In this section, we explore how posthumanism provides the ground for a critical humane education or a "post-humane education" that challenges anthropocentric education. Posthuman education is, of course, only one among a number of potentially non-anthropocentric learning possibilities. Others include education that introduces a deep ecology and animal rights perspective to environmental education, to "reposition students as responsible planetary citizens,"[89] and vegan education, which promotes abstention from exploiting nonhuman animals or consuming nonhuman animal products and which will "eventually shift the paradigm away from animals as property and toward animals as persons."[90] We turn our attention to posthumanism because we seek to return to the problematic similarity between, and frequent implicit conflation of, "humane" and "human," which posthumanism challenges. Posthumanism of any sort "fundamentally questions who or what gets to *matter*."[91] Our allusion to "any sort" draws us back to Martin Heidegger's observation, noted previously—there is no such thing as *the one* posthumanism.[92] Nathan Snaza and John A. Weaver show that what constitutes posthumanism is far from settled, and they suggest that, broadly, posthumanism is a perspective that seeks to consider the question "What if the human doesn't

[89] Helen Kopnina and Mickey Gjerris, "Are Some Animals More Equal than Others? Animal Rights and Deep Ecology in Environmental Education," *Canadian Journal of Environmental Education* 20 (2015): 108.

[90] Gary Francione, "Vegan Education Made Easy—Part 1," *Animal Rights: The Abolitionist Approach* (blog), March 25, 2008, https://www.abolitionistapproach.com/vegan-education-made-easy-part-i/.

[91] Annouchka Bayley, "How Did 'We' Become Human in the First Place? Entanglements of Posthumanism and Critical Pedagogy for the Twenty-First Century," in *Posthumanism and Higher Education: Reimagining Pedagogy, Practice and Research*, ed. Carol A. Taylor and Annouchka Bayley (Houndsmills: Palgrave Macmillan, 2019), 360 (emphasis in original).

[92] Heidegger, *The Basic Problems of Phenomenology*, 328.

have to be the measure?"[93] Francis Fukuyama is concerned that developments in biotechnology might affect human centrality because they could "alter human nature and thereby move us into a 'posthuman' stage of history."[94] This is not the sort of thinking about posthumanism that we have in mind. In the context of our discussion here, we are inclined toward Andy Miah's suggestion that posthumanism "should be an attempt to understand what has been omitted from an anthropocentric worldview."[95] In this regard, Miah argues that "the animal rights movement [is] a pre-cursor to a posthuman philosophy, since it similarly is concerned with the circle of moral concern and the relevance of locating such concern outside of specieist [*sic*] boundaries."[96] This ways of conceiving of posthumanism seeks to examine human–nonhuman entanglements from a non-hierarchical, non-anthropocentric perspective, and this can be achieved through education.[97]

In order for a non-anthropocentric posthuman approach to be achieved in formal education, posthumanism must be applied intentionally in education theory.[98] As Helena Pedersen notes, "Western pedagogy is firmly rooted in a 'humanist' tradition, where the human subject is considered both the instrument and the end product of education."[99] Decentering humanity is no easy task. It meets resistance in some quarters. Moreover, most of us—including those of us who are educators—have been educated anthropocentrically. In order to engage with posthumanism, we must think imaginatively. Because posthumanism is a "visionary stance,"[100] we must be experimental and creative in our thinking and ask different and inventive questions that step outside of

[93] Nathan Snaza and John A. Weaver, *Posthumanism and Educational Research* (New York: Routledge, 2015), 3.

[94] Francis Fukuyama, *Our Posthuman Future: Consequences of the Biotechnology Revolution* (London: Profile Books, 2003), 7.

[95] Andy Miah, "A Critical History of Posthumanism," in *Medical Enhancement and Posthumanity*, ed. Bert Gordijn and Ruth Chadwick, the International Library of Ethics, Law and Technology, vol 2. (Dordrecht: Springer, 2008), 77.

[96] Miah, "A Critical History," 95n16.

[97] Allesandro Ferrante and Daniele Sartori, "From Anthropocentrism to Post-Humanism in the Educational Debate," *Relations* 4, no. 2 (2016): 177.

[98] Ferrante and Sartori, "From Anthropocentrism to Post-Humanism," 176.

[99] Pedersen, "Is 'the Posthuman' Educable?," 241.

[100] Miah, "A Critical History," 96.

the human.[101] Karin Murris asks, "How can we find other, more tacit ways of experiencing the world that also account for more-than-human experiences—what to look out for and what to avoid when teaching and researching with a posthumanist orientation?"[102] How indeed! Drawing on the work of Karen Barad,[103] Murris suggests that posthuman education should seek to shorn us of "bodily boundaries" by asking searching and hypothetical questions about multispecies entanglements, by resisting becoming caught up in humanist discourses, and by "becoming sensitive to the always political and ethical nature of teaching and research."[104] This means posthumanism not only should be taught to students but also should be an essential part of teacher training. In this regard, Caitlin Howlett posits the possibility of including "speculative fiction" in the training of teachers.[105] Speculative fiction is "the domain wherein are expressed scientific and technological aspirations, prospects and possibilities."[106] Drawing on the work of Jessica Langer,[107] Howlett explains that speculative fiction "situates the reader in a complex relationship with the Other."[108] Thus, it can help students and student teachers "engage with systems of othering in ways that unsettle the seemingly inevitability of the categories of identity through which education operates."[109] Accordingly, rather than learning *about* imagination, it is important that student teachers *engage in* imagining in order to upset their "sense of self," their "relationship to others," and their "place in the world."[110]

[101] Karin Murris, "Review Essay: Propositions for Posthuman Teaching and Research—A Diffractive Re-View of Three Books," *Critical Studies in Teaching and Learning* 5, no. 1 (2017): 104.

[102] Murris, "Review Essay," 105.

[103] Karen Barad, "Diffracting Diffraction: Cutting Together-Apart," *Parallax* 20, no. 3 (2014): 168–87.

[104] Murris, "Review Essay," 107–8.

[105] Caitlin Howlett, "Teacher Education and Posthumanism," *Issues in Teacher Education* 27, no. 1 (2018): 106–18.

[106] Robbie B. H. Goh, "Engaging Future Asia: Techno-Orientalisms, Ethnography, Speculative Fiction," *Creative Industries Journal* 6, no. 1 (2013): 43.

[107] Jessica Langer, *Postcolonialism and Science Fiction* (New York: Palgrave Macmillan, 2011).

[108] Howlett, "Teacher Education and Posthumanism," 116.

[109] Howlett, "Teacher Education and Posthumanism," 116.

[110] Howlett, "Teacher Education and Posthumanism," 117.

This returns us to our point about "normal education." Rather than transmitting the known world by means of accepted knowledge that is characteristic of normal education, posthuman education offers the potential to uncover new understandings of human–nonhuman entanglements, understandings that are crucial to a critical form of humane education.

10.6 Concluding Remarks: Toward Post-Humane Education

Most usually, humane education teaches care and compassion toward humans and nonhumans. This is, of course, of central importance. Care and compassion are essential in our treatment of humans and nonhumans.[111] But humane education must go further. Being humane is often associated with a commitment to "animal welfare," which, though it offers improvements to the lives of nonhuman animals, fails to address—or fails to address adequately—anthropocentrism and the continuing consequences of anthropocentric assumptions and practices for nonhuman animals. The idea and practice of humane education do not necessarily counter anthropocentric values. Indeed, if humane education is regarded as an important part of the civilizing process that is discussed by Elias,[112] it can serve to cement rather than challenge anthropocentric notions about the superior status of human by *confirming* human distance from animality. In this respect, in the absence of a non-anthropocentric approach, humane education might simply bolster anthropocentric assumptions and distinctions—because, to use Tester's reflections on some elements of animal advocacy, "humans would be mere beasts" without such distinctions.[113] In short, frequently, the human remains central in humane education.

Posthuman education offers a possible redress or counter to anthropocentric tendencies in humane education, but only if applied critically. Thus, in direct contrast to Fukuyama's take on posthumanism,[114]

[111] For example, see Josephine Donovan and Carol J. Adams, *The Feminist Care Tradition in Animal Ethics: A Reader* (New York: Columbia University Press, 2007).

[112] Elias, *The Civilizing Process.*

[113] Tester, *Animals in Society*, 89.

[114] Fukuyama, *Our Posthuman Future.*

lamenting the possible decentering of humanity, for us intentionally decentering humanity is integral to challenging the "compulsory humanity" synonymous with anthropocentrism and its consequences for other species. Educating humans to question the human as the benchmark, to decenter and think beyond the human, to contest speciesist boundaries, and to use imaginative techniques to consider what is omitted when operating within an anthropocentric view of the world offers the potential for resistance to and displacement of humanist discourses.

In our view posthumanism is integral to the form of critical pedagogic practice required in education to expose the limits and limitations of anthropocentrism and the detrimental consequences inflicted upon other species and their habitats and on ecosystem services and the environment. Whether the approach is designated critical humane education or, our preference, *post-humane education* is of secondary importance to the radical changes in practice required in the conduct of "normal" education and, of course, the operation of other social institutions. The priority is educational practice that cultivates awareness of the frequently injurious effects of human–nonhuman animal entanglements on nonhuman animals and outlines ethical, ecological, and political grounds for radical policies to promote a progressive transformation of human conduct and lifestyles, in order to end exploitative use and abuse of nonhuman animals as property and resources to serve human ends and to engender positive relationships, non-relationships, or rewilding relationships with nonhuman animals. As David Shearman and Joseph Wayne Smith have argued, the vital responsibility of education is "preparation for the future," a responsibility that begins in the present and one that stands in stark contrast to "the form of education peddled by the higher education establishments of Western civilizations, for these have become the tools of our maladaptive economic culture."[115] In our judgment, a critical humane education, or post-humane education, is what is required to transform positively the lives and experiences of human and nonhuman animals by altering radically the ways in which humans conceive of, think about, treat, and relate to other species of animals.

[115] David Shearman and Joseph Wayne Smith, *The Climate Challenge and the Failure of Democracy* (London: Praeger, 2007), 166.

Bibliography

Almiron, Núria "Beyond Anthropocentrism: Critical Animal Studies and the Political Economy of Communication." *The Political Economy of Communication* 4, no. 2 (2016): 54–72.

Angell, George T. *Our Dumb Animals* Vol. 30, no. 3 (August 1897). Our Dumb Animals 1897–10: Vol 30 Iss 5 (archive.org)

Arbour, Rose, Tania Signal, and Nicola Taylor. "Teaching Kindness: The Promise of Humane Education." *Society & Animals* 17, no. 2 (2009): 136–48.

Aronowitz, Stanley, and Henry Giroux. "Radical Education and Transformative Intellectuals." *CTheory* 9, no. 3 (1985): 48–63.

Badmington, Neil. "Theorizing Posthumanism." *Cultural Critique* 53, no. 1 (2003): 10–27.

Barad, Karen. "Diffracting Diffraction: Cutting Together-Apart." *Parallax* 20, no. 3 (2014): 168–87.

Bayley, Annouchka. "How Did 'We' Become Human in the First Place? Entanglements of Posthumanism and Critical Pedagogy for the Twenty-First Century." In *Posthumanism and Higher Education: Reimagining Pedagogy, Practice and Research*, edited by Carol A. Taylor and Annouchka Bayley, 359–65. Houndsmills: Palgrave Macmillan, 2019.

Best, Steven. "Legally Blind: The Case for Granting Animals Legal Rights." Animal Rights Online. 2003 Legally Blind: The Case For Granting Animals Legal Rights - Animal Writes Newsletters From 15 June 2003 Issue From Animal Rights Online (all-creatures.org).

Best, Steven. *The Politics of Total Liberation: Revolution for the 21st Century*. New York: Palgrave-Macmillan, 2014.

Blakemore, Colin, Judy MacArthur Clark, Tino Nevalainen, Michael Oberdorfer, and Arthur. Sussman. "Implementing the 3Rs in Neuroscience Research: A Reasoned Approach." *Neuron* 75, no. 6 (2012): 948–50.

Boakye, Cecilia. "Climate Change Education." *SAGE Open* 5, no. 4 (December 2015).

Brown, H. A. "The Normal-School Curriculum." *The Elementary School Journal* 20, no. 4 (1919): 276–84.

Ceballos, Gerardo, Paul R. Ehrlich, and Rodolfo Dirzo. "Biological Annihilation via the Ongoing Sixth Mass Extinction Signalled by Vertebrate Population Losses and Declines." *PNAS* 114, no. 30 (2017): 6089–96. https://www.pnas.org/content/pnas/114/30/E6089.full.pdf.

Conty, Ariane. "The Politics of Nature: New Materialist Responses to the Anthropocene." *Theory, Culture & Society* 35, nos. 7–8 (2018): 73–96.

Cottrell Free, Ann. *Animals, Nature and Albert Schweitzer*. Washington, DC: Flying Fox Press, 1982.

Davis, Janet M. *The Gospel of Kindness: Animal Welfare and the Making of Modern America*. Oxford: Oxford University Press, 2016.

Davy, Barbara. "An Other Face of Ethics in Levinas." *Ethics & the Environment* 12, no. 1 (2007): 39–65.

DeLeon, Abraham. "What's That Nonhuman Doing on Your Lunch Tray: Disciplinary Spaces, School Cafeterias, and Possibilities for Resistance." In *School Food Politics: The Complex Ecology of Hunger and Feeding in Schools around the World*, edited by Sarah A. Robert and Marcus B. Weaver-Hightower, 183–200. New York: Peter Lang, 2011.

Derrida, Jacques. *The Animal That Therefore I Am*. New York: Fordham University Press, 2008.

Derrida, Jacques. "The Future of the Profession or the Unconditional University (Thanks to the 'Humanities,' What Could Take Place Tomorrow)." Translated by Peggy. Kamuf. In *Deconstructing Derrida*, edited by Peter Pericles Trifonas and Michael A. Peters, 11–24. New York: Palgrave Macmillan, 2005.

Derrida, Jacques, and Elisabeth Roudinesco. *For What Tomorrow: A Dialogue*. Translated by Jeff Fort. Stanford, CA: Stanford University Press, 2004.

Donaldson, Sue, and Will Kymlicka. *Zoopolis: A Political Theory of Animal Rights*. Oxford: Oxford University Press, 2011.

Donovan, Josephine, and Carol J. Adams. *The Feminist Care Tradition in Animal Ethics: A Reader*. New York: Columbia University Press, 2007.

Elias, Norbert. *The Civilizing Process: Sociogenetic and Psychogenetic Investigations*. Rev. ed. Edited by Eric Dunning, Johan Goudsblom, and Stephen Mennell. Oxford: Blackwell, 2000. First published 1939.

Elias, Norbert. "Problems of Involvement and Detachment." *The British Journal of Sociology* 7, no. 3 (1956): 226–52.

Ferrando, Francesca. "The Party of the Anthropocene: Post-Humanism, Environmentalism and the Post-anthropocentric Paradigm Shift." *Relations* 4, no. 2 (2016): 159–73.

Ferrante, Allesandro., and Daniele. Sartori. "From Anthropocentrism to Post-Humanism in the Educational Debate." *Relations* 4, no. 2 (2016): 175–94.

Flynn, Clifton P. "Examining the Links between Animal Abuse and Human Violence." *Crime, Law and Social Change* 55, no. 5 (2011): 453–68.

Francione, Gary. "Animal Welfare and the Moral Value of Nonhuman Animals." *Law, Culture and the Humanities* 6, no. 1 (2010): 24–36.

Francione, Gary. *Animals, Property, and the Law*. Philadelphia: Temple University Press, 1995.

Francione, Gary. "Vegan Education Made Easy—Part 1." *Animal Rights: The Abolitionist Approach* (blog), March 25, 2008. https://www.abolitionistapproach.com/vegan-education-made-easy-part-i/.

Fukuyama, Francis. *Our Posthuman Future: Consequences of the Biotechnology Revolution*. London: Profile Books, 2003.

Gibbs, Paul. “Higher Education: A Compassion Business or Edifying Experience.” In *The Pedagogy of Compassion at the Heart of Higher Education*, edited by Paul Gibbs, 1–18. Cham: Springer, 2017.

Giroux, Henry. *On Critical Pedagogy.* New York: Continuum International, 2011.

Goh, Robbie R. H. “Engaging Future Asia: Techno-Orientalisms, Ethnography, Speculative Fiction.” *Creative Industries Journal* 6, no. 1 (2013): 43–56.

Greenhough, Beth, and Emma Roe. “Exploring the Role of Animal Technologists in Implementing the 3Rs: An Ethnographic Investigation of the UK University Sector.” *Science, Technology, & Human Values* 43, no. 4 (2018): 694–722.

Hamilton, Clive. *Requiem for a Species: Why We Resist the Truth about Climate Change*. London: Earthscan, 2010.

Harris, Steven. B. “The Society for the Recovery of Persons Apparently Dead.” *Alcor Life Extension Foundation*. 1990. https://alcor.org/Library/html/PersonsApparentlyDead.htm.

Harrison, Brian. “Animals and the State in Nineteenth-Century England.” *The English Historical Review* 88, no. 349 (1973): 786–820.

Heidegger, Martin. *The Basic Problems of Phenomenology*. Translation, introduction, and lexicon by Albert Hofstadter. Bloomington: Indiana University Press, 1982.

Howlett, Caitlin. “Teacher Education and Posthumanism.” *Issues in Teacher Education* 27, no. 1 (2018): 106–18.

Humane Education Coalition. “What Is Humane Education?” Accessed August 2, 2019. https://www.hecoalition.org/what-is-humane-education.html.

Humane Education Coalition. “Who We Are.” Accessed August 2, 2019. https://www.hecoalition.org/about.html.

Humane Society International. “Protect All Animals.” Accessed July 31, 2019. https://www.hsi.org/.

Humane Society of the United States. “Our Mission.” Accessed July 30, 2019. https://www.humanesociety.org/our-mission.

Khurana, Muskan “The Wilderness Live” Albert Schweitzer Animal Welfare Fund (rjyrosesaksena.wixsite.com) 2021. Accessed July 30, 2021.

Kopnina, Helen, and Mickey Gjerris. “Are Some Animals More Equal than Others? Animal Rights and Deep Ecology in Environmental Education.” *Canadian Journal of Environmental Education* 20 (2015): 108–123.

Kuhn, Thomas. *The Structure of Scientific Revolutions: Fiftieth Anniversary Edition*. Chicago: University of Chicago Press, 2012. First published 1962.

Langer, Jessica. *Postcolonialism and Science Fiction*. New York: Palgrave Macmillan, 2011.

Levinas, Emmanuel, and Richard Kearney. "Dialogue with Emmanuel Levinas." In *Face to Face with Levinas*, edited by Richard A. Cohen, 13–34. New York: State University of New York, 1986.

Marshall, George and David Polling, *Albert Schweitzer: A Biography.* London and New York: John Hopkins Press, 1971

McKay, Robert. "'Identifying with the Animals': Language, Subjectivity, and the Animal Politics of Margaret Atwood's *Surfacing*." In *Figuring Animals: Essays on Animal Images in Art, Literature, Philosophy and Popular Culture*, edited by Mary Sanders Pollock and Catherine Rainwater, 207–27. New York: Palgrave Macmillan, 2005.

Miah, Andy. "A Critical History of Posthumanism." In *Medical Enhancement and Posthumanity*, edited by Bert Gordijn and Ruth Chadwick, 71–94. The International Library of Ethics, Law and Technology, vol 2. Dordrecht: Springer, 2008.

Murris, Karin. "Review Essay: Propositions for Posthuman Teaching and Research—A Diffractive Re-View of Three Books." *Critical Studies in Teaching and Learning* 5, no. 1 (2017): 103–9.

National Anti-Vivisection Society. "Laws That Recognize the Importance of Humane Education." Accessed August 2, 2019. https://www.navs.org/what-we-do/keep-you-informed/legal-arena/education/humane-education/#.XURGrvJKiM8.

National Museum of Animals & Society. "American Humane Education Society." *Be Kind: A Visual History of Humane Education 1880–1945*, curated by Keri Cronin. National Museum of Animals & Society, 2013. https://bekindexhibit.org/exhibition/american-humane-education-society/.

Nibert, David. *Animal Oppression and Human Violence: Domesecration, Capitalism, and Global Conflict.* New York: Columbia University Press, 2013.

Pedersen, Helena. "Is 'the Posthuman' Educable? On the Convergence of Educational Philosophy, Animal Studies, and Posthumanist Theory." *Discourse: Studies in the Cultural Politics of Education* 31, no. 2 (2010): 237–50.

Pedersen, Helena. "Schools, Speciesism, and Hidden Curricula: The Role of Critical Pedagogy for Humane Education Futures." *Journal of Futures Studies* 8, no. 4 (2004): 1–14.

Peggs, Kay. *Animals and Sociology*. Basingstoke: Palgrave Macmillan, 2012.

Peggs, Kay, and Barry Smart. "Nonhuman Animal Suffering: Critical Pedagogy and Practical Animal Ethics." *Society and Animals* 25, no. 2 (2017): 181–98.

Preece, Rod "Thoughts Out of Season on the History of Animal Ethics." *Society & Animals* 15, no. 4 (2007): 365–78.

Royal Humane Society. Home page. Accessed August 2, 2019. https://www.royalhumanesociety.org.uk/.

Russell, William, and Rex Burch. *The Principles of Humane Experimental Technique*. London: Methuen, 1959.

Schweitzer, Albert. *Reverence for Life. Sermons 1900–1919* Translated by Reginald. H. Fuller. New York: Irvington, 1969.

Scruton, Roger. "Humane Education." *The American Scholar* 49, no. 4 (1980): 489–98.

Shearman, David, and Joseph Wayne Smith. *The Climate Challenge and the Failure of Democracy*. London: Praeger, 2007.

Snaza, Nathan, and John A. Weaver, eds. *Posthumanism and Educational Research*. New York: Routledge, 2015.

Spannring, Reingard "Animals in Environmental Education Research." *Environmental Education Research* 23, no. 1 (2017): 63–74.

Taylor, Carol A., and Annouchka. Bayley, eds. *Posthumanism and Higher Education: Reimagining Pedagogy, Practice and Research*. Houndsmills: Palgrave Macmillan, 2019.

Taylor, Katy, Nicky Gordon, Gill Langley, and Wendy Higgins. "Estimates for Worldwide Laboratory Animal Use in 2005." *ATLA* 36 (2008): 327–42. Accessed April 15, 2014. http://www.buav.org/_lib/userfiles/files/Science_Reports/Estimates_For_Worldwide_Laboratory.pdf.

Tester, Keith *Animals in Society: The Humanity of Animal Rights*. London: Routledge, 1991.

Unti, Bernard, and Bill DeRosa. "Humane Education: Past, Present and Future." In *The State of the Animals II*, edited by Deborah. J. Salem and Andrew N. Rowan, 27–48. Washington, DC: Humane Society Press, 2003.

White, Richard. "Compassion in Philosophy and Education." In *The Pedagogy of Compassion at the Heart of Higher Education*, edited by Paul Gibbs, 19–34. Cham: Springer, 2017.

Williams, Raymond. *Keywords: A Vocabulary of Culture and Society*. Rev. ed. New York: Oxford University Press, 1983.

CHAPTER 11

The Kindness-to-Animals Ethic in Antebellum America

Robyn Hederman

In August 1841, New York City's annual dog slaughter had already begun. The author and social reformer Lydia Maria Child described this massacre in her "Letters from New York," writing that these "dog killers" had murdered "twelve to fifteen hundred" this summer—and "in the hottest of the weather at the rate of 300 a day."[1] They knocked down these "poor creatures" and beat them to death with "their bloody clubs." Sometimes, these dogs were not killed, but instead they were "horribly maimed, and [ran] howling and limping away." When the mob descended, Child fled from her window, trying to escape this "Reign of Terror."[2] Child, who wrote advice manuals for mothers, asked "whether

[1] Lydia Maria Child, *Letters from New York*, ed. Bruce Mills (Athens: University of Georgia Press, 1998), 12.

[2] Child, *Letters*, 12.

R. Hederman (✉)
New York State Supreme Court, Criminal Term, Brooklyn, NY, USA
e-mail: rhederman@yahoo.com

C. Linzey and A. Linzey (eds.), *The Palgrave Companion to Humane Education*, The Palgrave Macmillan Animal Ethics Series,
https://doi.org/10.1007/978-3-032-19795-5_11

such brutal scenes do not prepare the minds of the young to take place in bloody riots and revolutions."[3]

The creation of the American Society for the Prevention of Cruelty to Animals (ASPCA) by Henry Bergh in 1866 often symbolizes the start of the animal protection movement in the United States. Nonetheless, the kindness-to-animals ethic flourished in antebellum America. Americans, repulsed by the infliction of pain and suffering on sentient beings, questioned what it meant to be "humane."[4] As a result, magazines and advice books instructed parents to teach their children to be kind to nonhuman animals, acknowledging that cruelty to animals predicted their future character.[5]

For instance, in her biography of the abolitionist Isaac Hopper, Lydia Maria Child opined that Hopper's kind treatment of his childhood pet lamb foretold his later work on behalf of fugitive slaves.[6] From his earliest years, Hopper "evinced great fondness for animals and watched with lively interest all the little creatures of the woods and fields."[7] Hopper "took such excellent care of his pet" that it "trotted about after him like a dog."[8] During the American Revolutionary War, Hopper prevented British soldiers from seizing his lamb. Hopper used his "rusty knife" to cut the ties binding the lamb's legs, crying, "That's my lamb, and you shan't have it!"[9] A British officer intervened, reassuring him, "Well, they shan't have it. You'll make a fine soldier one of these days."[10]

[3] Child, *Letters*, 12.

[4] Bernard Oreste Unti, "The Quality of Mercy: Organized Animal Protection in the United States, 1866–1930" (PhD diss., American University, 2002), 27–28. Unti defines humane as "descriptive not only of personality type but also of a broad set of reform concerns." See also Norman S. Fiering, "Irresistible Compassion: An Aspect of Eighteenth-Century Sympathy and Humanitarianism," *Journal of the History of Ideas* 37, no. 2 (April–June 1976): 195–218.

[5] See Katherine C. Grier, "Childhood Socialization and Companion Animals: United States, 1820–1870," *Society and Animals* 7, no. 2 (1999): 95–120.

[6] Lydia Maria Child, *Isaac T. Hopper: A True Life* (Boston: John P. Jewett, 1853), 9, https://archive.org/details/isaacthoppertrue00chil/page/9.

[7] Child, *Isaac T. Hopper*, 18.

[8] Child, *Isaac T. Hopper*, 9.

[9] Child, *Isaac T. Hopper*, 9.

[10] Child, *Isaac T. Hopper*, 10.

Such literature promoted kindness to animals in postrevolutionary America and led to a flurry of antebellum reform efforts extending humanity's moral duty to other sentient beings. As a result, the kindness-to-animals ethic became emblematic of antebellum culture.

11.1 Postrevolutionary Americans

Postrevolutionary Americans hoped to create a "republican national identity"—namely, a society where "sovereign individuals joined in purely voluntary ties" and subordinated their "private interests to the public good."[11] In his article "Animals Made Americans Human," historian Bill Leon Smith examines several of the nation's founders who believed that kindness to nonhuman animals supported republican values and virtues.[12] According to Smith, as postrevolutionary Americans obtained greater access to British literature condemning animal cruelty, they contemplated whether kindness to all creatures reflected the values of the new nation.[13]

Middle- and upper-class Americans sought a growing number of children's books and advice manuals expanding on John Locke's principles of "enlightened child-rearing."[14] In *Some Thoughts on Education*, John

[11] Jacqueline S. Reinier, *From Virtue to Character: American Childhood, 1775–1850* (New York: Twayne, 1996), 32–33.

[12] Smith, however, makes clear that not all the founding fathers were disturbed by animal cruelty. According to Smith, Thomas Jefferson punished a dog for attacking his sheep by hanging the dog without "the slightest hint of compassion or magnanimity." See Bill Leon Smith, "Animals Made Americans Human: Sentient Creatures and the Creation of Early America's Moral Sensibility," *Journal of Animal Ethics* 2, no. 2 (Fall 2012): 127–40, 126. Jefferson's hostility to dogs was demonstrated in his September 24, 1811, letter to Peter Minor. Jefferson stated, "I participate in all your hostility to dogs, and would readily join in any plan for exterminating the whole race. I consider them as the most afflicting of all the follies for which men tax themselves." *The Papers of Thomas Jefferson, Retirement Series, Volume 4: June 1811 to April 1812*, ed. J. Jefferson Looney (Princeton: Princeton University Press, 2007), 170. Additionally, James Turner points out that the Federalists made use of this reputation in the 1800 election, calling Monticello "Dogs' Misery" because of his painful experiments on animals. See James Turner, *Animals, Pain, and Humanity in the Victorian Mind: Reckoning with the Beast* (Baltimore: Johns Hopkins University Press, 1980), 20.

[13] Smith, "Animals Made Americans Human," 130–32. For example, see Alexander Pope, "Essay on Man," in *Essay on Man and Other Poems* (1733/1734; repr., New York: Dover, 1994), 45–79.

[14] Reinier, *From Virtue to Character*, 18. See Jacqueline S. Renier, "Rearing the Republican Child: Attitudes and Practices in Post-Revolutionary Philadelphia," *William*

Locke warned that cruelty to animals should be watched in children because "the custom of tormenting and killing beasts, will, by degrees, harden their hearts even towards men."[15] Locke proposed that children should be raised with an "abhorrence of killing or tormenting any living creature" and cautioned that "they who delight in the suffering and destruction of inferior creatures will not be apt to be compassionate or benign to those of their own kind."[16]

The physician and educator Benjamin Rush, influenced by Locke, believed that the nation needed to pursue republican and Christian values. In an oration given before the American Philosophical Society in 1786 titled "The Influence of Physical Causes upon the Moral Faculty," Rush proposed that a Christian republic required its citizens to be virtuous and disciplined. To this end, Rush sought reforms including abolition, temperance, the elimination of capital punishment, and the humane treatment of nonhuman animals. Indeed, Rush believed that "cruelty to brute animals was another means of destroying moral sensibility."[17] Consequently, Rush, whose visions and ideals "in many ways refigured antebellum reform,"[18] helped to frame the kindness-to-animals ethic in antebellum America.

and Mary Quarterly, 3rd ser., 39, no. 1 (January 1982): 153–54. See also Richard L. Bushman, *The Refinement of America: Persons, Houses, Cities* (New York: Alfred A. Knopf, 1992).

[15] John Locke, *Some Thoughts on Education* (1693; repr., Boston: Gray and Bowen, 1830), 151–52, https://archive.org/details/somethoughtscon01lockuoft/page/151-152.

[16] Locke, *Some Thoughts*, 151–52.

[17] In this essay, Rush referred to William Hogarth's *Four Stages of Cruelty* prints, stating, "Mr. Hogarth's points out, in his ingenious prints, the connection between cruelty to brute animals in youth, and murder in manhood." Benjamin Rush, M.D., "An Enquiry into the Influence of Physical Causes upon the Moral Faculty," in *Two Essays on the Mind* (1786; New York: Brunner/Mazel, 1972), 32.

[18] Robert H. Abzug, *Cosmo Crumbling: American Reform and the Religious Imagination* (New York: Oxford University Press, 1994), 12. See also John L. Thomas, "Romantic Reform in America, 1815–1865," *American Quarterly* 17, no. 4 (Winter 1965): 656–81. For a full account of Republicanism ideology in the new nation, see Bernard Bailyn, *The Ideological Origins of the American Revolution* (Cambridge, MA: Belknap Press of Harvard University Press, 1967); Gordon S. Wood, *The Creation of the American Republic, 1776–1787* (Chapel Hill: University of North Carolina Press, 1969).

11.2 Republican Motherhood

Rush also believed that the kindness-to-animals ethic was consistent with republican motherhood and an essential component of childhood socialization. Rush demonstrated the link between republican government and methods of child-rearing in his essay "Of the Mode of Education Proper in a Republic."[19] According to Rush, republican mothers had the responsibility to produce virtuous citizens—specifically, to raise patriotic sons who would become the nation's future leaders. Rush proposed that women should be "taught the principles of liberty and government" because "the first impressions upon the minds of children are generally derived from women."[20] In his speech to the students of the Young Ladies' Academy in Philadelphia, Rush claimed that the "cultivation of reason in women" is "friendly to the order of nature, and to private as well as public happiness."[21]

The historian Linda Kerber claims that most colonial Americans believed that a "learned woman" was an "unenviable anomaly" who developed her mind at her own risk.[22] Although most postrevolutionary Americans were not willing to grant citizenship to women, they proposed an alternative model—namely, the "Republican Woman." The Republican Woman would not be a citizen, but "she might serve a political purpose," in that she would raise her sons to be the future citizens of the nation.[23] According to Sheila L. Skemp, in her study of the contemporary essayist

[19] Benjamin Rush, "Of the Mode of Education Proper in a Republic," in Essays, Literary, Moral, and Philosophical (Philadelphia: Thomas & William Bradford, 1798), 6–20. https://archive.org/details/essaysliterarymo00rush_0/page/6/mode/2up.

[20] Rush, "On Mode of Education," 19.

[21] Rush, "Thoughts upon Female Education, Accommodated to the Present State of Society, Manners, and Government, in the United States of America: Addressed to Visitors of the Young Ladies' Academy in Philadelphia, 28th July 1787," in *Essays: Literary, Moral and Philosophical* "92. For a further discussion, see David Freeman Hawke, *Benjamin Rush: The Revolutionary Gadfly* (Indianapolis: Bobbs-Merrill, 1971), 333–34. On the role of "republican motherhood," see Ruth H. Bloch, "American Feminine Ideals in Transition: The Rise of the Moral Mother, 1785–1815," *Feminist Studies* 4, no. 2 (June 1978): 100–26. See also Nancy F. Cott, *The Bonds of Womanhood: "Women's Sphere" in New England, 1780–1835* (New Haven: Yale University Press, 1977).

[22] Linda K. Kerber, *Women of the Republic: Intellect and Ideology in Revolutionary America* (New York: Norton, 1986), 190–91.

[23] Linda Kerber, "The Republican Mother: Women and the Enlightenment—An American Perspective," *American Quarterly* 28, no. 2 (Summer 1976): 204.

Judith Sargent Murray, Murray argued that women should be educated and that education would not destroy a women's femininity.[24] Similar to Rush, Murray believed that because women were primarily responsible for raising children, especially their sons, "the future of the Republic was in their hands."[25] Consequentially, educational opportunities for women expanded in the late eighteenth century.[26]

Rush, a proponent of female education, instructed mothers to teach their children to shun cruel amusements. In his essay "Thoughts upon the Amusements and Punishments Which Are Proper for Schools," Rush objected to hunting ("gunning"), claiming it "hardens the heart, by inflicting unnecessary pain and death upon animals."[27] According to Rush, hunting was "unnecessary in a civilized society"[28] and led "young men into low, and bad company."[29] Similarly, Rush objected to children receiving firearm training because it fostered "hostile ideas towards their fellow creatures."[30] Instead, Rush wished children to develop "sentiments of universal benevolence to men of all nations and colors."[31]

[24] Sheila L. Skemp, *Judith Sargent Murray: A Brief Biography with Documents*, Bedford Series in History and Culture (Boston: Bedford/St. Martin, 1998), 86–87.

[25] Skemp, *Judith Sargent Murray*, 88.

[26] However, for an alternative opinion, see Elaine F. Crane, "Dependence in the Era of Independence: The Role of Women in Republican Society," in *The American Revolution: Its Character and Limits*, ed. Jack Greene (New York: New York University Press, 1987), 253–75.

[27] Benjamin Rush, "Thoughts upon the Amusements and Punishments Which Are Proper for Schools. Addressed to George, Clymer, Esq.," in *Essays: Literary, Moral and Philosophical*, 60.

[28] Rush, "Thoughts on Amusements," 60.

[29] Rush, "Thoughts upon the Amusements," 61.

[30] Rush, "Thoughts upon the Amusements," 61.

[31] Rush, "Thoughts upon the Amusements," 61.

11.3 The Antebellum Era

Numerous studies have analyzed the causes of the burgeoning reform movements of the early nineteenth century.[32] Many of these studies either ignore or give cursory attention to the crusade to prevent animal cruelty. Nevertheless, the kindness-to-animals ethic became an ideal of what was described by historian Daniel Walker as "Victorian culture."[33]

Drastic economic and social changes took place in the postrevolutionary era and continued into the early nineteenth century. Industrial growth and the emergence of the market economy, as well as the role of women and the shifts in family dynamics, are beyond the scope of this chapter. Yet in *Moralists and Modernizers*, Steve Mintz claims that the "specter" of social decline "haunted early nineteenth-century Americans."[34] Americans feared the rapid growth of cities, the rise in homicides and other violent offenses, the widening social divisions, and the increase in immigration. Consequently, middle-class Americans developed a "cultural obsession with self-control" and became preoccupied with order and regulated self-improvement.[35] According to Howe, this preoccupation resulted in American Victorians practicing "humanistic cultivation, Protestant self-denial and/or bourgeois self-control."[36]

In her study of antebellum culture, Karen Halttunen examines nineteenth-century advice literature. Manuals and magazines directed

[32] Robert H. Abzug, *Cosmo Crumbling*; Steven Mintz, *Moralists and Modernizers: America's Pre–Civil War Reformers* (Baltimore: Johns Hopkins University Press, 1995); Lori D. Ginzberg, *Women and the Work of Benevolence: Morality, Politics, and Class in the Nineteenth-Century United States* (New Haven: Yale University Press, 1990).

[33] Daniel Walker Howe, "Victorian Culture in America," in *Victorian America*, ed. Daniel Walker Howe (Philadelphia: University of Pennsylvania Press, 1976), 3–28. Victorian American culture was defined by gentility, liberal evangelical Protestant religion, and domesticity. See Katherine C. Grier, *Pets in America: A History* (Orlando, FL: Harcourt, Inc., 2006), 166.

[34] Mintz, *Moralists and Modernizers*, 3.

[35] Mintz, *Moralists and Modernizers*, 11. For a detailed discussion of this period, see Charles Sellers, *The Market Revolution: Jacksonian America, 1815–1846* (New York: Oxford University Press, 1991); Gordon S. Wood, *The Radicalism of the American Revolution* (New York: Vintage Books, 1991).

[36] Howe, "Victorian Culture in America," 18.

Americans to conduct themselves with "perfect sincerity or 'transparency' of character."[37] Consequently, middle-class Americans, or those hoping to improve their social status, adopted a "sentimental ideal of social conduct" governed by social practices such as etiquette, dress, and bereavement practices.[38] Halttunen proposes that by adopting and adhering to these social norms, ambitious men and women could demonstrate that they "were true ladies and gentlemen deserving of the higher social place to which they aspired."[39]

Although not discussed by Halttunen, "pet-keeping" also became a social custom of middle-class America. According to Katherine C. Grier in *Pets in America: A History*, the domestic ethic of kindness to animals evolved from Victorian cultural ideals—namely, gentility, liberal evangelical Protestantism, and domesticity.[40] Americans believed that character development resulted from "conscious choice and self-control," and they understood animal cruelty to be a "problem both of individual or familial deficiency and of good and evil."[41] According to Grier, antebellum families were defined by how they treated animals within their care. As a result, the kindness-to-animals ethic became identified with what historians have described as "the cult of domesticity."[42]

Domesticity encouraged the separation of family life from the world of business. According to historian Nancy F. Cott, the "canon of domesticity" prescribed that women would create morally safe havens to act as

[37] Karen Halttunen, preface to *Confidence Men and Painted Women: A Study of Middle-Class Culture in America, 1830–1870* (New Haven: Yale University Press, 1982).

[38] Halttunen, *Confidence Men and Painted Women*, 193. These practices helped distinguish the true socially refined class from the ambitious pretenders, described as the confidence men and women.

[39] Halttunen, *Confidence Men and Painted Women*, xvii.

[40] Grier, *Pets in America*, 166.

[41] Grier, "Childhood Socialization," 95–96.

[42] Grier, *Pets in America*, 164. For works defining the "cult of domesticity," see Aileen S. Kraditor, introduction to *Up from the Pedestal: Selected Writings in the History of American Feminism* (Chicago: Quadrangle Books, 1968); Mary P. Ryan, *The Empire of the Mother: American Writing about Domesticity, 1830–1860* (New York, The Haworth Press, 1982); Barbara Welter, "The Cult of True Womanhood: 1820–1860," chap. 2 in *Dimity Convictions: The American Woman in the Nineteenth Century* (Athens: Ohio University Press,1976).

buffers against the "pecuniary standards in the work world."[43] Accordingly, a woman's "vocation"—that is, motherhood—was to educate and socialize her children to become virtuous, self-disciplined adults.[44]

The ethic of kindness, which provided children with these qualities, endowed pet-keeping with a "high moral purpose."[45] As a result, humane education became an essential method of educating children to pursue Victorian values.[46] Middle-class Americans believed that how children treated animals would predict how they, as adults, would treat other human beings. According to Grier, kindness to nonhuman animals "became a marker middle-class identity."[47]

11.4 Kindness-to-Animals Literature

By the mid-nineteenth century, there was an influx of children's literature and advice manuals teaching children to be kind to nonhuman animals. Popular fiction depicted animals as sharing emotional and physical attributes with other members of the family, including specifically the ability to feel pain.[48] Child-rearing literature instructed parents that childhood cruelty to nonhuman animals was a predictor of future violence.[49] For example, the author Lydia H. Sigourney proposed that mothers should teach their children to follow a moral code that promoted "kindness to all around."[50] In *Letters to Mothers*, she recommended that kindness to all was best taught by instructing children to treat nonhuman animals humanely. If the child "seizes a kitten by the back, or pulls its hair," Sigourney instructed, the mother must demonstrate the proper way

[43] Cott, *Bonds of Womanhood*, 69. See Barbara Leslie Epstein, *The Politics of Domesticity: Women, Evangelism, and Temperance in Nineteenth-Century America* (Middletown, CT: Wesleyan University Press, 1981).

[44] Cott, *Bonds of Womanhood*, 93–94.

[45] Grier, *Pets of America*, 164.

[46] Unti, "Quality of Mercy," 54–55.

[47] Grier, "Childhood Socialization," 100.

[48] Unti, "Quality of Mercy," 54. See, for example, Oliver Goldsmith, *The History of Little Goody Two Shoes* (1765; repr., New York: Macmillan, 1961).

[49] Grier, "Childhood Socialization," 101–2; Michelle C. Neely Yates, "The Antebellum Animal" (PhD diss., University of California, 2011), 22–24.

[50] Lydia H. Sigourney, *Letters to Mothers* (Hartford: Hudson and Skinner, 1838), 40, https://archive.org/details/letterstomother02sigogoog/page/n40.

to treat the kitten; for example, she should show "how it may be held properly, and soothed into confidence."[51]

Sigourney presented a twofold argument. Initially, she argued that cruelty to animals, no matter how insignificant, was a predictor of violent behavior in adulthood. For instance, Sigourney asserted that when the "traitor" Benedict Arnold was a boy, he had "loved to destroy insects, to mutilate toads, to steal the eggs of the mourning bird, and [to] torture quiet, domestic animals." When he became an adult, Arnold had "laid waste the shrinking, domestic charities, and would have drained the life-blood of his endangered country."[52]

Sigourney also claimed that nonhuman animals were moral beings who suffered physical pain and experienced joy. She advised mothers to "check the first buddings of those Domitian tastes" and to instruct their children that "the gift of life, to the poor beetle, or the crawling worm, is from the Great Father above, and not to be lightly trodden out."[53]

Similarly, in *The Mother's Book*, Lydia Maria Child emphasized that "kindness toward animals is of great importance" to cultivate affection in children.[54] If they are "guilty of any violent treatment toward them, they should see that you are grieved and displeased at such conduct."[55]

Child instructed mothers that their children must distinguish the kitten from the "brass andirons," which do not move or grow larger.[56] Child pondered, "Does George know what an animal is? It is a creature that grows and can move about from one place to another."[57] Child further

[51] Sigourney, *Letters to Mothers*, 40.

[52] Sigourney, *Letters to Mothers*, 39–40.

[53] Sigourney, *Letters to Mothers*, 38.

[54] Lydia Maria Child, *The Mother's Book* (1831; repr., New York: Francis, 1846), 6, https://archive.org/details/mothersbook1844chil/page/10.

[55] Child, *Mother's Book*, 6–7. Lydia Child uses the example of a mother who laughs when her two-year-old child pulls the tail of a kitten. When the kitten scratches the boy in self-defense, the mother then beats the kitten. Child describes the adverse effects such behavior has on the small child's character. Initially, the mother encourages the child's actions by her amusement. Further, she demonstrates injustice to the kitten and a lesson of tyranny to the boy. Finally, she teaches the boy the lesson of retaliation.

[56] Child, *Mother's Book*, 13. See also Yates, "Antebellum Animal," 23–24.

[57] Child, *Mother's Book*, 13.

explained that "young children are often cruel from the mere thoughtlessness of frolic; they strike an animal as they would strike a log of wood, without knowing that they occasion pain."[58]

Fictional stories and verse also dealt with the theme of animal cruelty.[59] The abolitionist Harriet Beecher Stowe wrote stories educating parents and children on animal welfare issues. In *Civilized Creatures*, Jennifer Mason asserted that the animal welfare agenda was often demonstrated in Stowe's stories.[60]

In *Queer Little People*, Stowe compiled childhood stories describing the habits and instincts of nonhuman animals. Stowe examined her aunt's rules for the proper treatment of animals in "Aunt Esther's Rules." For example, Aunt Esther proscribed giving animals to young children who did not know how to handle them. According to Aunt Esther, "one was never to frighten an animal for sport," nor ever to tease or play practical jokes on them, because "they have but one short life to live, they are dumb and cannot complain, and they are wholly in our power."[61]

Again, in Stowe's "A Talk about Birds," a boy learns why it is morally wrong to stone and kill birds.[62] "Two strange boys" tell Jamie, "We have our pockets full of stones, and we are going to kill birds with them; it's the

[58] Child, *Mother's Book*, 7.

[59] See, for example, Sarah Josepha Hale's "Mary Had a Little Lamb," published in 1830 and reprinted in Grier, *Pets in America*, 187. In the poem, Hale attempted to instruct children to treat their pets kindly, so that they would respond in kind. See also Ruth E. Finley, *The Lady of Godey's: Sarah Josepha Hale* (Philadelphia: J. B. Lippincott, 1931), 279–80.

[60] Jennifer Mason, *Civilized Creatures: Urban Animals, Sentimental Culture, and American Literature, 1850–1900* (Baltimore: Johns Hopkins University Press, 2005), 100.

[61] Harriet Beecher Stowe, "Aunt Esther's Rules," in *Queer Little People* (Boston: Houghton, Mifflin, 1867),152–57, https://archive.org/details/queerlittlepeopl00stow/page/n5. Although Aunt Esther promotes kind treatment toward animals, she also is an advocate of humane killing of unwanted animals. Stowe describes this act as "real brave humanity." "How much kinder and more really humane to take upon ourselves the momentary suffering of causing the death of an animal than to turn our back and leave it to drag out a life of torture and misery!" For a further discussion of "Aunt Esther's Rules," see Mason, *Civilized Creatures*, 100; see also Yates, "Antebellum Animal," 24–31.

[62] Harriet Elizabeth Beecher Stowe, "A Talk about Birds," in *Our Charlie, and What to Do with Him* (Philadelphia: J. B. Lippincott, 1859), 94–109, https://archive.org/details/in.ernet.dli.2015.350688/page/n1. Here, Jamie's mother explains that God must be displeased with children hurting and killing birds. Jamie responds, "And it is a great deal worse to kill little birds than it is to break looking glasses, and such things; because little birds can feel, you know."

best fun in the world."[63] Jamie, "a thoughtless little boy," joins them and fills "his pockets with stones."[64] Subsequently, the boys turn their attention to "a nice white cat." The frightened cat "scamper[s] with all her might" while "all three of the boys... chase after her."[65] Jamie's mother, "a very kind and tender-hearted woman," intervenes after "watching the whole affair."[66]

Stowe emphasizes that instead of "blaming him harshly," Jamie's mother "set out to make him *think*," believing that "when little boys did cruel things, it was more because they were *thoughtless* than because they at heart were cruel."[67] Initially, Jamie's mother informs him that he and birds share "nerves to feel with, and muscles to move with."[68] She questions whether God would be displeased that boys would amuse themselves with throwing stones at birds after God filled the "woods and fields" with "hundreds of little happy birds, all so curiously and beautifully made."[69]

To make her point, Jamie's mother reads him another story, "The Nest in the Orchard," which describes the slaughter of a family of bluebirds. Here, a mother bluebird and her mate set up a nest in a "snug little hole" inside the trunk of a "spreading apple tree."[70] The parents are delighted with the birth of their babies, and "while one would sit with wings outspread to keep the little creatures warm, the other would range about and get flies and worms to feed them."[71] Stowe brutally describes the birds' demise at the hands of a "man with a large bag tied about his neck, and a long gun in his hand," who "drew up his gun and fired" at the father bluebird.[72]

[63] Stowe, "A Talk About Birds," 95.

[64] Stowe, "A Talk About Birds," 95.

[65] Stowe, "A Talk About Birds," 96.

[66] Stowe, "A Talk About Birds," 96.

[67] Stowe, "A Talk About Birds," 97.

[68] Stowe, "A Talk About Birds," 103.

[69] Stowe, "A Talk About Birds," 108.

[70] Harriet Elizabeth Beecher Stowe, "The Nest in the Orchard," in *Our Charlie, and What to Do with Him* (Philadelphia: J. B. Lippincott, 1859), 112, https://archive.org/details/in.ernet.dli.2015.350688/page/n1.

[71] Stowe, "Nest in the Orchard," 115.

[72] Stowe, "Nest in the Orchard," 116.

After the death of the father bird, the "same man" aims his weapon at the mother bird, who is "picking up food for her young ones as well as she [is] able."[73] The mother bird is shot in her wing but makes it home to feed "her little ones once more."[74] But there she falls "down at the foot of the tree" and dies.[75] The five dead babies are later discovered in the nest. "The Nest in the Orchard" instructs Jamie and other children that killing bluebirds and other sentient creatures—who have strong emotions and a capacity to suffer—is morally wrong.[76]

11.5 Family Violence and the "Problem with Boys"

The "epidemic of multiple family murders," analyzed by Steven Mintz in *Moralists and Modernizers*, demonstrated to antebellum Americans that not only was there a breakdown of social order, but this violence also was moving into the domestic sphere.[77] Through contemporary social movements such as abolition, prison reform, animal protection, and opposition to capital and corporal punishment, humanitarian reformers worked to shift public attitudes towards society violence and the infliction of pain.[78] Nonetheless, reformers were reluctant to advocate in the area of domestic abuse.

In *Campaigns against Corporal Punishment*, Myra C. Glenn describes domestic violence as an unexamined problem in antebellum America.[79]

[73] Stowe, "Nest in the Orchard," 118.

[74] Stowe, "Nest in the Orchard," 119.

[75] Stowe, "Nest in the Orchard," 119.

[76] In this story, Jamie asks, "What could he want to shoot the little birds for?" Stowe, "Nest in the Orchard," 116.

[77] Steven Mintz, *Moralists and Modernizer*, 6. Mintz indicates that of the twelve most famous incidents of domestic murder between 1600 and 1900, ten occurred between 1770 and 1825.

[78] Karen Halttunen, "Humanitarianism and the Pornography of Pain in Anglo-American Culture," *American Historical Review* 100, no. 2 (April 1995): 303–34; Elizabeth B. Clark, "'The Sacred Rights of the Weak': Pain, Sympathy, and the Culture of Individual Rights in Antebellum America." *Journal of American History* 82, no. 2 (September 1995): 463–93.

[79] Myra C. Glenn, *Campaigns against Corporal Punishment: Prisoners, Sailors, Women, and Children in Antebellum America* (Albany: State University of New York Press, 1984), 82.

Middle-class Americans regarded domestic violence as a "tragic deviation from Victorian social ethic."[80] Nevertheless, their idealized view of marriage—as a sacred and permanent bond—imposed an "ideological barrier" to a public discussion of domestic abuse.[81] Instead, Katherine C. Grier suggests that antebellum literature warning parents about "the problem with boys," specifically violence toward animals, "was often a covert discourse about violence in human families."[82]

Accordingly, advice manuals focusing on violence committed by boys emphasized that they should develop empathy early in life. For example, Sara Josepha Hale advised parents that "pet-keeping" in the home would encourage boys to be kind. Pet-keeping, according to Hale, was "a great preventative against thoughtless cruelty and tyranny they are so apt to exercise toward all dependent beings!"[83] Likewise, Henry Ward Beecher, in his *Lectures to Young Men,* warned against the "intense thrill" obtained from "public executions, at pugilistic combats, at cock-fightings, horse-races, bear-baitings, bull-fights, [and] gladiatorial shows."[84] He described these "brutal amusements" as "the desperate excitements of debauched men."[85] Beecher described "wretches who have waded in all the depths of iniquity, and burned every passion to the socket, find rage and blows and blood, the only stimulus of which you are susceptible."[86] Beecher advised young men, "You are training yourself to be just such wretches, if you are exhausting your passions in illicit indulgences."[87]

Yet Grier points out that adults disagreed about what types of animal cruelty were permissible. For example, Beecher and many Americans accepted boyhood hunting as a "cultural rehearsal for manhood, even

[80] Glenn, *Campaigns against Corporal Punishment*, 77.

[81] Glenn, *Campaigns against Corporal Punishment*, 80.

[82] Grier, *Pets in America*, 186–91; Grier, "Childhood Socialization," 109.

[83] Sara Josepha Hale, "Pets and Their Uses," in *Manners; or, Happy Homes and Good Society All the Year Round* (1868; repr., New York: Arno Press, 1972), 244. See discussion in Grier, "Childhood Socialization and Companion Animals," 110–11.

[84] Henry Ward Beecher, *Lectures to Young Men, on Various Important Subjects* (Salem: D. B. Brooks and Brother, 1855), 148–49.

[85] Beecher, *Lectures to Young Men*, 223.

[86] Beecher, *Lectures to Young Men*, 223–24.

[87] Beecher, *Lectures to Young Men*, 224.

the manhood of white-collar work."[88] William A. Alcott, alternatively, worried that those who killed for amusement were "developing and cultivating the spirit of cruelty."[89] He advised "young ladies" to use their influence—"as sisters, daughters or mothers"—to prevent the "cruelties of infancy and youth" by "some of those who are beyond the family."[90] Alcott entrusted women to prevent the males in their lives from "decidedly and deeply unchristian" activities.[91] He warned ladies not to smile on these "juvenile murderers" or to eat "the fruit of their doings," because "a cruel young man will never make a delicate friend or a good husband."[92]

11.6 Conclusion

Despite these arguments, the kindness-to-animals ethic did not solely focus on the negative effects of animal cruelty upon humans. The efforts of antebellum Americans to prevent animal cruelty were also tied to the gruesome scenes they witnessed every day. Lydia Maria Child was not alone when she described the brutal killing of dogs on the streets of New York City in 1841. Numerous contemporary editorials and articles also condemned the extermination.[93]

In 1864, a *New York Times* editorial column commented that on "Thirteenth Street, near Second Avenue, in this *well governed* City, a dying horse has lain since yesterday morning." The writer lamented, "How are the ends of justice served by permitting a horse to linger in agony on a public street, in full view of our doors, tortured by wicked boys, and exposed to the sun and flies?" The author questioned, "Why

[88] Grier, "Childhood Socialization," 108.

[89] William Andrus Alcott, *Gift Book for Young Ladies; Familiar Letters on Their Acquaintances, Male and Female, Employments, Friendships, &C.* (Auburn: Derby and Miller, 1853), 273, https://archive.org/details/giftbookforyoung00alco/page/n4.

[90] Alcott, *Gift Book for Young Ladies*, 273.

[91] Alcott, *Gift Book for Young Ladies*, 273–74. See Grier, "Childhood Socialization," 108, for a discussion.

[92] Alcott, *Gift Book for Young Ladies*, 188, 274.

[93] "Where the Dogs Go To," *Frank Leslie's*, August 14, 1858, 166–67; "A Dog on Fire," *Harper's Weekly*, March 7, 1857, 150.

is there no law to punish the owner of the animal? Will someone answer these questions for the sake of HUMANITY?"[94]

In 1866, another *New York Times* editorial questioned why the United States had "no society to watch for and procure the punishment of such brutalities." The author noted that "the formation of societies for the prevention of cruelties to animals is not new in the Old World, and should long ago have been effected in the new."[95]

These hopes were soon realized. The kindness-to-animals ethic set the stage for the growth of American animal protection societies. In 1866, Henry Bergh formed the first animal protection organization in New York.[96] Other states followed New York's example; as a result, animal protection organizations emerged throughout the country.[97]

Post–Civil War animal protectionists advocated for humane education to be taught to children—adopting the kindness-to-animals ethic prevalent in antebellum America. Anna Sewell's hugely popular novel *Black Beauty* was first published in the United States in 1890 by the American Humane Society, at the behest of George T. Angell of the Massachusetts Society for the Prevention of Cruelty to Animals. Angell pronounced

[94] "Inhumane Cruelty," *New York Times*, July 29, 1864, 2. See also "Cruelty to Animals: Letters from the People," *New York Times*, January 10, 1864, 5.

[95] "Work for the Idle," *New York Times*, January 10, 1866, 4.

[96] The same year, Bergh drafted anti-cruelty legislation. See N.Y. Rev. Stat. Ch. Sec. 682.2 (1866). "Every owner, driver or possessor of an old, maimed, or diseased horse or mule, turned loose or left disabled in any street, lane or place of any street in this state, who shall allow such horse or mule to lie in any street, lane or public place for more than three hours after knowledge of such disability, shall, on conviction, be adjudged guilty of a misdemeanor." Animal Welfare Institute, *Animals and Their Legal Rights* (Washington, DC: AWI, 1990), 5.This law replaced the antiquated 1829 anti-cruelty law. See N.Y. Rev. Tit. 6, Sec. 26 (1829). In 1867, Bergh drafted "An Act for the More Effectual Prevention of Cruelty to Animals," which was passed April 12, 1867. This statute has served as an example for many anti-cruelty statutes throughout the United States. N.Y. Rev. Stat. Sec. 375.2–9. For a further discussion, see David Favre and Vivien Tsang, "The Development of Anti-Cruelty Laws during the 1800s," *Detroit College of Law Review* (Spring 1993): 16, appendix A: "The 1867 New York Anti-Cruelty Law."

[97] George Angel founded the Massachusetts Society for the Prevention of Cruelty to Animals in 1868. Richard Muckle founded the Pennsylvania Society for the Prevention of Cruelty to Animals in 1867, and Caroline Earle White founded the Women's Pennsylvania Society for the Prevention of Animals in 1869.

Black Beauty "the Uncle Tom's Cabin of the Horse," and by the time of his death, its circulation had reached three million copies.[98]

Bibliography

Abzug, Robert H. *Cosmos Crumbling: American Reform and the Religious Imagination*. New York: Oxford University Press, 1994.

Alcott, William Andrus. *Gift Book for Young Ladies; Familiar Letters on Their Acquaintances, Male and Female, Employments, Friendships, &C.* Auburn: Derby and Miller, 1853. https://archive.org/details/giftbookforyoung00alco/page/n4.

Animal Welfare Institute. *Animals and Their Legal Rights*. Washington, DC: AWI, 1990.

Bailyn, Bernard. *The Ideological Origins of the American Revolution*. Cambridge, MA: Belknap Press of Harvard University Press, 1967.

Beecher, Henry Ward. *Lectures to Young Men, on Various Important Subjects*. Salem: D. B. Brooks and Brother, 1855.

Bloch, Ruth H. "American Feminine Ideals in Transition: The Rise of the Moral Mother, 1785–1815." *Feminist Studies* 4 (2 June 1978): 100–126.

Bushman, Richard L. *The Refinement of America: Persons, Houses, Cities*. New York: Alfred A. Knopf, 1992.

Child, Lydia Maria. *Isaac T. Hopper: A True Life*. Boston: John P. Jewett, 1853. https://archive.org/details/isaacthoppertrue00chil/page/9.

Child, Lydia Maria. *Letters from New York*. Edited by Bruce Mills. Athens: University of Georgia Press, 1998.

Child, Lydia Maria. *The Mother's Book*. 1831. Reprint, New York: Francis, 1844. https://archive.org/details/mothersbook1844chil/page/10.

Clark, Elizabeth B. "'The Sacred Rights of the Weak': Pain, Sympathy, and the Culture of Individual Rights in Antebellum America." *Journal of American History* 82, no. 2 (September 1995): 463–93.

Coleman, Sydney H. *Humane Society Leaders in America, with a Sketch of the Early History of the Humane Movement in England*. Albany: American Humane Association, 1924.

Cott, Nancy F. *The Bonds of Womanhood: "Women's Sphere" in New England, 1780–1835*. New Haven: Yale University Press, 1977.

Crane, Elaine F. "Dependence in the Era of Independence: The Role of Women in Republican Society." In *The American Revolution: Its Character and*

[98] Sydney H. Coleman, *Humane Society Leaders in America, with a Sketch of the Early History of the Humane Movement in England* (Albany: American Humane Association, 1924), 106.

Limits, edited by Jack P. Greene, 253–75. New York: New York University, 1987.

Epstein, Barbara Leslie. *The Politics of Domesticity: Women, Evangelism, and Temperance in Nineteenth-Century America, 1820–1860.* Middletown, CT: Wesleyan University Press, 1981.

Favre, David, and Vivien Tsang. "The Development of Anti-Cruelty Laws During the 1800s." *Detroit College of Law Review* (Spring 1993): 1–35.

Fiering, Norman S. "Irresistible Compassion: An Aspect of Eighteenth-Century Sympathy and Humanitarianism." *Journal of the History of Ideas* 37 (2 April–June 1976): 195–218.

Finley, Ruth E. *The Lady of Godey's: Sara Josepha Hale*. Philadelphia: J. B. Lippincott, 1931.

Frank Leslie's. "Where the Dogs Go To." August 14, 1858.

Ginzberg, Lori D. *Women and the Work of Benevolence: Morality, Politics, and Class in the Nineteenth-Century United States*. New Haven: Yale University Press, 1990.

Glenn, Myra C. *Campaigns against Corporal Punishment: Prisoners, Sailors, Women, and Children in Antebellum America*. Albany: State University of New York Press, 1984.

Goldsmith, Oliver. 1961. *The History of Little Goody Two Shoes*. 1765. Repr., New York: Macmillan, 1961.

Grier, Katherine C. "Childhood Socialization and Companion Animals: United States, 1820–1870." *Society and Animals* 7, no. 2 (1999): 95–120.

Grier, Katherine C. *Pets in America: A History*. Orlando, FL: Harcourt Inc., 2006.

Hale, Sara Josepha. "Pets and Their Uses." In *Manners; or, Happy Homes and Good Society All the Year Round*. 1868. Reprint, New York: Arno Press, 1972.

Halttunen, Karen. *Confidence Men and Painted Women: A Study of Middle-Class Culture in America, 1830–1870*. New Haven: Yale University Press, 1982.

Halttunen, Karen. "Humanitarianism and the Pornography of Pain in Anglo-American Culture." *American Historical Review* 100, no. 2 (April 1995): 303–34.

Harper's Weekly. "A Dog on Fire." March 7, 1857.

Hawke, David Freeman. *Benjamin Rush: The Revolutionary Gadfly*. Indianapolis: Bobbs-Merrill, 1971.

Howe, Daniel Walker. "Victorian Culture in America." In *Victorian America*, edited by Daniel Walker Howe, 3–28. Philadelphia: University of Pennsylvania Press, 1976.

Jefferson, Thomas. *The Papers of Thomas Jefferson, Retirement Series, Volume 4: June 1811 to April 1812*. Edited by J. Jefferson Looney. Princeton: Princeton University Press, 2007.

Kerber, Linda K. "The Republican Mother: Women and the Enlightenment—An American Perspective." *American Quarterly* 28, no. 2 (Summer 1976): 187–205.

Kerber, Linda K. *Women of the Republic: Intellect and Ideology in Revolutionary America*. New York: Norton, 1986.

Kraditor, Aileen S. *Up from the Pedestal: Selected Writings in the History of American Feminism*. Chicago: Quadrangle Books, 1968.

Locke, John. *Some Thoughts on Education*. 1693. Reprint, Boston: Gray and Bowen, 1830. https://archive.org/details/somethoughtscon01lockuoft.

Mason, Jennifer. *Civilized Creatures: Urban Animals, Sentimental Culture, and American Literature, 1850-1900*. Baltimore: John Hopkins University, 2005.

Mintz, Steven. *Moralists and Modernizers: America's Pre–Civil War Reformers*. Baltimore: Johns Hopkins University Press, 1995.

New York Times. "Cruelty to Animals: Letters from the People." January 10, 1864.

New York Times. "Inhumane Cruelty." July 29, 1864.

New York Times. "Work for the Idle." January 10, 1866.

Pope, Alexander. "Essay on Man." In *Essay on Man and Other Poems*, 45–79. 1733/1734. Reprint, New York: Dover, 1994.

Reinier, Jacqueline S. *From Virtue to Character: American Childhood, 1775–1850*. New York: Twayne, 1996.

Reinier, Jacqueline S. "Rearing the Republican Child: Attitudes and Practices in Post-Revolutionary Philadelphia." *William and Mary Quarterly*, 3rd ser., 39, no. 1 (January 1982): 150–63.

Rush, Benjamin, M. D. "An Enquiry into the Influence of Physical Causes upon Moral Faculty." In *Two Essays on the Mind*, 1–40. 1786. New York: Brunner/Mazel, 1972.

Rush, Benjamin, M. D. "Thoughts upon the Amusements and Punishments Which Are Proper for Schools. Addressed to George Clymer, Esq." In *Essays: Literary, Moral and Philosophical*, 57–74. Philadelphia: Thomas &William Bradford, 1798. https://archive.org/details/essaysliterarymo00rush_0/page/56/mode/2up.

Rush, Benjamin, M. D. "Of Mode of Education Proper in a Republic, In *Essays: Literary, Moral and Philosophical*, 6–19. Philadelphia: Thomas &William Bradford, 1798. https://archive.org/details/essaysliterarymo00rush_0/page/6/mode/2up.

Rush, Benjamin, M. D. "Thoughts upon Female Education, Accommodated to the Present State of Society, Manners, and Government in the United States: Addressed to Visitors of the Young Ladies' Academy in Philadelphia, 28th July 1787." In *Essays: Literary, Moral and Philosophical*, 75–92. Philadelphia: Thomas & William Bradford, 1798. https://archive.org/details/essaysliterarymo00rush_0/page/74/mode/2up.

Ryan, Mary P. *The Empire of the Mother: American Writing about Domesticity, 1830–1860*. New York: The Haworth Press, 1982.
Sellers, Charles. *The Market Revolution: Jacksonian America, 1815–1846*. New York: Oxford University Press, 1991.
Sigourney, Lydia H. *Letters to Mothers*. Hartford: Hudson and Skinner, 1838. https://archive.org/details/letterstomother02sigogoog/page/n40.
Skemp, Sheila L. *Judith Sargent Murray: A Brief Biography with Documents*. Bedford Series in History and Culture. Boston: Bedford/St. Martin, 1998.
Smith, Bill Leon. "Animals Made Americans Human: Sentient Creatures and the Creation of Early America's Moral Sensibility." *Journal of Animal Ethics* 2, no. 2 (Fall 2012): 127–40.
Stowe, Harriet Beecher. "Aunt Esther's Rules." In *Queer Little People*, 152–57. Boston: Houghton, Mifflin, 1867. https://archive.org/details/queerlittlepeopl00stow/page/n5.
Stowe, Harriet Beecher. "The Nest in the Orchard." In *Our Charlie, and What to Do with Him*, 108–17. https://archive.org/details/in.ernet.dli.2015.350688/page/n1.
Stowe, Harriet Beecher. "A Talk About Birds." In *Our Charlie, and What to Do with Him*, 94–108. https://archive.org/details/in.ernet.dli.2015.350688/page/n1.
Thomas, John L. "Romantic Reform in America, 1815–1865." *American Quarterly* 17, no. 4 (Winter 1965): 656–81.
Turner, James. *Animals, Pain, and Humanity in the Victorian Mind: Reckoning with the Beast*. Baltimore: Johns Hopkins University Press, 1980.
Unti, Bernard Oreste. "The Quality of Mercy: Organized Animal Protection in the United States, 1866–1930." PhD diss., American University, 2002.
Welter, Barbara. "The Cult of True Womanhood: 1820–1860." In *Dimity Convictions: The American Woman in the Nineteenth Century*, 21–41. Athens: Ohio University Press, 1976.
Wood, Gordon. *The Creation of the American Republic, 1776–1787*. Chapel Hill: University of North Carolina Press, 1969.
Wood, Gordon. *The Radicalism of the American Revolution*. New York: Vintage Books, 1991.
Yates, Michelle C. Neely. "The Antebellum Animal." PhD diss., University of California, 2011.

CHAPTER 12

“Habits of Mercy”: RSPCA Humane Education in Nineteenth-Century Britain

Isabel Barber

12.1 Introduction

Since Harriet Ritvo’s pioneering history *The Animal Estate*, which explored social and symbolic relationships between Victorians and animals,[1] her work has informed what is often termed the “animal turn” in social and cultural history. Historians have argued that major changes in the urban landscape transformed the relationships between humans and other animals in nineteenth-century Britain.[2] This included how animals were perceived, how animals were treated, and how charitable organizations sought to protect them.

[1] Harriet Ritvo, *The Animal Estate* (Cambridge: Harvard University Press, 1987).

[2] Chien-hui Li, “A Union of Christianity, Humanity, and Philanthropy: The Christian Tradition and the Prevention of Cruelty to Animals in Nineteenth-Century England,” *Society and Animals* 8, no. 3 (2000): 266.

I. Barber (✉)
London, UK
e-mail: isabelbarber98@gmail.com

C. Linzey and A. Linzey (eds.), *The Palgrave Companion to Humane Education*, The Palgrave Macmillan Animal Ethics Series,
https://doi.org/10.1007/978-3-032-19795-5_12

The Royal Society for the Prevention of Cruelty to Animals (RSPCA) believed education was a vital means of protecting animals from human cruelty. The year 1870 marked a drastic shift within the organization's direction as the RSPCA reignited its humane education efforts with two main initiatives. The first was the creation of a new monthly journal, *The Animal World*, first published October 1869. The second enterprise was the formation of a new Ladies' Humane Education Committee in London. Tasked by the General Committee to oversee the management of *The Animal World*, the Ladies' Committee was responsible for leading RSPCA education efforts across the country. In the first part of this chapter, I will provide a brief history of these events, while aiming to situate the educational initiatives of the RSPCA in a wider context of non-punitive philanthropy.

In the second part of the chapter, I will turn to examine the specifics of RSPCA humane education. The promoted rhetoric of "kindness" in *The Animal World* and notable textbooks distributed to schools was inconsistent in its application. I will explore why children were taught to cherish certain animals while others were presented instead through the discussion of humane slaughter. In my following paragraphs and conclusion, I will consider how humane education was influenced by sociopolitical pressures and suggest these may still present problems today.

12.2 Historical Overview

The Ladies' Committee

The catalyst for formation of the Ladies' Humane Education Committee was a meeting in 1869 between George Angell, an American lawyer, philanthropist, and president of the Massachusetts Society for the Prevention of Cruelty to Animals, and Angela Burdett-Coutts, one of the wealthiest British philanthropists of the period. Burdett-Coutts had written to the *Times* of London, appealing to educators to adopt humane education, and John Colam, Liberal minister and RSPCA secretary, had responded by requesting public help to form a ladies' society for the purposes of education.[3] The committee first met in London in July 1870. Its membership is first listed in the RSPCA's 1875 annual report, by which

[3] "Systematic Education for the Humane Treatment of Animals," *Times* (London), September 14, 1869, p. 6, col a, Times Digital Archive.

time it had become a well-established auxiliary to the General Committee, with several sister branches across the country.[4]

The Animal World

In the 1870s, children became a focal audience for the RSPCA's charity funds. Previously, in 1861, the *Annual Report* had shown that of forty-five educational pamphlets and books, only six were aimed at children, and overall educational spending remained limited.[5] The catalyst for the RSPCA noticing the value of education was the appointment of John Colam as RSPCA secretary in 1861. Colam began a program of lectures and school visits,[6] and it was this intensification of humane education endeavors that culminated in the publication of *The Animal World* in 1869. The journal was printed on toned paper, ran sixteen pages in length, was illustrated throughout, and could be purchased at the cost of two pennies.

In an 1870 circular issued by the Ladies' Committee, Burdett-Coutts was explicit in wanting the involvement and education of "all classes."[7] To target the working class and avoid reliance solely on middle-class parents who could afford journal subscription fees, the committee distributed free copies of *The Animal World* to schools. This amounted to 310,000 copies printed during the publication's twelve-month infancy. The committee remained adamant during the following years that *The Animal World* was "not to realize a pecuniary profit, but a moral one; and it will fail in its mission until it is placed in all schools."[8]

Alongside *The Animal World*, the RSPCA organized an annual essay competition, promoted within the journal. This was a typical nineteenth-century phenomenon popular with numerous temperance and "nature

[4] *RSPCA Annual Report 1875*, 4, RSPCA Archive, Horsham.

[5] Frederick Stephen Milton, "Taking the Pledge: A Study of Children's Societies for the Prevention of Cruelty to Birds and Animals in Britain, c.1870–1914" (PhD diss., University of Newcastle, 2009), 73.

[6] Milton, "Taking the Pledge," 161.

[7] "Circular Issued by the Society of the Prevention of Cruelty to Animals Ladies' Humane Education Committee," December 1, 1870, City of Westminster Archives Centre, London, BUR/L5, BUR/2704/7/4.

[8] *Ladies' Committee Minute Books: July 1870–May 1905*, 20, RSPCA Archive, Horsham.

societies," although the efforts of the RSPCA dwarfed all other competitions.[9] By 1878, the London branch had invited "every school... whose address is published in the London Directory" to take part.[10] Covering a radius of twenty miles, over five thousand schools received invites.[11] Each year, a prestigious prize ceremony was held to reward winning submissions, and the contest's scale and reputation grew throughout the 1870s.[12]

The Wider Cause

The clear shift in RSPCA policy toward humane education mirrored wider reform movements. Many organizations changed their approach from protection through legislation to protection through morally informed prevention. No longer did the RSPCA's mission end at prosecutions; in the 1870s it embraced a non-punitive educational approach. In 1869, the *Annual Report* clearly acknowledged this change in ethos: "the Society, although actively engaged for many years past in the prosecution of offenders, has not hitherto sufficiently devoted its energies and funds to the promotion of humane education."[13]

This preventive focus is explicit in the first issue of *The Animal World*. "Prosecutions" alone were described as "insufficient," and the editor stressed that "measures should be devised for training the mind when it is young."[14] Otherwise, it was feared that a youth would develop into a "hardened man of evil communications, with fixed ideas and habits," without a means of "soften[ing] him down to tenderness."[15] This appears to illustrate a Victorian concern that immorality was harder to combat in adults, due to a propensity for acting cruelly developing in individuals at a younger age.

[9] Milton, "Taking the Pledge," 188.

[10] RSPCA Annual Report 1878, 67, RSPCA Archive, Horsham.

[11] Milton, "Taking the Pledge," 291.

[12] *RSPCA Annual Report 1871*, 25, RSPCA Archive, Horsham; RSPCA Annual Report 1873, 24, RSPCA Archive, Horsham.

[13] *RSPCA Annual Report 1869*, 7, RSPCA Archive, Horsham.

[14] *The Animal World: A Monthly Advocate for Humanity* 1, no. 1 (October 1869): 8.

[15] *The Animal World* 1, no. 1 (October 1869): 8.

The Animal World indicates that humane education targeting animals was therefore as much an act motivated by a wish to reform human society. It stated that teaching children "kindness to animals" would tackle the individual who is a "burden to society, a terror to everybody—a selfish, sensual, woman-beating, child-neglecting, animal-torturing animal."[16] The notion of an ideological correlation between violence to humans and violence to animals was not uncommon.[17] This rhetoric permeates the introductory focus of the new committee and its education endeavors. An early circular of the Ladies' Committee includes supporting input from teachers and academics. One contribution, from Harold Browne, theologian and bishop of Ely, stressed the need to create "habits of mercy" not only toward animals, but "both to man and beast."[18] Similarly, at its 1870 annual meeting, the RSPCA proclaimed that "while [*The Animal World*'s] primary object is the protection of creatures who minister to man's wants, it is obvious that in no less degree it seeks to elevate human nature."[19] The two objects were from the offset seen as parallel, complementary aims.

The work of the Ladies' Committee was internally viewed as part of a larger philanthropic mission in society. The individuals involved believed themselves agents of progressive reform and drew parallels not just between their present undertakings but also with the past legacy of British activism. Speaking five years into the new humane education program on the meaning of "civilised" society, the RSPCA chairman, K.G. Earl of Harrowby, identified the "real triumphs" of the Ladies' Committee as "softening and regulating human passions," before proceeding to discuss their work in relation to slavery. He declared that "not long ago we had a great struggle for the emancipation of slaves" and expressed hope "that the name of England will be hereafter associated with the protection of the lower animals as well as of our own kind."[20]

Many of the initial proponents of legislating against cruelty to animals also had been active in the antislavery movement. Renowned British

[16] *The Animal World* 1, no. 1 (October 1869): 8.

[17] Ritvo, *Animal Estate*, 131–32.

[18] "Circular Issued by... Ladies' Humane Education Committee," City of Westminster Archives Centre.

[19] RSPCA Annual Report 1869, 7.

[20] RSPCA Annual Report 1874, 36, RSPCA Archive, Horsham.

politician and philanthropist William Wilberforce not only led the British abolitionist campaign but also in 1800 was one of the first politicians to initiate legislating against animal cruelty; he was subsequently one of the SPCA founders in 1824 (the organization became the RSPCA in 1840). Liverpool, previously the slavery capital of Europe, witnessed the pinnacle of antislavery activism, and similar organizing occurred on the animal front.[21] That the city was one of the earliest to create an RSPCA sister branch to the London humane education committee should come as no surprise. The chairman's allusion to this history tells us the RSPCA was aware of its members' past involvement in the "emancipation of slaves" and prided itself on continuing to fight similar "great struggle[s]."

12.3 Kindness to Animals

Utility

RSPCA humane education in the 1870s differed in its treatment of different animals. Although sometimes the RSPCA's rhetoric claimed that kindness to animals was universal and automatic, in much of the literature it produced, some animals were granted consideration only in terms of their utility to provide meat or clothing. Varying degrees and forms of violence against animals played a crucial role in shaping how society distinguished between different species and different persons.[22] Although each species already shared its own relationship with humans, these relationships were not fixed. The status of the companion animal grew; meat-eating was reevaluated against new attitudes to violence; and the morals of vivisection and hunting were challenged.

The Dog

Bradshaw argues that the nineteenth century saw a growing distinction between "outdoor" and "household" animals, as attitudes toward domesticated animals changed.[23] In particular, Bradshaw explores how

[21] Joanna Bourke, *What It Means to Be Human: Reflections from 1791 to the Present* (London: Virago, 2011), 129.

[22] Bourke, *What It Means*, 128.

[23] John Bradshaw, *The Animals among Us: The New Science of Anthrozoology* (London: Allen Lane, 2017), 52.

dogs became caricatured as "selfless heroes, exemplars of fidelity and loyalty."[24] RSPCA humane education enlightens this case. Visual depictions in *The Animal World* appear to encourage children to feel a stronger emotional connection with "pets" over free-roaming animals and domesticated livestock. Illustrations by renowned Victorian natural history artist Harrison Weir, an active participant within the RSPCA, feature prominently. Although they are predominantly sentimental and provide detailed depictions of both "household" and "outdoor" animals, the nature of the empathy evoked in each image is variable, depending on the type of animal in question.

Occasionally depicted with their young, cows and sheep most often appear heterogeneously in herds (see Figs. 12.1 and 12.2).[25] Only the dog appears as a heroic individual. He is often a distinct character within a narrative, and his relationship with humans is always explicit. In three of the illustrations that appear throughout volume 1 of *The Animal World*, the dog is depicted as "A Churchyard Mourner" (Figs. 12.3, 12.4, and 12.5).[26] In a bolder example, another dog can be seen at the forefront of a stormy shipwreck, pulling a boat in danger, alongside the caption "The Dog that Saved the Crew" (Fig. 12.6).[27]

Children were encouraged to develop sympathy with an array of animals throughout the illustrations, but the importance of an emotional connection with dogs took precedence over connection with all other species. Tague summarizes, "The nineteenth century saw the increasing bifurcation of animals into two distinct categories: there were pets, and then there were all other animals."[28] The greater emphasis on dogs throughout RSPCA publications was a reflection of the public's growing consideration toward their welfare at large. Statistics from 1857 to 1860 showed that although canine and feline victims accounted for only 2

24 Bradshaw, *Animals among Us*, 53.

25 *The Animal World* 1, no. 5 (February 1870): 89; *The Animal World* 1, no. 3 (December 1869): 77.

26 *The Animal World* 1, no. 2 (November 1869): 37; *The Animal World* 1, no. 8 (May 1870): 1.

27 *The Animal World* 1, no. 5 (February 1870): 93.

28 Ingrid H. Tague, "The History of Emotional Attachment to Animals," in *The Routledge Companion to Animal–Human History*, ed. Hilda Kean and Philip Howell (London: Routledge, 2019), 358.

Fig. 12.1 Highland cattle. *The Animal World* 1, no. 5 (February 1870): 89

percent of animal cruelty convictions, cats and dogs were disproportionately the main focus of 13 percent of the RSPCA's published reports on cruelty.[29]

[29] Ritvo, *Animal Estate*, 141.

Fig. 12.2 Sheep. *The Animal World* 1, no. 3 (December 1869): 77

One explanation for this preference is that the image and reality of the dog became a new symbol of family life. Dogs now functioned less as useful guards for the house and more as emotional beings within the home. Intelligence was assigned to them based on their expression of "sagacity," which Milton clarifies was less about "intelligent mental power" and more about "the ability to adapt to human surroundings and

Fig. 12.3 "A Churchyard Mourner," by Harrison Weir, 1861. *The Animal World* 1, no. 2 (November 1869): 37

to please people."[30] As Howell argues, dogs became a vital part of the home for the middle class's "domestic family life," alongside women and children. Within this new domestic sphere, the "urban bourgeoisie" could

[30] Milton, "Taking the Pledge," 267; Hilda Kean, *Animal Rights: Political and Social Change in Britain Since 1800* (London: Reaktion Books, 1998), 84–88.

Fig. 12.4 Grey friar's bobby. *The Animal World* 1, no. 8 (May 1870): 1

adopt pet-keeping in order to contrast themselves against the working classes' relationship with their animals; the practice was "emblematic of their values" of "domestication."[31]

[31] Philip Howell, *At Home and Astray: The Domestic Dog in Victorian Britain* (Charlottesville: University of Virginia Press, 2015), 14.

Fig. 12.5 "Poor Shelvey is dead and buried, but his dog and children remain." *The Animal World* 2, no. 16 (January 1871): 1

Fig. 12.6 "The Dog that Saved the Crew," by Harrison Weir, 1863. *The Animal World* 1, no. 5 (February 1870): 93

Attitudes to Violence

The early nineteenth century was a period subject to a "heightened sensitivity about pain."[32] Nonhuman animals were often viewed as able to suffer, yet the association of deliberate "pain" with "cruelty" was subject to the experience of the animal within urban geography and culture. Activists sought to protect dogs through public displays, such as an "exhibition of painted and cautionary notices in the public parks to prevent cruelty," while for animals who were traditionally eaten, kindness was promoted through the "erection of public slaughterhouses."[33] Not only did killing continue, but it also evaded, despite increased "sensitivities" to "pain," the label of "cruelty." Two strands of thought facilitated this: perceiving meat-eating as necessary and redefining slaughter as "humane."

RSPCA humane education provided justifications for the consumption of animals for food. In one widely distributed textbook, *Our Duty to Animals*, Caroline Bray, writer of school textbooks and avid supporter of the RSPCA in Coventry, teaches,

> When we are eating meat we sometimes think how sad it is that we cannot take this wholesome food without having caused pain and death to some poor animal! … Some persons feel this so strongly that they refuse to eat meat all their lives. But let us see what would happen if animals never fed upon one another. All that must live must eat … all that live must die; therefore, if animals were never killed or eaten, the earth would soon become covered in carcasses of the dead.[34]

It is interesting to consider whether this rhetoric was truly believed. Perhaps Bray's argument served to articulate a narrative of necessity in order to soothe perturbed morals. Bray goes on to argue that "all that die a natural death from old age or disease, must generally pass through a time of weakness, helplessness, and suffering" and that it is crueler to leave animals to die on their own accord than it is to kill them first.[35] This

[32] James Turner, *Reckoning with the Beast: Animals, Pain, and Humanity in the Victorian Mind* (London: Johns Hopkins University Press, 1980), 81.

[33] RSPCA Annual Report 1871, 21.

[34] Caroline Bray, *Our Duty to Animals* (London: S. W. Partridge, 1870), 3.

[35] Bray, *Our Duty to Animals*, 4.

promotion of meat-eating was indicative of society at large. Vegetarianism did not proliferate, and meat consumption grew per person throughout the century from the 1840s.[36] The RSPCA never expressed any official vegetarian sympathies.

The focus on producing justifications for meat-eating within humane education literature suggests that violence to animals became a predicament for those Victorians who congratulated themselves on moral improvement. A stoic and civilized character was idealized as distinct from the barbaric violence that moralists identified in those involved in the 1830 and 1848 French Revolutions.[37] But civilized behavior was not by default synonymous with the practice of killing. To overcome this problem, humane education avoided a focus on the animal subject and instead chose to wholly stage the matter around the method of death.

Street butchers and market poulterers were "branded as savages" and viewed as the exception to otherwise acceptable slaughter practice.[38] Humanitarian concerns for animals were in part raised by sanitary reformers who wanted to improve conditions and promote welfare on the basis of public health.[39] Clara Balfour, temperance activist and advocate for women's advancement, wrote for *The Animal World* on the subject of humane slaughter. She praised "the Abattoirs of Paris" for "their cleanliness" and "merciful preparations for swift, and nearly painless death."[40] By contrast, she declared, in England butchers would "coldly calculate" "blows on the head and feet" of the animal, "the most sensitive parts of his own frame."[41] Balfour was intrigued by the idea of adopting these new slaughter methods from Paris, convinced that butchery could become "a merciful trade."[42] McDonnell further discusses how attitudes to killing shifted in this period, as the ideology of cruelty to animals was seen to

[36] Christopher Otter, "Civilising Slaughter: The Development of the British Public Abattoir, 1850–1910," in *Meat, Modernity, and the Rise of the Slaughterhouse*, ed. Paula Young Lee (Durham: University of New Hampshire Press, 2008), 89.

[37] Colin Matthew, "Public Life and Politics," in *The Nineteenth Century: The British Isles 1815–1901*, ed. Colin Matthew (Oxford: Oxford University Press, 2000), 132.

[38] *RSPCA Annual Report 1869*, 8.

[39] Kean, *Animal Rights*, 103.

[40] *The Animal World* 1, no. 10 (July 1870): 170.

[41] *The Animal World* 1, no. 10 (July 1870): 170.

[42] *The Animal World* 1, no. 10 (July 1870): 171.

corrupt and dehumanize humans. Engaging with the idea of the "politics of sight," a proposed slaughterhouse bill in 1857, to prevent children under fourteen from witnessing slaughterhouse deaths, was one of many motions concerned with how the sight could brutalize people.[43] Meat might not have been regarded as murder, but it did need to become a commodity that resembled something clean and palatable to a "civilized" society. Blood and death had to be made invisible, reflected in the rise in this period of the public abattoir housed on the perimeter of the city.[44]

The determination of kindness versus cruelty often depended more on who was involved in the animal–human interaction than on what harm was being done. In Bray's textbook, addressing the oxen killed to supply "beef and suet, milk, and stout leather," Bray instructs children to "[treat] them with kindness," a term that is left undefined.[45] The book instead ends by exploring the question "What is cruelty?" Bray argues, "When a surgeon cuts off a limb, or a dentist pulls out a tooth, he gives a great deal of pain. But he is not cruel. When a butcher kills the animal for food, he cannot help giving him pain; but this is not cruelty … cruelty is the giving of needless pain; and the worst kind of pain is to have pleasure in giving it."[46] In this way, "cruelty" was determined less by the threshold of pain itself than by the social authority of those administering it, reflecting a classed distribution of "habits of mercy."

Vivisection

The first 1869 issue of *The Animal World* declared an interest and "duty" to provide arguments "for and against vivisections."[47] This was due to conflicting pressures from intellectual middle-class scientists against public opinion. Howell argues that scientists were viewed as "dog stealers" and vivisection as the "antithesis of domestic values" for the cruelty it brought

[43] Jennifer McDonnell, "Representing Animals in the Literature of Victorian Britain," in *The Routledge Companion to Animal–Human History*, ed. Hilda Kean and Philip Howell (London: Routledge, 2019), 231.

[44] Otter, "Civilising Slaughter," 96.

[45] Bray, *Our Duty to Animals*, 9.

[46] Bray, *Our Duty to Animals*, 151.

[47] *The Animal World* 1, no. 1 (October 1869): 14.

against the beloved dog.[48] Despite this, the RSPCA was tolerant of vivisection once mandatory anesthesia was debated and won.[49]

A year later, the tone from the editorial team changed. When the Ladies' Committee took responsibility for *The Animal World* in 1870, they expressed a wish to "thank the General Committee for the blow they have given to excessive experimentation," after they refused to "make further grants of money for the promotion of experiments upon animals."[50] However, the ambiguous phrasing of "excessive" rather than "all" leaves considerable room for interpretation. In this context, the categories of cruelty and kindness within humane education could be reconfigured to accommodate the infliction of pain on nonhuman animals in the name of scientific progress. This reflects Ritvo's argument that there was an "exclusive prerogative" afforded to educated scientists who believed they "automatically belonged to a different moral category from drovers and omnibus drivers,"[51] reinforcing the role of class in shaping "habits of mercy."

Hunting

Social-class conflict continued to exert tensions on the RSPCA's ethical stance regarding hunting and field sports. The RSPCA as an organization did not provide a clear condemnation of hunting, but the messages within its humane education publications were more complex. In the letter columns of *The Animal World*, various readers expressed an opinion that hunting was an instance of "cruelty to the lower animal" that "disgraces the humanity of the upper classes" who so eagerly denigrated against the "plebeian cruelties" of "cattle-dealers" and "butchers."[52] The editor of *The Animal World* did not respond to each letter individually but published a statement that declared the RSPCA's "limits." He wrote that although it was "flattering to the conductors of this Journal to receive letters every day of the most earnest entreaty to the Editor to attack this

[48] Howell, *At Home and Astray*, 114.

[49] Mark N. Ozer, "The British Vivisection Controversy," *Bulletin of the History of Medicine* 40, no. 2 (1966): 161.

[50] *The Animal World* 2, no. 13 (October 1870): 6.

[51] Ritvo, *Animal Estate*, 157–58.

[52] *The Animal World* 1, no. 5 (February 1870): 94.

and that abuse … the entire base of our supporters are, perhaps, equally divided in opinion as to the cruelty of Field Sports."[53]

Debates within *The Animal World* coincided with a national debate on hunting between historian and Liberal politician E. A. Freeman and novelist Anthony Trollope in newspaper columns during 1869 and 1870.[54] Freeman, who attacked the RSPCA for its nonchalant statement on hunting in *The Animal World*, argued that fox hunting was a form of "savage amusement," whereas Trollope argued that hunting was necessary to one's ability to be "manly." Freeman argued that "cruelty" was a subjective concept that depended on different stages of sensibility; something acceptable in one era could be perceived as barbaric in the next. Violence was "invented," or "continually discovered" in new forms.[55] When redefined in Victorian Britain, the new parameters of unacceptable human violence to animals did not include the "sport" of the gentry. Just as in the case of vivisection and the middle-class professional, the British gentleman was made an exception to the rule of deliberate cruelty to animals and shrouded from official criticism for breaching "habits of mercy."

The Animal World's neutral statement on field sports did not, however, fully represent the moral arguments made more subtly by some of its journalists. The power of visual imagery was utilized effectively, as seen in the May 1870 issue's inclusion of a startling illustration of an injured pigeon. The bird lies suffering in the fore frame, while in the background gentry are depicted shooting birds.[56] The aim was clear: to evoke sympathy with the wounded bird in children. Such instances supporting anti-hunting lobbyists within *The Animal World* were not mirrored within the RSPCA's annual reports, suggesting a distinction between editorial advocacy and official institutional discourse. Ritvo notes that in one of the early essay competitions, the society "felt obligated to exempt hunting from their strictures" to avoid controversy,[57] and none of the remaining archival essay extracts provides any mention of the "sport." Thus hunting

[53] *The Animal World* 1, no. 5 (February 1870): 88.

[54] Rob Boddice, "Manliness and the 'Morality of Field Sports': E. A. Freeman and Anthony Trollope, 1869–71," *Historian* 70, no.1 (2008): 20.

[55] Boddice, "Manliness," 15–16.

[56] *The Animal World* 1, no. 8 (May 1870): 133.

[57] Ritvo, *Animal Estate*, 134.

emerged as a category of violence simultaneously visible and unspeakable within the RSPCA educational framework.

12.4 Conclusion

The revival of RSPCA humane education in the nineteenth century was part of a wider endeavor to eliminate the suffering of animals and humans. Members of the Ladies' Committee had their roots in the campaign to abolish slavery and saw themselves as inhabiting a place in a larger social movement seeking to morally reform society. Animals were considered, alongside a plethora of human causes, as members of a vulnerable class worthy of consideration. Notably, charities striving to protect animals from harm preceded any national organization attempting the same for children; the National Society for the Prevention of Cruelty to Children was founded in 1884, sixty years after the RSPCA.

Changing conceptions of the civilized human self both encouraged pet-keeping and produced new anxieties around killing, requiring the negotiation of attitudes toward violence. The notion of "kindness" associated with humane slaughter established in the 1870s increasingly operated within a framework in which violence was not eliminated but differentially classified according to social legitimacy. Within the RSPCA, this structure persists into the present. As of May 2026, the RSPCA still promotes "RSPCA Assured" meat according to "strict welfare standards." Although "habits of mercy" associated with vivisection and hunting have faced widespread scrutiny and condemnation within the organization, the premise of vegetarianism remains absent on the humane education agenda. The Victorian era legislation, which, despite recognizing that animals were worthy of moral consideration and protection, worked to regulate and systematize violence, remains visible in the organization. For the majority of animal species, the nineteenth century should be characterized as a period of increased suffering.[58] This pattern persists into the present day where nineteenth- and twentieth-century moral assumptions about acceptable violence towards animals continue to shape humane education, often in tension to anti-speciesist critiques. Contemporary humane education advocates therefore

[58] McDonnell, "Representing Animals," 230.

face the challenge of arguing to schools, teachers, and government officials alike that nonhuman animals deserve their place on the curriculum alongside other human and environmental causes.

Acknowledgment Thank you to the RSPCA Archive at Horsham for permission to reproduce images from *The Animal World*.

Bibliography

Primary Sources

The Animal World: A Monthly Advocate for Humanity 1, no. 1 (October 1869), no. 2 (November 1869), no. 3 (December 1869), no. 5 (February 1870), no. 8 (May 1870), no. 10 (July 1870); 2, no. 13 (October 1870); 2, no. 16 (January 1871).

Bray, Caroline. *Our Duty to Animals*. London: S. W. Partridge, 1870.

"Circular Issued by the Society of the Prevention of Cruelty to Animals Ladies' Humane Education Committee." December 1, 1870. City of Westminster Archives Centre, London, BUR/L5, BUR/2704/7/4.

Ladies' Committee Minute Books: July 1870–May 1905. RSPCA Archive, Horsham.

RSPCA Annual Report 1869. RSPCA Archive, Horsham.

RSPCA Annual Report 1871. RSPCA Archive, Horsham.

RSPCA Annual Report 1873. RSPCA Archive, Horsham.

RSPCA Annual Report 1874. RSPCA Archive, Horsham.

RSPCA Annual Report 1875. RSPCA Archive, Horsham.

RSPCA Annual Report 1878. RSPCA Archive, Horsham.

Times (London). "Systematic Education for the Humane Treatment of Animals." September 14, 1869, p. 6, col. a. Times Digital Archive.

Secondary Sources

Boddice, Rob. "Manliness and the 'Morality of Field Sports': E.A. Freeman and Anthony Trollope, 1869–71." *Historian* 70, no. 1 (2008): 1–29.

Bourke, Joanna. *What It Means to Be Human: Reflections from 1791 to the Present*. London: Virago, 2011.

Bradshaw, John. *The Animals Among Us: The New Science of Anthrozoology*. London: Allen Lane, 2017.

Howell, Philip. *At Home and Astray: The Domestic Dog in Victorian Britain*. Charlottesville: University of Virginia Press, 2015.

Kean, Hilda. *Animal Rights: Political and Social Change in Britain Since 1800*. London: Reaktion Books, 1998.

Li, Chien-hui. "A Union of Christianity, Humanity, and Philanthropy: The Christian Tradition and the Prevention of Cruelty to Animals in Nineteenth-Century England." *Society and Animals* 8, no. 3 (2000): 265–85.

Matthew, Colin. "Public Life and Politics." In *The Nineteenth Century: The British Isles 1815–1901*, edited by Colin Matthew, 85–134. Oxford: Oxford University Press, 2000.

McDonnell, Jennifer. "Representing Animals in the Literature of Victorian Britain." In *The Routledge Companion to Animal–Human History*, edited by Hilda Kean and Philip Howell, 222–50. London: Routledge, 2019.

Milton, Frederick Stephen. "Taking the Pledge: A Study of Children's Societies for the Prevention of Cruelty to Birds and Animals in Britain, c.1870–1914." PhD diss., University of Newcastle, 2009.

Otter, Christopher. "Civilising Slaughter: The Development of the British Public Abattoir, 1850–1910." In *Meat, Modernity, and the Rise of the Slaughterhouse*, edited by Paula Young Lee, 89–106. Durham: University of New Hampshire Press, 2008.

Ozer, Mark N. "The British Vivisection Controversy." *Bulletin of the History of Medicine* 40, no. 2 (1966): 158–67.

Ritvo, Harriet. *The Animal Estate*. Cambridge: Harvard University Press, 1987.

Tague, Ingrid H. "The History of Emotional Attachment to Animals." In *The Routledge Companion to Animal–Human History*, edited by Hilda Kean and Philip Howell, 345–66. London: Routledge, 2019.

Turner, James. *Reckoning with the Beast: Animals, Pain, and Humanity in the Victorian Mind*. London: Johns Hopkins University Press, 1980.

CHAPTER 13

Deep in the Heartland: Humane Education in Oklahoma

Bob Blackburn, Louisa McCune, and Nancy E. Snow

Endeavoring toward an ideal—Plato's "Form of the Good"—is an actionable goal that has existed since humans began engaging with one another, most famously so in the Socratic dialogues. These ideals constitute the basis of American life and, throughout the nation's 245 years, have percolated in all fifty states in varying ways and at myriad times.

One place may serve as a watchtower of kindness to animals, with the holy grail of humane education discovered in its founding legislative documents. In the United States, as proponents of humane education gained momentum during the opening decades of the twentieth century, the reform movement—a period dedicated to American "spiritual and

B. Blackburn · N. E. Snow (✉)
Department of Philosphy, The University of Kansas, Lawrence, KS, USA
e-mail: nsnow@ku.edu

L. McCune
Kirkpatrick Foundation, Oklahoma City, OK, USA
e-mail: susan@kirkpatrickfoundation.com

C. Linzey and A. Linzey (eds.), *The Palgrave Companion to Humane Education*, The Palgrave Macmillan Animal Ethics Series,
https://doi.org/10.1007/978-3-032-19795-5_13

secular uplift"[1]—achieved some of its earliest victories in an unlikely part of the country that was only slightly removed from the frontier era of American Indian reservations, cowboys, and land runs. There, deep in the American heartland, laws were enacted in 1905 and 1914 mandating that humane education be considered core curriculum in all public schools, with stiff penalties for noncompliance. That unlikely beacon of hope was Oklahoma.

The roots of that success story are found in the twin movements known as populism and progressivism, which dominated the politics of Oklahoma Territory after 1890 and in the first decade of statehood after 1907. Whereas populism is often simplified as a romantic uprising of rural farmers fighting railroads and special interest groups, it is most easily understood as a movement based on evangelical Christianity with a strong belief that corporate elites, corrupt elected officials, and urban culture were destroying what was sacred to common folks in America.

Populism, both as a grassroots movement and as a political party, overlapped the progressive movement that rose out of the urban middle class as a response to monopolies, unregulated exploitation of resources both human and natural, and the belief that society could be improved through legislation, regulation, and strong government. Although progressives such as Theodore Roosevelt and populists such as William Jennings Bryan had conflicting solutions to perceived problems, they had one thing in common—a passion for reform at both the personal and public levels.

The leading citizens of the territory and early state of Oklahoma, drawn from all corners of the nation and world by free land and oil booms, largely embraced reform. The Women's Christian Temperance Union, led in Oklahoma by forceful personalities such as the axe-swinging Carrie Nation, succeeded in embedding prohibition of alcoholic beverages in the state constitution. The campaign to protect children, led in Oklahoma by crusading Kate Barnard, also enshrined child labor laws and compulsory education in the state constitution. Although the right of women to vote was not initially secured in the state constitution, the Federation of Women's Clubs, led in Oklahoma by Mrs. Selwyn Douglas, achieved its goal of suffrage in Oklahoma a year earlier than national suffrage.

The pages of Oklahoma history from the turn of the twentieth century include other victories for reform. From 1902 to 1905, with strong

[1] Brittanica.com, s.v. "An Age of Reform," accessed September 10, 2021, https://www.britannica.com/place/United-States/An-age-of-reform.

support across the state, Congress set aside public land and created Platt National Park and the Wichita Wildlife Refuge to conserve natural resources. In 1907, after Kate Barnard exposed the brutal exploitation of state prisoners housed in Kansas, the Oklahoma Legislature funded a new state prison founded on the principles of rehabilitation and education rather than punishment alone. In 1909, the people of Oklahoma City overwhelmingly approved a bond issue to create four expansive parks far beyond the city limits, connected by a scenic parkway called Grand Boulevard. Each of these achievements relied on the urge to reform both the individual and society in general.

It was an easy jump to the cause of humane treatment of animals as another tool to fight immorality, criminal behavior, and the abuse of wives and children, all threats to the social fabric. "Cruelty to animals jeopardizes a society's claim to civilization," wrote one national reformer. "It is the responsibility of those who understand that to instruct and influence those who do not."[2] Linked with a growing concern for the natural environment, humane education promised "ethical benefits, moral guidance, spiritual inspiration, and healthy exercise."[3] At a time when Horatio Alger's books were runaway bestsellers as a moral guide to a good and healthy life, and when the Boy Scouts movement was gaining strength with rules such as "A Scout is kind... he is a friend to animals," humane education struck a chord with many Oklahomans.[4]

The Bena Dea Club in the small town of Tecumseh, Oklahoma, consisting of sixteen self-described "active and enthusiastic women," supported the crusade through their local newspaper. "The object of our club is self-advancement and social improvement," they wrote. "The club has done all it could to support humane education... we have placed in the public schools of each district in our county a scrapbook pertaining to cruelty to dumb animals, a work in which our late President of the Women's Clubs Federation, Mrs. Selwyn Douglas, was so deeply interested."[5]

[2] Quoted in Bernard Oreste Unti, "The Quality of Mercy: Organized Animal Protection in the United States 1866-1930" (PhD diss., American University, 2002), 264.

[3] Unti, "The Quality of Mercy," 575.

[4] Robert Baden-Powell, "Scout Law," in *Scouting for Boys* (London: C. Arthur Pearson, 1908).

[5] "News from the Clubs," *Daily Oklahoman*, April 12, 1903, 11.

The Reverend C. J. Christianson, pastor of the Harrison Avenue Congregational Church in Oklahoma City, launched his church into the battle with a fiery sermon that was covered in the newspaper. "Do unto animals as you would have them do unto you if you were an animal," he began. "Crack, crack, crack goes the whip upon the poor tired horse.... Is the driver a Christian? Is he a man?"[6] With mounting passion, the reverend continued the assault. "The man who maltreats a poor helpless animal is more of a brute than the animal he maltreats... he may say he loves his fellow man, his church, his God, but there must be something decidedly wrong somewhere, for God loves these animals."[7]

Responding to this chorus of voices in the grassroots community, Leslie Niblack, a newspaper publisher in Guthrie, Oklahoma, and outspoken leader in the Democratic Party, filed a humane education bill for consideration during the 1905 legislative session. Only one other such law existed in the country, in Pennsylvania. Seeking to amend the laws of the territory, it started with a summary: "An Act to provide for moral and humane education in the public schools of the Territory of Oklahoma." The body of the text read,

> Section 8233: Teach Children Ethics: In each and every public school it shall be the duty of each and every teacher to teach morality in the broadest meaning of the word, for the purpose of elevating and refining the character of school children up to the highest plane in life; that they may know how to conduct themselves as social beings in relation to each other, as respects right and wrong and rectitude of life.
> Section 8234: Humanity: In each and every public school, in addition to the branches of study now prescribed, not less than one-half hour of each week, during the whole of each term of school, shall be devoted to teaching the school children attending said school, kindness to and humane treatment and protection of dumb animals and birds; their lives, habits and usefulness, and the important part they are intended to fulfill in the economy of nature, and such studies on the subject as the board of public education may adopt.[8]

[6] Quoted in "Makes Plea for Beasts of Burden," *Daily Oklahoman*, June 5, 1911, 5.

[7] Quoted in "Makes Plea," 5.

[8] "A Lesson from Oklahoma," *The Leader* (Guthrie, OK), April 25, 1905, 4.

Advocates for humane education, standing shoulder to shoulder with crusaders for child welfare and moral behavior, celebrated when the bill was passed and signed into law. Even the *Topeka Journal* newspaper in Kansas weighed in on the law, in a piece titled "A Lesson from Oklahoma": "Although a comparatively recently settled country and with many of the social and political characteristics of the border, Oklahoma territory has taken a step for civilization and progress which many older states might follow with credit to themselves."[9]

Still, the battle was not over. On November 16, 1907, Oklahoma became the forty-sixth American state, and the territorial statutes soon were replaced by state laws. Although many sections of progressive legislation were eliminated or weakened by special interest groups, the statute on "humane education" was resurrected and passed again in 1914. As in 1905, the law included an important provision with teeth to enforce what otherwise might have been taken as a mere suggestion. The new law included this provision: "No teacher in the public schools shall be entitled to receive any portion of the public school moneys as compensation for services unless such teacher shall have complied with the provisions of this Act."[10]

By 1922, after two decades of progressive legislation nationwide, twenty states had a law for humane education, but only three states included punishment for failure to fulfill the purpose of the law. Oklahoma was one of the three, along with New York and Illinois. Deep in the heartland, not far removed from a territorial frontier, the humane treatment of animals was viewed as an important part of a person's moral character.

13.1 Eudaimonia as a Model for a Humane Life in Oklahoma

So what is humane education, and how does the concept apply to the philosophical ideal of virtue ethics and creating ethical, virtuous people, which, after all, is the goal of implementing such a policy in all of public education? Flash forward one hundred years later to Norman, Oklahoma,

[9] "A Lesson from Oklahoma," 4.

[10] William R. Hood, "State Laws Relating to Education Enacted in 1915, 1916 and 1917," *Department of the Interior Bulletin*, 1918, 23 (Washington: Government Printing Office, 1919).

where the idea of educating students to be responsible and caring citizens of the world is getting a renewed examination.

In early 2013, a team of faculty at the University of Oklahoma, in a conversation that has spanned across the university and included external partners, began developing the idea for an interdisciplinary center for the study and promotion of flourishing. The idea blossomed in 2014 when the team received a planning grant. The goal was to develop a project that would define the place of virtues in a public university, deploy sustainable and replicable programs that cultivate a specified set of virtues in students, and develop a rigorous social-science platform to measure the impact of virtue education on behaviors and outcomes. The John Templeton Foundation approved the proposal for $2.95 million, enabling the Institute for the Study of Human Flourishing to open its Oklahoma doors on September 1, 2015.

Since then, the institute has focused on nine key virtues: open-mindedness, intellectual humility, love of learning, honesty, self-regulation, perseverance, civility, fairness, and compassion. It has broadened its scope to include community and education outreach, and its focus on virtue ethics is clearly compatible with humane education, demonstrated by its similar aims for both individuals and society. As the institute moves forward, the cross-section of doctoral fellows will expand the outlook to examine how the welfare of animals and the environment is essential for human flourishing, thereby aligning even more completely with the goals of humane education.

The Humane Education Coalition, a global nonprofit (based in Pennsylvania, a fellow humane-minded state) that acts as a Collective Impact field catalyst, provides this definition:

> Humane education is a comprehensive and necessary methodology that equips learners with the tools to think critically and identify creative and impactful solutions to interconnected global challenges. Humane education encourages many character values, including compassion, respect, empathy, and inclusion. Learners develop sensitivity to other living beings and the abiotic world within which we all survive, and they learn to appreciate the diversity of life on our planet. Humane education also provides opportunities for individuals to develop a sense of responsibility, stewardship, and self-esteem through its solutions-oriented approach, empowering learners

to make choices that align with their values and take action as confident, capable citizens.[11]

The avowed aim of humane education is to form a certain kind of person—one who is able and ready to address the problems that prevent people, animals, and the natural world from fully thriving or flourishing. Three features of this educational agenda are especially noteworthy. The first is the cultivation of character. Humane education seeks to instill virtues. Some are listed in the preceding quotation, but we can add others, such as kindness, generosity, and integrity. Second, through instilling these virtues, humane education aims to equip people with the capacities for positive and effective choices and action. It addresses not only the prosocial and affective domains but also the cognitive domain. Finally, humane education wishes to influence society by forming socially responsible citizens.

Virtue ethics is a type of ethical theory that has recently enjoyed a resurgence in academic philosophy. It takes the concept of "virtue" to be of central importance and boasts an ancient pedigree, harking to philosophers such as Plato and Aristotle. Both philosophers were concerned with a seminal question: "How should we live?" Both took virtues to be entrenched dispositions of character and intellect. In Greek, the term for "virtue" is *arête*—which means, broadly, "excellence." The virtues are dispositions such as generosity, honesty, and courage. For the Greeks these dispositions were the stable and controlling elements that humans need to flourish, or to live a good life, as expressed by the Greek term *eudaimonia*.

Virtues such as compassion, kindness, generosity, and integrity are included in a central aim of humane education. Humane education seeks to transform schooling to form more virtuous people—people who are able to view the world and societal problems from a certain kind of perspective and are able to approach those problems with a certain kind of motivation, which is highly consistent with the Aristotelian approach to virtue ethics.

Aristotle believed that virtuous people have the capacity to perceive situations as calling for virtue and to respond with appropriately virtuous motivation. A compassionate person, for example, will be able to notice

[11] "What Is Humane Education?," Humane Education Coalition, accessed September 13, 2021, https://www.hecoalition.org/what-is-humane-education.html/

the facts of situations as calling for a compassionate response—able to perceive a suffering animal, for example, and be moved to help the animal. Additionally, Aristotle thought that the virtuous possess a certain kind of deliberative capacity—practical wisdom, or *phronêsis* in Greek. Practical wisdom enables the virtuous person to correctly deliberate about whether and how to act in any given situation calling for virtue. It also enables the virtuous person to adjudicate cases in which virtues can conflict. This is highly congruent with humane education's emphasis on cultivating the cognitive abilities required for prosocial choices and action.

Finally, Aristotle thought that a virtuous person should act with appropriate affect. A generous person, for example, should give wholeheartedly and not grudgingly.

All of these features of the virtuous person reflect the same approach taken to schooling by humane education. Educating people to have the aforementioned virtues will empower and motivate them to see problems that need to be corrected, seek solutions, enact those solutions in practice, and do so with the appropriate affect—namely, the generosity of spirit needed to effect positive societal change.

Aristotelian virtue ethics, like humane education, is concerned also with creating a certain type of society. This is a society characterized by *eudaimonia*—one in which virtuous people can thrive. Like proponents of humane education, Aristotle thought that people are social or political animals. They can flourish only in the company of others—in the context of the *polis*, or Greek city-state. The virtues are those qualities of character that enable citizens to contribute to their city and to thrive, both as individuals and collectively.

These similarities notwithstanding, humane education and Aristotelian virtue ethics have important differences. Aristotle was a man of his time and therefore was not concerned about the impact that people have on the planet; lacking this global perspective, he did not recognize the importance of how we interact with the environment. He also neglected to recognize the significance of animal well-being for human well-being. Contemporary virtue ethicists have begun to address both deficits, but much work remains to be done. Here, perhaps, proponents of this ancient ethical tradition can learn from strides made by humane education.

13.2 Promise Meets the Impact of Globalization

While philanthropic and nonprofit leaders in Oklahoma continue work in these ethical trenches to make the state one of "the safest and most humane places to be an animal,"[12] the tides of globalization require new efforts at collaborations and deep dives in economics, social science, theology, culture, education, and plain old-fashioned marketing. Oklahoma is most known for the Rodgers and Hammerstein Broadway show and later movie, both blockbusters that intrinsically yoke the cowman with his cattle, the farmer with his horse—profound interspecies relationships between individuals who rely on one another for comfort. The inherent tension depicted between farmers and ranchers in the story creates a sort of balance in the ecosystem, a "brand new state" that represents a promised land for the characters, among them Curley, Laurie, Will, and Ado Annie. "He rides for days on end, with just a pony for a friend," says Aunt Eller. In the very opening lines, rendered silk by Gordon MacRae's lyrical meditation on the morning light across the meadow, Curley seems to say that animals are our friends and deserve to be treated with dignity.[13]

In much of the animal protection movement, important attention has been placed on companion animals, with expert focus and financial resources dedicated to addressing the dog and cat overpopulation crisis, to dramatic success. From a high of twenty million animals killed annually in municipal shelters twenty-five years ago to fewer than 700,000 today, the focus and network of SPCAs and rescue groups have resulted in nothing short of a success, with the goal of ending needless euthanasia in America by 2025.[14]

A shadow, however, lies over the profound suffering of farmed animals in extreme confinement, away from urban centers and far removed from the main highways. In a development simultaneous to the erosion of

[12] Kirkpatrick Foundation, *Safe & Humane: A New Vision* (Oklahoma City: Kirkpatrick Foundation, 2012).

[13] Richard Rodgers and Oscar Hammerstein II, "Oh, What a Beautiful Morning," *Oklahoma!* (Williamson Music, 1941).

[14] Best Friends Animal Society, "Animal Welfare Coalition-Building Action Kit," *Best Friends Animal Society*, accessed September 20, 2024, https://bestfriends.org/pet-care-resources/animal-welfare-coalition-building-action-kit.

explicit humane education laws in the state code, Oklahoma has experienced a dramatic rise in industrialized animal agriculture. In 1991, the state legislature approved historic exceptions to its long-standing, populist-made, anti-corporate farming laws, a watershed moment for out-of-state pig and poultry "producers." Within days of the near unanimous vote—the sole nay voice in the Oklahoma Senate was Larry Lawler, a schoolteacher and rancher from rural Stephens County—announcements were made by multinational companies for new state-of-the-art facilities designed to house and "process" millions of hogs and chickens.

Today, Oklahoma is home to 185,000,000 farmed animals, the vast majority brought indoors, with approximately three million female sows and hens living the duration of their lives in extreme confinement. Both farmer and rancher—once nervous, tenuous friends—are now referred to as "ag producers" by industry groups, a corporatized term so friendless and faceless that such fundamentals as animals' ability to lie down, turn around, stand up, spread their wings, and tend to their young have faded from humane practices of husbandry.

Although the sands of time have removed Oklahoma's landmark humane education requirements, the passion remains for teaching future generations about the value of compassion, justice, and courage toward—and on behalf of—all living creatures. Programs such as Red Rover and Muttigrees (developed by Yale University in partnership with the North Shore Animal League) have found their way into a handful of schools around the state over the last decade. Local farmers and ranchers engaging in gentler or less intensive practices are returning to rural Oklahoma, proudly embracing the best of sustainable agriculture. A new, transformative $38 million municipal animal shelter in the capital city is slated to begin construction in 2024. With the impact of commercial globalization on real people, animals, and communities, in Oklahoma and elsewhere, a global perspective on virtuous living for the environment and for each other is slowing returning to the cauldrons of social progress, the classroom, and the civic conversation.

Bibliography

Baden-Powell, Robert. "Scout Law." In *Scouting for Boys*. London: C. Arthur Pearson, 1908.

Best Friends Animal Society. "Animal Welfare Coalition-Building Action Kit." *Best Friends Animal Society*. Accessed September 20, 2024. https://bestfriends.org/pet-care-resources/animal-welfare-coalition-building-action-kit

Brittanica.com. s.v. "An Age of Reform." Accessed September 10, 2021. https://www.britannica.com/place/United-States/An-age-of-reform.

Daily Oklahoman. "Makes Plea for Beasts of Burden." June 5, 1911.

Daily Oklahoman. "News from the Clubs." April 12, 1903.

Hood, William R. "State Laws Relating to Education Enacted in 1915, 1916 and 1917." *Department of the Interior Bulletin*, 1918, 23. Washington: Government Printing Office, 1919.

Humane Education Coalition. "What Is Humane Education?" Accessed September 13, 2021. https://www.hecoalition.org/what-is-humane-education.html.

The Leader (Guthrie, OK). "A Lesson from Oklahoma." April 25, 1905.

Kirkpatrick Foundation. *Safe & Humane: A New Vision*. Oklahoma City: Kirkpatrick Foundation, 2012.

Rodgers, Richard, and Oscar Hammerstein II. "Oh, What a Beautiful Morning." *Oklahoma!* Williamson Music, 1941.

Unti, Bernard Oreste. "The Quality of Mercy: Organized Animal Protection in the United States, 1866–1930." PhD diss., American University, 2002.

CHAPTER 14

Is the Singapore Education System Ready to Introduce Humane Education?

Low Wei Ling

14.1 Introduction

Humane education remains a relatively underexplored concept in Singapore, with the most comparable framework being the nation's moral education system. As Singapore continues to develop as a progressive state, the government and civil society have increasingly prioritised initiatives addressing environmental degradation, social inequalities, and other societal challenges. This chapter examines how humane education can be contextualised within Singapore and considers its potential role in promoting harmonious coexistence between humans and the natural environment, as well as fostering a more compassionate urban society.

L. W. Ling (✉)
Keep C.A.T.S., Singapore, Singapore
e-mail: wei-ling@hotmail.co.jp

C. Linzey and A. Linzey (eds.), *The Palgrave Companion to Humane Education*, The Palgrave Macmillan Animal Ethics Series,
https://doi.org/10.1007/978-3-032-19795-5_14

14.2 Character and Citizen Education in Relation to Humane Education

In Singapore, the most comparable educational framework to humane education is the Character and Citizenship Education (also known as "CCE") syllabus, launched in 2015 to cultivate ethical values and civic responsibility.[1] This syllabus replaced its predecessors—Civics and Moral Education, National Education, and 好 公 民 (pronounced as *Hao Gong Min*; literally, "Good Citizen") for Mandarin-speaking students. In response to emerging challenges associated with technological advancement and globalisation,[2] CCE was introduced to equip students with the skills and dispositions necessary for responsible citizenship and to foster character development in preparation for the future. Importantly, CCE provides a foundation for integrating humane education by encouraging empathy, respect for life, and a sense of responsibility towards the natural environment and wildlife in Singapore, thus promoting a more compassionate and ecologically aware society.

The implementation of CCE differs between primary and secondary schools and pre-university institutions. In primary and secondary schools, the syllabus is delivered through weekly lessons, amounting to approximately two to two and a half hours per week.[3] In contrast, pre-university CCE spans two to three years, comprising forty hours of classroom instruction and an additional forty hours of cohort-based education, supplemented by customised experiential learning opportunities tailored to students' developmental needs.[4]

[1] S. Sreedharan, "Syllabus on Character, Citizenship Education Launched," *TODAY* (Singapore), January 3, 2014, https://www.todayonline.com/singapore/syllabus-character-citizenship-education-launched.

[2] G. Ng, "New Focus on Values in Key Education Shift," *AsiaOne*, September 23, 2011, https://www.asiaone.com/News/AsiaOne%2BNews/Singapore/Story/A1Story20110923-300971.html.

[3] Sreedharan, "Syllabus on Character."

[4] Ministry of Education, *Character and Citizenship Education Syllabus: Pre-University* (Student Development Curriculum Division, 2016), accessed September 16, 2019, https://www.moe.gov.sg/docs/default-source/document/education/syllabuses/character-citizenship-education/files/character_and_citizenship_education_preu_syllabus.pdf.

14.3 Phases of the Singapore Education System and Its Economic Development

To comprehend the evolution of Singapore's education system, it is essential to consider the educational imperatives that emerged in response to the nation's challenging economic conditions following the Second World War (Fig. 14.1).

Singapore's education system emerged from the urgent need to develop an economically robust nation following independence.[6] At the time, the national literacy rate was approximately 60%,[7] and the

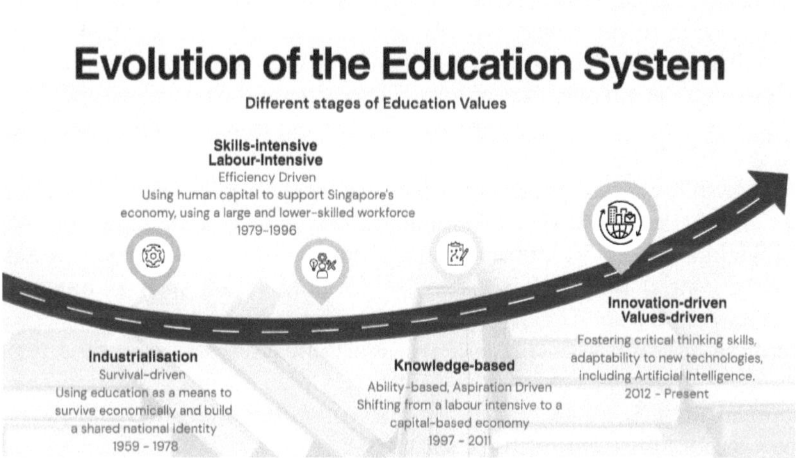

Fig. 14.1 Evolution of the Singapore education system with the economic development[5]

[5] Adapted with reference to Ministry of Education, "Singapore's Education System: Presentation to Hayes Academy Bromley" (PowerPoint presentation, 2015), slide 5, https://slideplayer.com/slide/6169299/.

[6] National Institute of Education, Nanyang Technological University, PowerPoint presentation to ICSEI on Singapore's education system, January 10, 2017, accessed September 16, 2019, https://www.nie.edu.sg/docs/default-source/oer/presentation-to-icsei-on-singapore-39-s-education-system5e0005ffdeb0605693ab4b9265f5641c.pdf?sfvrsn=0.

[7] Civil Service College, Singapore Public Service, "Infrastructure And Economic Development—Singapore Case" (PowerPoint presentation, June 26, 2009), accessed September

majority of the population was employed in low-skilled occupations. Consequently, the education system was primarily designed to support economic development, facilitating the transition of the workforce into manufacturing and industrial roles.[8] Over subsequent decades, sustained economic growth transformed Singapore into a prosperous, high-income country.

While Singapore's early education system was primarily oriented towards economic development, contemporary educational initiatives reflect a broader vision that extends beyond workforce preparation. With increasing recognition of social and environmental challenges, the Ministry of Education has sought to cultivate values such as empathy, responsibility, and ethical reasoning among students. This gave rise to the introduction of the Civics and Moral Education (known as "CME") on 23 February 1991[9] in primary schools and lower secondary schools.

Prior to the introduction of formal character and citizenship initiatives, ethical education in Singapore was largely imparted through Religious Knowledge[10] classes which were later phased out.

This shift reflects Singapore's national agenda to position education not merely as a driver of economic success but as a strategic tool for cultivating socially responsible and ethically conscious citizens. CME plays a central role in this vision by instilling values such as integrity, resilience, and responsibility, which underpin the nation's goal of fostering a cohesive, forward-looking society. By helping students understand their place in Singapore's history and future, CME nurtures pride in their national identity while reinforcing the communal obligations necessary for a harmonious and inclusive society.

16, 2019, http://siteresources.worldbank.org/EDUCATION/Resources/278200-1121703274255/1439264-1242337549970/PSA_Infrastructure_Economic_Devt.pdf.

[8] K. Blackburn, *Education, Industrialization and the End of Empire in Singapore* (Milton: Taylor and Francis, 2016), 86.

[9] The Sunday Times, *Younger pupils will be taught new civics course The Straits Times, February 24, 1991, Page 13*, https://eresources.nlb.gov.sg/newspapers/digitised/article/straitstimes19910224-1.2.20.1.

[10] Tan, June, The Straits Times education reporter, *Religion To Be A Compulsory Subject The Straits Times*, January 17, 1982, Page 1, https://eresources.nlb.gov.sg/newspapers/digitised/page/straitstimes19820117-1.1.1.

14.4 The New Character and Citizenship Framework

In 2011, then Minister for Education, Heng Swee Keat, announced a new Character and Citizenship Education (known as "CCE") framework to emphasise character development.[11] Three key areas were identified: personal values, moral values, and values of citizenship. This represented a paradigm shift from nation-building towards the cultivation of individual character and personal development, marking the latest phase of Singapore's education system as innovation-driven, student-centric, and values-oriented.

The new CCE framework integrates elements from three preceding curricula: Civics and Moral Education, which emphasises civic-mindedness and moral development; Social Studies, which explores Singaporean society and its interconnections with the wider world,[12] and National Education, aimed at fostering national identity and social cohesion (Fig. 14.2).[13]

At the time of writing, four years have passed since the launch of the CCE framework. An examination of the CCE curriculum, however, indicates that its content is predominantly individual-focused rather than community-oriented, suggesting that the framework prioritises the development of interpersonal skills.

Each chapter in the CCE textbooks incorporates a narrative designed to convey a specific moral lesson, reminiscent of Aesop's Fables. However, an analysis of textbooks for primary grades one to four reveals that only three of ninety-two chapters address issues related to environmental and animal welfare (see Annex A).[14] The three chapters address, respectively,

[11] S. K. Heng, "Work Plan Seminar Opening Address," September 21, 2011, National Archives of Singapore, Ministry of Education, http://www.nas.gov.sg/archivesonline/data/pdfdoc/20110929001/wps_opening_address_(media)(checked).pdf.

[12] Ministry of Education, "Why Study Social Studies?," *Schoolbag*, August 30, 2017, accessed September 16, 2019. http://www.schoolbag.edu.sg/story/why-study-social-studies.

[13] C. T. Goh, "Prepare Our Children for the New Century: Teach Them Well," *Speeches: A Bimonthly Selection of Ministerial Speeches* (Publicity Division, Singapore: Ministry of Culture), September–October 1996, 1–13.

[14] Ministry of Education, *好品德好公民. 三年级, 三年级* (Student Development Curriculum Division, 2015), 78–82; Ministry of Education, *好品德好公民. 四年级, 四年级* (Student Development Curriculum Division, 2015), 64–75.

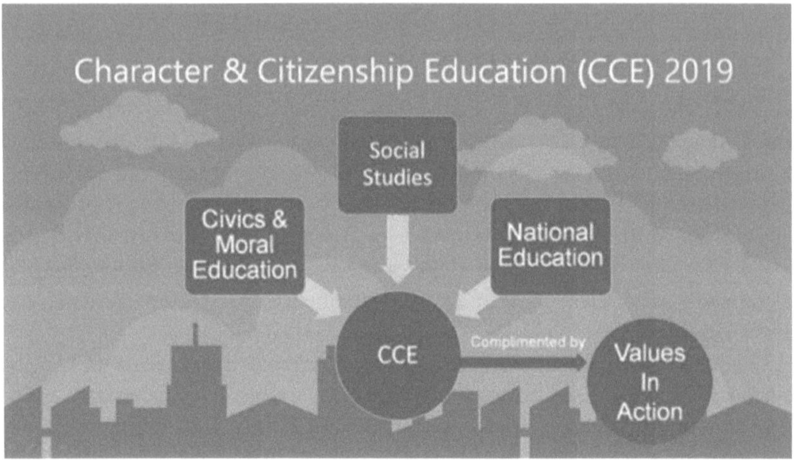

Fig. 14.2 Components of Character and Citizenship Education 2019[15]

responsible care of companion animals, conservation and recycling initiatives, and the urban ecosystem. While the inclusion of these "humane education" chapters represents a commendable effort to raise awareness, their coverage appears superficial and limited, offering insufficient opportunity for students to engage in deeper discussion or to internalise these principles in daily life. Moreover, the textbooks tend to avoid more complex or challenging topics, such as wars and social justice issues, which could provide important context for ethical reasoning and civic understanding.

An interview with Edwina Lim, a humane educator with the Singapore-based animal protection charity Animal Concerns Research and Education Society (ACRES), suggests that introducing simple and accessible topics can serve as an effective starting point. According to Lim, "Judging from the educational materials provided in primary schools, the light content allows students in grades 3 and 4 (ages 9 and 10, respectively) to comprehend the concepts more easily. It also lays the

[15] W. L. Low, "Is the Singapore Education System Ready to Introduce Humane Education?" (PowerPoint presentation at Sixth Annual Oxford Summer School on Animal Ethics, July 24, 2019).

groundwork for deeper discussions, whether initiated by the teacher or by the students themselves".[16]

Lim further emphasises the importance of incorporating experiential learning and daily practices to ensure the effectiveness of humane education. To some extent, this approach has been introduced through the Values in Action (VIA) Programme, the successor to the Community Involvement Programme (CIP)[17] which operates in conjunction with the CCE curriculum.

Previously, the Community Involvement Programme (CIP) was a mandatory initiative requiring students to complete at least six hours of community service annually, spanning from primary school through to pre-university education. In 2005, the requirement was revised to apply solely to the primary and secondary levels. Completed service hours could be converted into higher-education admission points, providing an advantage to students who had fulfilled the maximum one hundred hours compared with those who had not.

The Values in Action (VIA) Programme[18] has since eliminated the requirement to log service hours, instead prioritising the development of intended value outcomes for students. This represents a significant shift in emphasis, moving from an "hours-based" approach to one centred on experiential learning. In 2017, the local newspaper *The Straits Times* reported on the experiences of students at three schools. VIA coordinator Soon Suet Peng explained the programme's objectives: "Rather than one-off programmes, we want students to be consistently engaged in long-term projects, which is better for cultivating the values we want in them".[19]

In addition to VIA, schools have the flexibility to incorporate humane education through the Social and Emotional Learning (SEL)

[16] Edwina Lim, personal interview by author, April 12, 2019.

[17] Ministry of Education, National Education Branch, "National Education in Schools," in *Securing Our Future: Sourcebook for Infusing National Education into the Primary School Curriculum*, ed. S. K. S. Tan and C. B. Goh (Singapore: Prentice Hall, 2003), 14.

[18] Ministry of Education, "Values in Action," https://www.moe.gov.sg/education/secondary/values-in-action.

[19] Quoted in T. Fan Xi and L. Min Zhang, "Values in Action Programme: Instilling a Sense of Duty, Empathy," *Straits Times*, July 24, 2017, https://www.straitstimes.com/singapore/education/chung-cheng-high-main-lessons-from-butterflies.

Programme[20], an external elective that can be tailored to the needs of each student population and delivered as either long-term projects or one-off sessions. SEL is typically facilitated by external providers engaged by the schools, providing a platform for animal welfare organisations to introduce humane education to both students and educators, thereby complementing the experiential learning opportunities emphasised in VIA.

For example, ACRES staff deliver assembly talks under the SEL programme at both schools and corporate organisations, presenting their projects in areas such as human–wildlife conflict, the illegal keeping of free-living animals, ecotourism, and the promotion of cruelty-free lifestyles. ACRES has also collaborated with students on VIA projects, including initiatives such as Meat-Free Monday or fundraising and awareness campaigns for vulnerable animals. Similarly, the Singapore SPCA implements its own humane education programmes, conducted both in schools and at its shelter, further contributing to the development of empathy and ethical awareness among students.[21]

One of the primary challenges in integrating humane education into Singapore's school system lies in persuading schools of its significance for students' emotional and social development. Schools often prioritise programmes perceived as offering greater value or engagement, such as excursions to the zoo, which are considered more "exciting" and hence yield higher perceived educational or economic value. Consequently, animal welfare organisations face considerable difficulty in securing school participation, owing to the relative unfamiliarity of its concept in Singapore, resulting in limited demand. In practice, school programmes tend to focus on more relatable content, such as the welfare of domesticated companion animals, which can then serve as an entry point for exploring broader humane education topics.

Confucianism has also influenced the limited presence of humane education in Singapore schools. The CCE curriculum is strongly shaped

[20] Ministry of Education, *Social and Emotional Learning*, last updated October 27, 2025, accessed on October 30, 2025 https://www.moe.gov.sg/education-in-sg/our-programmes/social-and-emotional-learning.

[21] Society for the Prevention of Cruelty to Animals, "Animal Kind" (Singapore: SPCA, 2005), accessed September 16, 2019, http://www.spca.org.sg/pdf/animalkind.pdf.

by Confucian principles,[22] which prioritise societal interests over individualistic concerns. While Confucianism does not entirely disregard humane education or animal ethics,[23] its primary focus is on social harmony and the interdependence of human relationships.[24] A self-regulated hierarchical system governs interpersonal relations within families and broader societal units, positioning humane education as a peripheral yet important element in maintaining societal morality. From a Confucian perspective, the mistreatment of animals is discouraged because it exposes individuals to violent or uncivilised behaviours, which may threaten social stability and harmony. Confucius recognised that the natural equilibrium depends on benevolence and respect for one's environment, encompassing both human and non-human inhabitants.[25]

14.5 Singapore's Urban Animals Conflict and Management System

Animal welfare awareness in Singapore remains in its infancy. As in many other Asian countries,[26] issues related to animal welfare are often subordinated to societal concerns considered more urgent. This is somewhat paradoxical, given that human–animal interactions occur daily, frequently

[22] C. Tan, "'Our Shared Values' in Singapore: A Confucian Perspective," *Educational Theory* 62, no. 4 (2012): 449–63, https://doi.org/10.1111/j.1741-5446.2012.00456.x.

[23] D. N. Blakeley, "Listening to the Animals: The Confucian View of Animal Welfare," *Journal of Chinese Philosophy* 30, no. 2 (May 6, 2003): 137–57, https://www.blackwellpublishing.com/content/BPL_Images/Journal_Samples/JOCP0301-8121~30~2/111.PDF.

[24] A. Teon, "Confucianism and the Law in Singapore and Taiwan," *Greater China Journal*, June 2, 2016, https://china-journal.org/2016/06/02/confucianism-and-the-law-in-singapore-and-taiwan.

[25] A. Bosworth, N. Chaipradikul, M. M. Cheng, K. Junmookda, P. Kadam, D. Macer, et al., *Ethics and Biodiversity*, Ethics and Climate Change in Asia and the Pacific (ECCAP) Project Working Group 16 report (Bangkok: UNESCO Bangkok), https://citeseerx.ist.psu.edu/viewdoc/download?doi=10.1.1.472.6731&rep=rep1&type=pdf.

[26] Animals Asia, "Facts about Cruelty to Animals in Asia," accessed September 16, 2019, https://www.animalsasia.org/us/facts-about-cruelty-to-animals-in-asia.html; *Business Times*, "No Dog's Life for Elderly Animals in Singapore Zoo," October 25, 2017, https://www.businesstimes.com.sg/lifestyle/no-dogs-life-elderly-animals-singapore-zoo.

giving rise to conflicts between humans and animals.[27] Such interactions can be categorised into those involving domesticated animals and those involving free-living animals, or "wildlife".

It is important to note that traditional relationships with companion animals, or experiences of "pet ownership", differ markedly between Asia and the West although the Asian experience is gradually becoming more Westernised. In this context, Westernisation refers to the tendency to treat companion animals as surrogate children. Consequently, animal welfare laws in Western countries recognise animals as sentient beings and provide legal protection against mistreatment and neglect. In contrast, socioeconomic constraints in many Asian families limit the keeping of animals primarily to livestock or working animals, rather than as companions.[28] These conditions, combined with comparatively less stringent animal welfare legislation, result in both limited deterrence against animal mistreatment and minimal formal education on humane treatment.

Furthermore, animals are more widely recognised as integral members of society in the West than in Singapore. For example, visually impaired individuals and their guide dogs in Singapore continue to encounter significant discrimination in public spaces.[29] Singapore's animal rights and welfare legislation remains comparatively archaic relative to countries such as Germany and New Zealand,[30] although amendments to the Animals and Birds Act 2002 indicate a commitment to strengthening legal protections for animals.[31]

[27] B. Lim, "Man versus Wildlife: Experts Say Education Is Key for People and Wild Animals to Co-exist," *The Pride*, February 22, 2018, https://pride.kindness.sg/man-versus-wildlife/.

[28] D. Levin, "Q. & A.: Jill Robinson on Animals and Empathy in Asia," *Sinosphere* (blog), *New York Times*, December 26, 2014, https://sinosphere.blogs.nytimes.com/2014/12/26/q-and-a-jill-robinson-on-animals-and-empathy-in-asia/.

[29] M. Lin and N. Kuan, "Ignorance Keeps Guide Dogs Out at Singapore Businesses," *Straits Times*, June 17, 2013, https://www.straitstimes.com/singapore/ignorance-keeps-guide-dogs-out-at-singapore-businesses.

[30] R. Madasamy, "Animal Rights in Singapore—Time to Give Animals a Voice in Court," *Independent Singapore*, February 16, 2017, http://theindependent.sg/animal-rights-in-singapore-time-to-give-animals-a-voice-in-court/.

[31] Parliament of Singapore. "Animals and Birds (Amendment) Bill," Bill No. 14/2014 (October 7, 2014), https://www.parliament.gov.sg/docs/default-source/Bills-Introduced/animals-and-birds-(amendment)-bill-40-2014.pdf.

This limited recognition and protection of animals extends beyond domesticated species to Singapore's wildlife. Otters, monitor lizards, and long-tailed macaques are among the species inhabiting Singapore's natural environment, with the more charismatic animals often receiving greater attention and protection.[32] Belicia Teo, writing for RICE Media, observed that Singapore exhibits "both a contradiction and an inconsistency in wildlife management", noting that "an appreciation for wildlife ... founded only on its potential economic and symbolic value, the commitment to co-existence is tenuous at best".[33]

Prior to the enactment of the National Parks Board (Amendment) Bill 2019, oversight of animal-related matters in Singapore fell under the Agri-Food and Veterinary Authority (AVA). Following the bill's passage on 1 April 2019, responsibilities pertaining to plants and animals not used for food were transferred to the National Parks Board (NParks), an agency previously focused primarily on conservation, ecology, and horticultural science.

The government employs several methods to manage animal populations,[34] including culling, poison, repellents, contraceptive measures, and the removal of roosting or nesting sites. Culling, in particular, was historically a preferred approach due to its immediate results, which often appeased dissatisfied human residents. Such measures were typically executed by pest control companies engaged by the government. Given that stray cats are among the most visible and frequently encountered animals in urban environments, the Singapore cat management system will be examined in the following paragraphs.

The Agri-Food and Veterinary Authority of Singapore (AVA), in collaboration with the Cat Welfare Society, initiated efforts to advance more humane approaches to cat sterilisation and population management. These initiatives were implemented through the Stray Cat Rehabilitation

[32] Belicia Teo, "Singapore's Love for Otters Reveals Its Double-Standards towards Wildlife in the City," *RICE*, June 28, 2019, https://www.ricemedia.co/current-affairs-commentary-singapore-otters-double-standards-wildlife/.

[33] Teo, "Singapore's Love for Otters."

[34] Neo, Chai Chin, Today newspaper, page 13, Culling of healthy animals hard to justify: Researchers: TODAY, published on March 14, accessed on https://www.nus.edu.sg/newshub/news/2017/2017-03/2017-03-14/ANIMALS-tdy-14mar-p13.pdf.

Scheme (SCRS)[35] launched in 1998. However, the scheme was discontinued in 2003 amid the Severe Acute Respiratory Syndrome (SARS) epidemic, primarily due to heightened concerns regarding hygiene and sanitation.[36] Despite the limited long-term efficacy of culling as a population-control measure, governmental authorities continued to rely on this method until 2011.

Trap–Neuter–Release–Manage (TNRM) initiatives in Singapore have largely been driven by grassroots efforts within civil society. During the implementation of the Stray Cat Rehabilitation Scheme (SCRS), the Agri-Food and Veterinary Authority of Singapore (AVA) collaborated extensively with volunteers from the Cat Welfare Society. Following the termination of the SCRS, the Cat Welfare Society continued to operate its own TNRM programme, under which volunteers trap free-roaming cats and convey them to veterinarians for sterilisation, thereby filling the operational gap left by the discontinued scheme.[37] It was only after an eight-year interval that sustained reductions in the stray cat population persuaded the AVA to introduce a successor to the SCRS, namely the Stray Cats Sterilisation Programme. This programme was subsequently expanded in 2018 to encompass the management of stray dog populations as well.[38]

What the authorities have not adequately acknowledged is that issues concerning free-roaming animals are, at their core, consequences of human behaviour and human failings. In its press statement announcing the discontinuation of the SCRS, the Agri-Food and Veterinary Authority (AVA) cited several reasons for the programme's termination[39]: (1) pest

[35] Agri-Food and Veterinary Authority, "Discontinuation of the Stray Cat Rehabilitation Scheme (SCRS) in Public Housing Estates" (press statement, Ministry of National Development, October 8, 2003), National Archives of Singapore, http://www.nas.gov.sg/archivesonline/data/pdfdoc/AVA20031008001.pdf.

[36] I. Rochlitz (Ed.), *The Welfare of Cats* (Berlin: Springer Science & Business Media, 2007).

[37] Cat Welfare Society, *Report for the Year Ended 31 December 2001* (Singapore: Cat Welfare Society, 2001), http://catwelfare.org/wp-content/uploads/2017/08/CWS%20Annual%20Report%20YE2001.pdf.

[38] T. W. Toh, "Indonesian Officials Conduct Training for S'pore Animal Welfare Groups Ahead of Dog Sterilisation Scheme," *Straits Times*, October 9, 2018, https://www.straitstimes.com/singapore/environment/indonesian-officials-conduct-training-for-local-animal-welfare-groups-ahead-of.

[39] Hood, "Pet Ownership and Asian Multiculturalism."

problems arising from food that had not been properly cleared by feeders; (2) instances of cats defecating in common areas; and (3) an increase, or at least an absence of significant reduction, in the stray cat population and the volume of related public complaints.

With respect to the first issue, the pest problems in question were not caused by the cats themselves but by irresponsible feeders who failed to clear food waste. Even in the absence of cats, human littering continues to contribute to such hygiene concerns. Regarding the second issue, stray cats are frequently used as convenient scapegoats for faecal matter found in common areas, although untrained free-roaming pet cats or other animals may equally be responsible.[40] Finally, increases in the stray cat population are often attributable to the abandonment of pet cats rather than to any inherent behaviour of the animals.[41] Taken together, these points demonstrate that the underlying causes of animal–human conflict can be traced to irresponsible human behaviour, which has artificially created and perpetuated this social problem.

Living within a densely urbanised environment presents significant challenges for fostering harmony and tolerance, as humans and animals necessarily coexist in close proximity. The authorities have, at times, been criticised for responding to residents' complaints with what appears to be a knee-jerk resort to culling free-roaming animals.[42] The official position, however, is that such measures are undertaken to manage free-ranging animal populations, particularly in cases where species such as wild boars and primates lack natural predators and may therefore proliferate unchecked.

The substantial volume of complaints received by the AVA explains why culling has historically been employed as a means of placating a vocal minority of residents.

[40] Cat Welfare Society, "Say it again, for the person in the back! This is our response to Mr Peter Khaw's letter in the link below. Point 1: 'Cats can be a nuisance and are known to wander...,'" Facebook, January 24, 2017, https://www.facebook.com/catwelfaresg/posts/10153965370991065.

[41] Agri-Food and Veterinary Authority, "Dog Shooting Accident on 2th Nov 1998 at Lorong Halus" (Ministry of National Development, September 11, 2000), National Archives of Singapore, http://www.nas.gov.sg/archivesonline/data/pdfdoc/AVA20000911001.pdf.

[42] Siau Ming En, "The Big Read: As Animal Encounters Hit the Headlines, a Divide Opens Up," *TODAY* (Singapore), February 18, 2017, https://www.todayonline.com/singapore/big-read-animal-encounters-hit-headlines-divide-opens-0.

Moreover, the continued encroachment of human developments into nature reserves and green spaces has further heightened tensions between humans and wildlife.[43] In a country defined by acute land scarcity and high population density, human interests are almost invariably privileged over non-human needs. Yet, ironically, there remains little acknowledgement that human activity is the principal force disrupting natural habitats, effectively casting humans as the inadvertent invaders and colonisers of these environments.

According to the website *Statistics Times*,[44] Singapore's population density stands at 8,292 people per square kilometer, a figure that underscores the city-state's extreme land-pressure, making it the third-densest country in the world. Consequently, large-scale deforestation has been undertaken to accommodate residential, industrial, and commercial development. As a result, it is estimated that only about 1% of mangrove forests[45] and 3% of tropical rainforests remain today.[46]

Moreover, in an urbanised city-state such as Singapore, residents have limited opportunities to interact with or develop familiarity with the local biodiversity.[47] As urbanisation intensifies and everyday encounters with native biodiversity diminish, city dwellers gradually lose both familiarity with animals and an understanding of the ecological dynamics that shape coexistence in densely populated environments. This disconnect can

[43] Lim, Zachary and Begum, Shabana, *From shooting to sterilising: How human-wildlife conflict in Singapore has evolved*, The Straits Time, accessed on November 10, 2025, https://www.straitstimes.com/singapore/environment/shooting-to-sterilising-how-human-wildlife-conflict-in-spore-has-evolved.

[44] "List of Countries by Population Density," StatisticsTimes, August 8, 2019, accessed September 16, 2019, http://statisticstimes.com/demographics/countries-by-population-density.php.

[45] Yang Shufen, Rachel L. F. Lim, Sheue Chiou-Ron, and Jean W. H. Yong, "The Current Status of Mangrove Forests in Singapore," in *Proceedings of Nature Society, Singapore's Conference on "Nature Conservation for a Sustainable Singapore"—16th October 2011*, 99–120, accessed September 16, 2019, https://www.nss.org.sg/documents/Pages%2099-120.%20Yang%20et%20al.,%202013.%20Singapore%20Mangroves.pdf.

[46] G. Davison, "Urban Forest Rehabilitation—A Case Study from Singapore," in *Keep Asia Green Volume 1: Southeast Asia*, ed. Don Koo Lee, IUFRO World Series Vol. 20-1 (Vienna: IUFRO Headquarters, 2007), 171–82, https://www.iufro.org/download/file/7383/5120/Singapore_pdf/.

[47] National Parks Board Singapore, *5th National Report to the Convention on Biological Diversity (2010–2014)* (National Parks Board Singapore, 2015), accessed September 16, 2019, https://www.cbd.int/doc/world/sg/sg-nr-05-en.pdf.

influence public perception, often resulting in fear, misunderstanding, or intolerance towards urban wildlife and free-roaming animals. In such a context, humane education becomes an essential framework for shaping responsible attitudes and behaviours towards animals. The next section that follows will examine how humane education can serve as a foundational tool in promoting sustainable and compassionate approaches to urban animal management.

14.6 Urban Animal Management Through Humane Education

While it may not be feasible to reduce population density or significantly expand green spaces, there remains substantial scope to strengthen public understanding of the importance of respecting biodiversity and recognising that animals behave according to their natural instincts. Humane education plays a critical role in this regard. By fostering empathy, ecological literacy, and informed decision-making, humane education equips individuals and communities with the capacity to respond constructively to challenges such as human–animal conflict, population management, responsible pet ownership, and the ethical treatment of free-living species.

Humane education is a key mechanism for nurturing empathy and ethical awareness in human–animal relationships. It addresses immediate manifestations of conflict while simultaneously targeting the deeper structural and attitudinal roots from which these issues arise.[48] Unless Singaporeans understand the underlying reasons why issues involving free-roaming animals persist, the continued reliance on the same management approaches will yield little meaningful progress.

A similar critique had been articulated by *Straits Times* environmental correspondent Feng Zengkun:

> Rather than being quick to cull, or sticking with piecemeal reactions to individual incidents, Singapore needs a more holistic wildlife management plan. In the same way that the country needs a long-term infrastructure plan for transport or housing, it needs to develop a long-term plan for managing wildlife that balances animal welfare with public safety concerns. But to do this, it needs to do two things better: study more options other

[48] Sustainable Jersey, "Animals in the Community Education," accessed September 16, 2019, http://sj-site-legacy-editor-doc.s3.amazonaws.com/p16!41.pdf.

than culling to manage animals, and collect better data on Singapore's wildlife.[49]

To bridge the awareness gap and gather data, the National Parks Board (NParks)—Singapore's lead agency for greenery, biodiversity conservation, and the health, welfare, and management of wildlife and animals—has employed targeted strategies to engage diverse demographic groups. For students, NParks has collaborated with the Ministry of Education to implement programmes such as the Programme for Active Learning[50] and Ecolife by the River Plains[51] at the primary and secondary levels. Additionally, information on Singapore's biodiversity has been incorporated into school assessment books and textbooks, including the CCE materials.[52] NParks has also conducted educational outreach through assembly talks, exhibitions, roadshows, competitions, and project grants.

To engage the general public, NParks has actively embraced technological platforms such as social media, mobile applications, and online information portals. The agency encourages an open-source approach to knowledge sharing, enabling individuals to report animal sightings via the SGBioAtlas mobile application or to interact with followers on NParks' social media accounts.[53] Similar collaborative behaviours have also been observed in grassroots social media accounts and nature community groups, fostering a sense of shared stewardship of Singapore's biodiversity.[54]

[49] Z. Feng, "Let's Not Monkey Around with Wildlife Plan," *Straits Times*, October 5, 2013, https://www.straitstimes.com/singapore/lets-not-monkey-around-with-wildlife-plan.

[50] National Parks Board, Community in Nature Schools Award 2022, accessed on November 4, 2025 https://www.nparks.gov.sg/docs/default-source/learn-docs/programme/community-in-nature-schools-award/community-in-nature-schools-award2022-eBooklet.pdf.

[51] National Parks Board, 2015, 5th National Report to the Convention on Biological Diversity (2010–2014), page 45, accessed on October 31, 2025 https://www.cbd.int/doc/world/sg/sg-nr-05-en.pdf.

[52] National Parks Board, *Happy Pets, Happy People*, accessed October 31, 2025 https://www.nparks.gov.sg/learn/programmes/happy-pets-happy-people.

[53] National Parks Board, Facebook profile, accessed September 16, 2019, https://www.facebook.com/nparksbuzz/.

[54] National Parks Board, "Community in Nature Initiative," accessed September 16, 2019, https://www.nparks.gov.sg/biodiversity/community-in-nature-initiative; A. Tan,

However, these efforts have, to date, achieved only limited traction. Mayur Singh, founder of the Green Collective Singapore, observed, "I don't see the pie [people who are environmentally aware] growing".[55] Similarly, environmental advocate Melissa Chong noted that environmentalism has yet to attain mainstream recognition in Singapore.[56] These observations suggest that, despite the availability of technological platforms, educational programmes, and community initiatives, changing societal attitudes and behaviours remains a slow and incremental process.

A major contributing factor is the pragmatism deeply embedded in Singaporean culture. While Singapore experienced a rapid rise in economic growth, reaching first-world gross domestic product status, ecological consciousness did not develop at the same pace. The transition of the education system from an industrialisation-focused model to a knowledge-based framework, guided by political priorities and policies, has cultivated an outcome-oriented mindset among Singaporeans.[57] Consequently, the success of environmental movements in Singapore often hinges on the perceived economic or material benefits for participants, rather than on intrinsic ecological or ethical motivations.[58]

Another factor contributing to the slow adoption of pro-environmental behaviours in Singapore is the presence of competing societal priorities. In a fast-paced, highly urbanised society, issues such as economic growth, career advancement, and education often take precedence over environmental and animal welfare concerns.[59] As a result, individuals

"Putting Nature Back in Nature Photography," *Straits Times*, August 12, 2016, https://www.straitstimes.com/singapore/putting-nature-back-in-nature-photography.

[55] Cited in R. Hicks, "Is Singapore's Green Movement for Real?," *Eco-Business*, February 4, 2019, http://www.eco-business.com/news/is-singapores-green-movement-for-real/.

[56] Cited in Hicks, "Is Singapore's Green Movement for Real?".

[57] N. Ghosh, "Singapore Focuses on Outcomes, Not Ideology, to Foster Inclusive Society, Says Masagos Zulkifli in New York," *Straits Times*, July 18, 2018, https://www.straitstimes.com/singapore/singapore-focuses-on-outcomes-not-ideology-says-minister-for-the-environment-and-water.

[58] Ng and Sivasothi, *Guide to the Mangroves*, Vol. 1: *Ecosystem and Plant Diversity*.

[59] Euston, Quah and Jun Rui, Tan. Pursuing Growth And Managing The Environment: The Singapore Model, Journal of Business and Economic Analysis Volume 05, Issue, April 1, 2022, page 3, accessed on October 28, 2025, https://www.worldscientific.com/doi/epdf/10.1142/S2737566822500013.

and institutions may allocate limited time, attention, and resources to sustainability initiatives, even when they recognise their importance.

In addition, despite programmes such as the Character and Citizenship Education (CCE), Values in Action (VIA), and Social and Emotional Learning (SEL) which provide some opportunities for engagement, the coverage of these topics is often brief or peripheral. Due to the limited exposure to environmental and animal welfare issues in formal education, students may encounter animal welfare, biodiversity, and environmental issues only in short textbook chapters or one-off projects, restricting the depth of understanding and internalisation of pro-environmental values.

Finally, the perception of environmental and animal welfare issues as peripheral or non-essential further hinders behavioural change. When these concerns are seen as secondary to economic, academic, or social imperatives, both students and the general public may struggle to view them as integral to daily decision-making. This perception reinforces a pragmatic, outcome-oriented mindset, where actions are valued primarily if they yield tangible benefits, providing less incentives for them to make voluntary engagement with conservation and humane education initiatives.

There is an urgent need for sustained and targeted initiatives that embed environmental and humane education into everyday experiences, rendering them more relatable and actionable for both students and the wider public, thereby maximising their potential to generate meaningful impact.

14.7 Is Singapore Ready to Introduce Humane Education in Its Education System?

With the advent of the "student-centric, values-driven" phase of education, the Singaporean government appears increasingly willing to invest in curricula that extend beyond purely economic objectives. The Animals and Birds (Amendment) Act 2014 further codified this awareness into statute, reflecting a growing societal recognition of animal welfare. Rising levels of education and active citizen engagement have contributed to a shift in Singapore's moral compass, resulting in heightened sensitivity to the ethical treatment of animals. In this conservative context, subtle interventions or "soft nudges" are generally more effective in influencing public attitudes than confrontational or radical approaches, such as those sometimes employed by organisations like People for the Ethical Treatment of Animals (PETA)[60] which can provoke legal repercussions or public disapproval.[61]

On 18 August 2019, Prime Minister Lee Hsien Loong delivered a public address highlighting the detrimental impact of climate change emphasising the need for all stakeholders to act urgently to safeguard the environment and its biodiversity.[62] The resolute stance of Singapore's political leadership is likely to serve as a key catalyst for the broader acceptance of humane education and the advancement of environmental protection initiatives.

While strong political leadership and increasing public awareness create a favourable environment for humane and environmental education, practical considerations remain a key determinant of how such programmes can be implemented. Even so, should Singapore wish to introduce humane education, it will likely be designed with economic pragmatism in mind, whether targeting working adults or students.

The existing Character and Citizenship Education (CCE) curriculum provides an excellent foundation, although there remains scope for expanding topics and facilitating more in-depth discussion with students. Given the budgetary constraints faced by individual schools, a cost-effective approach would be for the Ministry of Education to collaborate with humane educators, for instance with ACRES, to co-develop the curriculum and implement it in a structured, top-down manner, rather than relying on schools to hire private instructors independently.

[60] "Singapore Squashes PETA Chicken Protest at KFC," *Hospitality Business News*, June 11, 2010, accessed September 16, 2019, http://hospitalitybusinessnews.com/2010063768/singapore-squashes-peta-chicken-protest-kfc; S. Tay, "Protesters, Ready to Strip, Told to Pack Up, Go Home," *Channel NewsAsia*, September 9, 2006, accessed September 16, 2019, Wild Singapore, http://www.wildsingapore.com/news/20060910/060908-7.htm.

[61] J. Ong, "Curb Radical Animal Rights Activists," *Straits Times*, January 16, 2019, https://www.straitstimes.com/forum/letters-in-print/curb-radical-animal-rights-activists; J. S. Gill, "Claims Misrepresent Singapore's Animal Protection Movement" (response from SPCA to forum letter "Curb Radical Animal Rights Activists"), *Straits Times*, January 19, 2018, last updated January 22, 2018, https://www.straitstimes.com/forum/letters-in-print/claims-misrepresent-singapores-animal-protection-movement.

[62] M. Matthew, "NDR 2019: Climate Change One of the 'Gravest Challenges Facing Mankind,' Impact on Singapore to Worsen, Says PM Lee," *Channel NewsAsia*, August 18, 2019, https://www.channelnewsasia.com/news/singapore/ndr-2019-climate-change-impact- singapore-greatest-threat-sea-11819382.

Humane education should extend beyond the school system and into the broader community, including the business sector. Corporate social responsibility (CSR)[63] provides a framework through which humane education can be integrated into organisational practices. By engaging businesses and economic actors, a ripple effect may be generated, influencing consumer behaviour and promoting ethical consumption. For example, retailers can market products made from sustainable materials in a manner that appeals to consumers, encouraging environmentally responsible purchasing decisions.

Furthermore, the Minister for the Environment and Water Resources, Masagos Zulkifli, has publicly identified environmental issues as a "pressing priority" and an "existential challenge".[64] This statement signals that the government is prioritising environmental concerns and actively engaging in multipartite partnerships to mitigate the detrimental effects of climate change. While incentive-based or regulatory measures can encourage environmentally responsible behaviour, such interventions are insufficient without accompanying education. Without understanding the intricate interdependence of ecosystems, individuals are unlikely to appreciate the crucial roles that animals play in maintaining ecological balance and sustaining human life.

In conclusion, Singapore's education system is well-positioned to introduce humane education, building upon existing initiatives such as the CCE curriculum and VIA programmes. However, the implementation of humane education in Singapore is likely to be guided more by pragmatic, needs-based considerations than by ideological motivations. To maximise its impact, the current education model could be adapted to broaden the range of topics covered and extend the duration of VIA projects, allowing students to engage more deeply and internalise key lessons. Additionally, demonstrating the potential economic, social, or tangible benefits of humane education would provide a compelling rationale for its adoption, helping humane educators and animal welfare organisations to popularise these programmes effectively. By aligning ethical and ecological learning

[63] J. Cheam, "The Evolution of CSR in Singapore," *Straits Times*, September 1, 2015, https://www.straitstimes.com/opinion/the-evolution-of-csr-in-singapore.

[64] Navene Elangovan, New research office to boost Singapore's climate science capabilities: Masagos, Today, published July 17, 2019, updated December 24, 2019, https://www.todayonline.com/singapore/new-research-office-boost-singapores-climate-science-capabilities.

with practical outcomes, Singapore can foster a generation that is both socially conscious and environmentally responsible, paving the way for a more compassionate and sustainable society.

Annex A: Interview with Edwina Lim, Humane Educator at ACRES[65]

1. Can you explain your role at ACRES?

I am a part-time educator executive at ACRES, where I helped launch the organisation's humane education efforts through the Values in Action programme. My role includes conducting outreach training for volunteers and sharing ACRES's background with both the public and volunteers through talks at the wildlife centre and school assemblies. I also mentor four groups—two from the "Youth for Causes" programme and two corporate teams—guiding them as they support our initiatives. In many ways, I also nurture students' entrepreneurial skills, as their spirit of enterprise emerges naturally when they plan and run fundraising projects to raise awareness for ACRES.

2. As a humane education officer, what are some of the projects you have spearheaded?

As the only humane education officer, I have spearheaded many projects such as the Youth for Causes and the Values in Action programme. Students first learn about ACRES and what we do before selecting which programme they want to be involved in. Current selection choices are human–wildlife conflict, illegal [keeping of] wildlife, living a cruelty-free lifestyle (non-animal food and products), and ecotourism that is animal-friendly.

3. How is the average Singaporean's understanding of Humane Education (HE)? Does he/she see it as another Character and Civics Education (CCE) (Note to readers, CCE is a curriculum that aims to nurture values and social-emotional competencies so that

[65] This interview has been lightly edited for clarity.

students grow up to be morally upright, active and contributing citizens)?

Many see it as an extra subject, and not many people are aware of HE unless they come from an international school or from a local school where the teacher has an interest in HE. In Eton House, an international kindergarten, the kindergarten–grade 2 students were brought on walks around the river, fed fish, and picked up rubbish. International schools have a higher level of awareness and more flexible teaching methods. In Singapore, local schools have many things to cover in their curriculum, and HE is often placed on the backburner. Flexibility plays a large role in education as it gives rise to more interactivity. Furthermore, with smaller and more intimate class sizes, it is easier to manage and impart knowledge—a luxury that local government schools do not have.

Schools usually seek vendors based on their price and thus choose the "most value-for-money" programmes, such as visits to the zoo, which are favoured by many primary schools. ACRES's site is perceived as less exciting than a zoo, which impedes our education outreach to these schools.

4. What are the common misconceptions about humane education?

At my time in ACRES, I have found that many people think that HE is solely focused on animal conservation (broad-based, organisational-level work), but HE also covers animal protection (individual-oriented efforts). For instance, to protect sharks, a single person can refuse to eat shark fin or help lay the groundwork for more effective petitions.

Another misconception is that HE is focused mostly on animals, when it actually covers other issues as well. There is the flexibility to include other topics, and since education in general can and should be flexible, we can easily incorporate other ethical lessons. The results can be similar to a row of dominos—we can explain the relevance of human baby abandonment and compare it to animal abandonment, and in doing so, hope that people will understand the moral implications of being unkind or cruel to animals and reflect on their human consciousness.

5. How lacking is humane education within the current CCE syllabus provided by the Ministry of Education?

Judging from the education materials provided in primary school, the concise content makes it easy for primary-school Primary 3 and 4 students (ages nine and ten, respectively) to comprehend. It also paves the way for deeper discussion initiated by the teacher or by the students themselves. ACRES is generally well known among Singaporean schools and works to educate students through different methods. For instance, through the SGA Green Awards, primary-school students run their own projects and can be nominated for awards. One project is "Meat-Free Mondays" for a year. We also speak to teachers about human–wildlife conflict and educate people on what they can do if they encounter wildlife.

6. Which aspect of humane education is missing from our CCE?

Wildlife, especially Singapore's native wildlife, is often unfamiliar to students. Students are unfamiliar with local animals and the wildlife found in our rainforests and nature reserves—most would say that animals can be found only in the zoo! In fact, they often seem unaware that our urban animals, such as birds and rats, are part of Singapore's urban wildlife.

7. What are the difficulties in getting schools, or even corporate or public groups, to participate in humane education programmes run by ACRES?

For schools, we have to work around their schedules. The peak period for engagement is from March to April, when we have engagements every day. The quiet periods are at the beginning and middle of the year and during exam seasons. Primary and secondary schools usually want us to wrap up programmes within the first two terms (January to June) or finish before October, as the end-of-year examinations start at the end of October. Primary schools may also participate in Earth Day on April 22. For corporate groups, they usually want to customise their own workshops and do not want to follow a fixed programme.

8. After participants go through the HE programme, do many of them just "forget" the lessons?

I hope not. I try to insert real-life experiences and connections so that the content becomes more relatable, and even if they forget the more "theory" part of it, I hope they remain more aware of the plight of animals and the impact of their actions on these animals.

9. How long has ACRES been running the humane education programme, and how has the programme changed over time?

ACRES has been running the HE programme since 2001. We stick to the same topics but use different approaches. In the past, when we had sufficient manpower, we performed skits with the participants, but now we let them explore how an animal feels. For example, we conduct a demonstration on how to catch a snake or let them experience the sensory features of fruit bats—their ears, eyes, and noses, and how the bats' sense of smell is affected by different scents. In addition, four years ago, we adopted a more general touch-and-go approach, but we now focus on illegal bat trade and human–wildlife conflict.

10. Does ACRES work with other organisations to craft its humane education?

No, we develop it independently. In the past, a former colleague was interested in the HEAL programme, but this programme is no longer in use. Instead, we now portray animals as sentient beings.

We try to correct misconceptions about animals. It is commonplace to see teachers telling children that animals have no feelings. Thus, we use HE to acknowledge animals' capacity to feel and educate students that animals have feelings and how we can lead a "cruelty-free" lifestyle, whether directly or indirectly.

ACRES offers a range of topics, and one session focuses on a single topic. Schools all have a budget, so usually they only pick one topic, which lasts between one and one and a half hours per session. Usually, only international schools select all the topics. The topics we offer are endangered animals, rainforest animals (nature wildlife), and how we can make a difference.

11. Which country is the "gold standard" for humane education, and how did they achieve it?

There isn't a single best country, but the UK comes close, as it has good animal welfare practices. The United States also performs well in this regard. I also believe that the education system and its focus are crucial to the teaching of humane education. Finland has the strongest education system overall, and Denmark does not focus on examinations but on interpersonal skills. For example, Denmark has a childcare centre in an elderly nursing home, helping to foster empathy towards the elderly, since children can sometimes be fearful of them. Germany has forest schools, which allow students to connect closely with nature.

12. How can we go about pushing Singapore to the same awareness level as Europe? How much longer does Singapore need to reach the same enlightened level as the aforementioned countries?

We should stop emphasising examinations, as the narrow focus on academic grades impedes our attention to world issues and discourages students from engaging in outdoor play. Also, in Singapore, parents tend to have the "Asian mindset", which is grades-centric and prioritises societal conformity, where alternative career paths or lifestyles are often not accepted. Only with a shift in focus away from textbooks will we be able to bring students back to world issues and the world around them.

13. In a Singapore-specific context, how can humane education help to resolve social ills, for instance environmental sustainability or animal welfare cases?

The premise of HE is about demonstrating kindness and empathy to the beings around us. As humans are intrinsically linked to one another, this connection also extends to our relationship with the Earth. In the grand scheme of things, actions of selfishness can compound and escalate into the slow destruction of the Earth, potentially resulting in the death of many humans or even humanity as a whole.

Thus, tying back to HE, it allows for sustainability in Singapore by considering the role of animal welfare. A good example would be farm animals in Singapore. If humane education is applied to animal husbandry, we can have more animal welfare-oriented practices, leading to healthier livestock.

Another utility of HE is changing perceptions and promoting greater understanding of misunderstood creatures. For instance, rats are more than just disease carriers; they have remarkable abilities to aid humans, such as in search and rescue efforts. By being educated and more open-minded, we are able to see the inherent worth of animals and act more ethically in our management of animals in society.

14. In Singapore, the urban wildlife conflict has repeatedly surfaced. Apart from the usual lectures and talks, how do you think we can use humane education to maintain harmony between humans and wildlife?

A campaign officer can conduct social engagement with residents and observe where places are reported to have conflict. This campaign officer would also observe and learn about the issue before engaging with residents and follow up to see if the issue is completely resolved.

We also have road shows, but not many people will come for them. For kids, games are the best way to attract their attention. This will in turn attract parents and educate them. We enable kids to create their own posters and can do cutouts of photos—through this process, kids gain knowledge when they collect information.

We also conduct demonstrations to show how animals are smuggled, and kids can role-play as officers, mimicking real life. Adults generally prefer to read short pointers.

15. Let's talk about innovation—how can we use technology to get humane education to the masses?

We can use live, high-definition video for presentations or incorporate interactive technology to gamify learning.

16. Are there any potential partnerships you are looking at to further promote humane education?

I am seeking a long-term relationship with schools and regular follow-ups. As for companies, I am trying to work with vegetarian and vegan shops. I am also advocating for all F&B (food and beverage) outlets to offer vegan options, while aiming to prevent the rise of militant vegans, who tend to be over-zealous in their efforts to promote veganism, to the point that their methods damage the public image of vegans.

Bibliography

Agri-Food and Veterinary Authority. "Discontinuation of the Stray Cat Rehabilitation Scheme (SCRS) in Public Housing Estates." Press statement. Ministry of National Development, October 8, 2003. National Archives of Singapore, http://www.nas.gov.sg/archivesonline/data/pdfdoc/AVA20031008001.pdf.

Agri-Food and Veterinary Authority. "Dog Shooting Accident on 2th Nov 1998 at Lorong Halus." Ministry of National Development, September 11, 2000. National Archives of Singapore, http://www.nas.gov.sg/archivesonline/data/pdfdoc/AVA20000911001.pdf.

Animals Asia. "Facts About Cruelty to Animals in Asia." Accessed September 16, 2019. https://www.animalsasia.org/us/facts-about-cruelty-to-animals-in-asia.html.

Blackburn, K. *Education, Industrialization and the End of Empire in Singapore.* Milton: Taylor and Francis, 2016.

Blakeley, D. N. "Listening to the Animals: The Confucian View of Animal Welfare." *Journal of Chinese Philosophy* 30, no. 2 (May 6, 2003): 137–57. https://www.blackwellpublishing.com/content/BPL_Images/Journal_Samples/JOCP0301- 8121~30~2/111.PDF.

Bosworth, A., N. Chaipradikul, M. M. Cheng, K. Junmookda, P. Kadam, D. Macer, et al. *Ethics and Biodiversity.* Ethics and Climate Change in Asia and the Pacific (ECCAP) Project Working Group 16 Report. Bangkok: UNESCO Bangkok. https://citeseerx.ist.psu.edu/viewdoc/download?doi=10.1.1.472.6731&rep=rep1&type=pdf.

Business Times. "No Dog's Life for Elderly Animals in Singapore Zoo." October 25, 2017. https://www.businesstimes.com.sg/lifestyle/no-dogs-life-elderly-animals-singapore-zoo.

Cat Welfare Society. *Report for the Year Ended 31 December 2001.* Singapore: Cat Welfare Society, 2001. http://catwelfare.org/wp-content/uploads/2017/08/CWS%20Annual%20Report%20YE2001.pdf.

Cat Welfare Society. "Say It Again, for the Person in the Back! This Is Our Response to Mr Peter Khaw's Letter in the Link Below. Point 1: 'Cats Can Be a Nuisance and Are Known to Wander …'" Facebook, January 24, 2017. https://www.facebook.com/catwelfaresg/posts/10153965370991065.

Cheam, J. "The Evolution of CSR in Singapore." *Straits Times*, September 1, 2015. https://www.straitstimes.com/opinion/the-evolution-of-csr-in-singapore.

Civil Service College, Singapore Public Service. "Infrastructure and Economic Development—Singapore Case." PowerPoint presentation, June 26, 2009.

Accessed September 16, 2019. http://siteresources.worldbank.org/EDUCATION/Resources/278200-1121703274255/1439264-1242337549970/PSA_Infrastructure_Economic_Devt.pdf.

Davison, G. "Urban Forest Rehabilitation—A Case Study from Singapore." In *Keep Asia Green Volume 1: Southeast Asia*, edited by Don Koo Lee, 171–82. IUFRO World Series Vol. 20–1. Vienna: IUFRO Headquarters, 2007. https://www.iufro.org/download/file/7383/5120/Singapore_pdf/.

Fan Xi, T., and L. Min Zhang. "Values in Action Programme: Instilling a Sense of Duty, Empathy." *Straits Times*, July 24, 2017. https://www.straitstimes.com/singapore/education/chung-cheng-high-main-lessons-from-butterflies.

Feng, Z. "Let's Not Monkey Around with Wildlife Plan." *Straits Times*, October 5, 2013. https://www.straitstimes.com/singapore/lets-not-monkey-around-with- wildlife-plan.

Ghosh, N. "Singapore Focuses on Outcomes, Not Ideology, to Foster Inclusive Society, Says Masagos Zulkifli in New York." *Straits Times*, July 18, 2018. https://www.straitstimes.com/singapore/singapore-focuses-on-outcomes-not-ideology-says-minister-for-the-environment-and-water.

Gill, J. S. "Claims Misrepresent Singapore's Animal Protection Movement." Response from SPCA to forum letter "Curb Radical Animal Rights Activists." *Straits Times*, January 19, 2018, last updated January 22, 2018. https://www.straitstimes.com/forum/letters-in-print/claims-misrepresent-singapores-animal-protection-movement.

Goh, C. T. "Prepare Our Children for the New Century: Teach Them Well." *Speeches: A Bimonthly Selection of Ministerial Speeches* (Publicity Division, Singapore: Ministry of Culture), September–October 1996, 1–13.

Heng, S. K. "Work Plan Seminar Opening Address." September 21, 2011. National Archives of Singapore, Ministry of Education. http://www.nas.gov.sg/archivesonline/data/pdfdoc/20110929001/wps_opening_address_(media)(ch ecked).pdf.

Hicks, R. "Is Singapore's Green Movement for Real?" *Eco-Business*, February 4, 2019. http://www.eco-business.com/news/is-singapores-green-movement-for-real/.

Hood, J. "Pet Ownership and Asian Multiculturalism." Presentation at the Urban Animal Management Conference, Perth, Australia, October 7–9, 1998. Accessed September 16, 2019. http://aiam.org.au/resources/Documents/1998%20UAM/PUB_Pro98_JeniHood.pdf.

Hospitality Business News. "Singapore Squashes PETA Chicken Protest at KFC." June 11, 2010. Accessed September 16, 2019. http://hospitalitybusinessnews.com/2010063768/singapore-squashes-peta-chicken-protest-kfc.

"Land Use Plan." *If Only Singaporeans Stopped to Think* (blog), February 2, 2013. Accessed September 16, 2019. https://ifonlysingaporeans.blogspot.com/2013/02/land-use-plan.html.

Levin, D. "Q. & A.: Jill Robinson on Animals and Empathy in Asia." *Sinosphere* (blog). *New York Times*, December 26, 2014. https://sinosphere.blogs.nytimes.com/2014/12/26/q-and-a-jill-robinson-on-animals-and-empathy-in-asia/.

Lim, B. "Man Versus Wildlife: Experts Say Education Is Key for People and Wild Animals to Co-exist." *The Pride*, February 22, 2018. https://pride.kindness.sg/man-versus-wildlife/.

Lim, Edwina. Personal interview by author. April 12, 2019.

Lin, M., and N. Kuan. "Ignorance Keeps Guide Dogs Out at Singapore Businesses." *Straits Times*, June 17, 2013. https://www.straitstimes.com/singapore/ignorance-keeps-guide-dogs-out-at-singapore-businesses.

"List of Countries by Population Density." Statistics Times, August 8, 2019. Accessed September 16, 2019. http://statisticstimes.com/demographics/countries-by-population-density.php.

Low, W. L. "Is the Singapore Education System Ready to Introduce Humane Education?" PowerPoint presentation at the Sixth Annual Oxford Summer School on Animal Ethics, July 24, 2019.

Madasamy, R. "Animal Rights in Singapore—Time to Give Animals a Voice in Court." *Independent Singapore*, February 16, 2017. http://theindependent.sg/animal-rights-in-singapore-time-to-give-animals-a-voice-in-court/.

Matthew, M. "NDR 2019: Climate Change One of the 'Gravest Challenges Facing Mankind,' Impact on Singapore to Worsen, Says PM Lee." *Channel NewsAsia*, August 18, 2019. https://www.channelnewsasia.com/news/singapore/ndr-2019-climate-change-impact- singapore-greatest-threat-sea-11819382.

Ming En, Siau. "The Big Read: As Animal Encounters Hit the Headlines, a Divide Opens Up." *TODAY* (Singapore), February 18, 2017. https://www.todayonline.com/singapore/big-read-animal-encounters-hit-headlines-divide-opens-0.

Ministry of Education. *Character and Citizenship Education Syllabus: Pre-University*. Student Development Curriculum Division, 2016. Accessed September 16, 2019. https://www.moe.gov.sg/docs/default-source/document/education/syllabuses/character-citizenship-education/files/character_and_citizenship_education_preu_syllabus.pdf.

Ministry of Education. "Singapore's Education System Presentation to Hayes Academy Bromley." PowerPoint presentation, 2015, slide 5. https://slideplayer.com/slide/6169299/

Ministry of Education. "Values in Action." https://www.moe.gov.sg/education/secondary/values-in-action.

Ministry of Education. "Why Study Social Studies?" *Schoolbag*, August 30, 2017. Accessed September 16, 2019. http://www.schoolbag.edu.sg/story/why-study-social-studies.

Ministry of Education. 好品德好公民. 三年级, 三年级. Student Development Curriculum Division, 2015.

Ministry of Education. 好品德好公民. 四年级, 四年级. Student Development Curriculum Division, 2015.

Ministry of Education, National Education Branch. "National Education in Schools." In *Securing Our Future: Sourcebook for Infusing National Education into the Primary School Curriculum*, edited by S. K. S. Tan and C. B. Goh, 14. Singapore: Prentice Hall, 2003.

National Institute of Education, Nanyang Technological University. PowerPoint Presentation to ICSEI on Singapore's Education System. January 10, 2017. Accessed September 16, 2019. https://www.nie.edu.sg/docs/default-source/oer/presentation-to-icsei-on-singapore-39-s-education-system5e0005ffdeb0605693ab4b9265f5641c.pdf?sfvrsn=0.

National Parks Board. "Community in Nature Initiative." Accessed September 16, 2019. https://www.nparks.gov.sg/biodiversity/community-in-nature-initiative.

National Parks Board. *5th National Report to the Convention on Biological Diversity (2010–2014)*. National Parks Board Singapore, 2015. Accessed September 16, 2019. https://www.cbd.int/doc/world/sg/sg-nr-05-en.pdf.

National Parks Board. (@NParks). Facebook profile. Accessed September 16, 2019. https://www.facebook.com/nparksbuzz/.

Ng, G. "New Focus on Values in Key Education Shift." *AsiaOne*, September 23, 2011. https://www.asiaone.com/News/AsiaOne%2BNews/Singapore/Story/A1Story20110923-300971.html.

Ng, Peter K. L., and N. Sivasothi, eds. *A Guide to the Mangroves of Singapore*. Vol. 1: *The Ecosystem and Plant Diversity*, Vol. 2: *Animal Diversity*. Singapore: Singapore Science Centre, 1999. Accessed September 16, 2019. http://mangrove.nus.edu.sg/guidebooks/text/1004.htm.

Ong, J. "Curb Radical Animal Rights Activists." *Straits Times*, January 16, 2019. https://www.straitstimes.com/forum/letters-in-print/curb-radical-animal-rights-activists.

Parliament of Singapore. "Animals and Birds (Amendment) Bill." Bill No. 14/2014. October 7, 2014. https://www.parliament.gov.sg/docs/default-source/Bills-Introduced/animals-and-birds-(amendment)-bill-40-2014.pdf.

Rochlitz, I. (Ed.). *The Welfare of Cats*. Berlin: Springer Science & Business Media, 2007.

Shufen, Yang, Rachel L. F. Lim, Sheue Chiou-Ron, and Jean W. H. Yong. "The Current Status of Mangrove Forests in Singapore." In *Proceedings of Nature Society, Singapore's Conference on "Nature Conservation for a Sustainable Singapore"—16th October 2011*, 99–120. Accessed September 16, 2019. https://www.nss.org.sg/documents/Pages%2099-120.%20Yang%20et%20al.,%202013.%20Singapore%20Mangroves.pdf.

Singapore Department of Statistics. "Population and Population Structure—Latest Data." Accessed September 21, 2019. http://www.singstat.gov.sg/find-data/search-by-theme/population/population-and-population-structure/latest-data.

Society for the Prevention of Cruelty to Animals. "Animal Kind." Singapore: SPCA, 2005. Accessed September 16, 2019. http://www.spca.org.sg/pdf/animalkind.pdf.

Sreedharan, S. "Syllabus on Character, Citizenship Education Launched." *TODAY* (Singapore), January 3, 2014. https://www.todayonline.com/singapore/syllabus-character-citizenship-education-launched.

Sustainable Jersey. "Animals in the Community Education." Accessed September 16, 2019. http://sj-site-legacy-editor-doc.s3.amazonaws.com/p16!41.pdf.

Tan, A. "Putting Nature Back in Nature Photography." *Straits Times*, August 12, 2016. https://www.straitstimes.com/singapore/putting-nature-back-in-nature-photography.

Tan, C. "'Our Shared Values' in Singapore: A Confucian Perspective." *Educational Theory* 62, no. 4 (2012): 449–63. https://doi.org/10.1111/j.1741-5446.2012.00456.x.

Tay, S. "Protesters, Ready to Strip, Told to Pack Up, Go Home." *Channel NewsAsia*, September 9, 2006. Accessed September 16, 2019. Wild Singapore, http://www.wildsingapore.com/news/20060910/060908-7.htm.

Teo, Belicia. "Singapore's Love for Otters Reveals Its Double-Standards towards Wildlife in the City." *RICE*, June 28, 2019. https://www.ricemedia.co/current-affairs-commentary-singapore-otters-double-standards-wildlife/.

Teon, A. "Confucianism and the Law in Singapore and Taiwan." *Greater China Journal*, June 2, 2016. https://china-journal.org/2016/06/02/confucianism-and- the-law-in-singapore-and-taiwan.

Toh, T. W. "Indonesian Officials Conduct Training for S'pore Animal Welfare Groups Ahead of Dog Sterilisation Scheme." *Straits Times*, October 9, 2018. https://www.straitstimes.com/singapore/environment/indonesian-officials-conduct-training-for-local-animal-welfare-groups-ahead-of.

Yee, A. T., R. T. Corlett, S. C. Liew, and H. T. Tan. "The Vegetation of Singapore—An Updated Map." *Gardens' Bulletin Singapore* 63, nos. 1–2 (2011): 205–12. Accessed September 16, 2019. https://www.nparks.gov.sg/sbg/research/publications/gardens-bulletin-singapore/-/media/sbg/gardens-bulletin/gbs_63_01,-a-,02_y2011_v63_01,-a-,02/63_01,-a-,02_205_y2011_v63p1,-a-,p2_gbs_pg_205.pdf.

CHAPTER 15

The Educational Role of Croatian Children's Magazines in Promoting Animal Rights

Boris Bakota and Lidija Bakota

Contents

B. Bakota (✉) · L. Bakota
Faculty of Law Osijek, Josip Juraj Strossmayer University of Osijek, Osijek, Croatia
e-mail: bbakota@pravos.hr

C. Linzey and A. Linzey (eds.), *The Palgrave Companion to Humane Education*, The Palgrave Macmillan Animal Ethics Series,
https://doi.org/10.1007/978-3-032-19795-5_15

15.1 The Beginnings of Croatian Children's Literature

The publication of *Mali tobolac raznog cvetja za dobru i pomnjivu mladež* (A little pouch of assorted flowers for good and attentive youngsters) by Ivan Filipović (1823–95) in 1950[1] marks the birth of children's literature in Croatia.[2] Children's literature in Croatia developed at a relatively late date compared to other European countries for several reasons. Children's literature historians Milan Crnković and Dubravka Težak believe the late appearance of children's literature in Croatia to be the result of the absence of notable children's authors in all Mediterranean countries, including Croatia, at the peak of the Renaissance and Baroque periods in literature.[3] The Illyrian Movement[4] was an important period in the development of Croatian children's literature. It was a time of strong development of national awareness, linguistic unification of the Croatian people, and a political and cultural renaissance that

L. Bakota
e-mail: lbakota@foozos.hr

[1] Ivan Filipović, *Mali tobolac raznog cvetja za dobru i pomnjivu mladež* [A little pouch of assorted flowers for good and attentive youngsters] (Zagreb: Tiskom i troškom Franje Župana, 1850).

[2] Crnković argues that we must not forget that children's literature indeed existed in Croatia for centuries before, but in form of oral folk literature. Milan Crnković, *Hrvatska dječja književnost* [Croatian children's literature] (Zagreb: Školska knjiga, 1978). Still, systematic and continuous development of Croatian children's literature began with the publication of Ivan Filipović's (1823–95) book *Mali tobolac raznog cvetja za dobru i pomnjivu mladež* in 1850. Ivan Filipović can therefore be considered the first Croatian children's author. He wrote children's poems and translated the works of other children's authors (most notably, Christoph Schmid and Franz Hoffman). Filipović was also the editor of the first Croatian children's magazine (1864–68). As an organizer of Croatian teachers, he promoted children's literature in schools and established the first institutions that would work on the expansion and development of Croatian children's literature. For more information about Ivan Filipović, the first Croatian children's author, see M. Ogrizović, *Ivan Filipović, učitelj učitelja* [Ivan Filipović, teachers' teacher] (Zagreb: HPKZ, 1974).

[3] Milan Crnković and Dubravka Težak, *Povijest hrvatske dječje književnosti od početaka do 1955. godine* [The history of Croatian children's literature from the beginning to 1955] (Zagreb: Znanje, 2002).

[4] The name Illyrian Movement (1830–43) stems from the belief that Croats are descended from Illyrians.

went hand in hand with an enthusiasm for culture and literature.[5] The first Croatian children's magazine, *Bosiljak* (1864–68), started coming out in the mid-nineteenth century. Its contributors were teachers who promoted children's literature. Apart from *Bosiljak*, two other children's magazines played a vital role in the development of Croatian children's literature in the early period: *Smilje* (1873–1945) and *Bršljan* (1873–76, 1889–1903). "All Croatian children's authors started their careers in these magazines and collaborated with them for many years."[6]

15.2 Periodization of Croatian Children's Literature and Its Magazine Production

We analyze the development of Croatian children's literature in this chapter in the context of major historical and political events in Europe and their impact on the political, economic, and social circumstances in Croatia. Throughout history, Croatia was a part of several different states (e.g., the Austro-Hungarian monarchy, the Kingdom of Yugoslavia, and the Socialist Federal Republic of Yugoslavia). Today, it is an independent, democratic country. In line with social and political developments from the mid-nineteenth century to the present, the development of Croatian children's literature can be divided into the following four periods:

1. The first period, 1848–1918: a time of strong development of national awareness, linguistic unification of the Croatian people, and a political and cultural renaissance that went hand in hand with an enthusiasm for culture and literature. Croatia was a part of the Austro-Hungarian monarchy during this time.[7]

[5] The year 1843, when Ivan Kukuljević Sakscinski delivered the first speech in the Croatian language in the Croatian Parliament, is a very important year for Croatian political and linguistic history.

[6] Crnković and Težak, *Povijest Hrvatske*, 146. All translations from the children's magazines are the authors.

[7] This is an important period for Croatia, among other reasons, because the first Croatian education law was adopted on October 14, 1874.

2. The second period, 1918–41: the period from the end of World War I to the beginning of World War II, when Croatia was a part of the Kingdom of Yugoslavia.[8]
3. The third period, 1941–91: from World War II to the beginning of democratic changes in the Republic of Croatia. Croatia was independent for a brief time (Independent State of Croatia, 1941–45), and after World War II, in 1945, it became one of the six socialist federal republics in Yugoslavia, ruled by a communist regime.
4. The fourth period, 1991–present: the period of the first democratic changes in the Republic of Croatia and its official recognition as an independent state on January 15, 1992. Croatia was freed from half a century of communist rule, and the period of democracy began.

Croatian Children's Magazines in the First Period (1848–1918)

The advent of the first Croatian children's magazines, *Bosiljak*, *Smilje* (1873–1945), and *Bršljan* (1873–76, 1889–1903), is the most notable event in the first period of development of Croatian children's literature. In addition to these three most important magazines, another six appeared by the end of the nineteenth century, "with almost the entire children's literary production of the time written in their collective pages."[9] Compared to *Bosiljak*, *Smilje*, and *Bršljan*, the other magazines came out for only a short period of time: *Zlatni orasi* (1869), *Ljiljan* (1884), *Hrvatska omladina* (1885–87), *Pobratim* (1891), and *Mali dobrotvor* (1895).[10]

[8] Two events are considered the beginning of World War II in Croatia: the bombing of Belgrade, the capital of Yugoslavia, by the Germans on April 6, 1941, and the declaration of the Independent State of Croatia on April 10, 1941.

[9] B. Pašagić, "Hrvatska dječja književnost od 1964. do 1950. Godine" [Croatian children's literature 1964–1950], *Zaprešićki godišnjak* 9 (1999): 170.

[10] For more information about Croatian children's magazines, please see the following reference works: Pašagić, "Hrvatska dječja književnost"; B. Pašagić, "Umjetnička proza u hrvatskim dječjim časopisima od 1864. do 1945. g." [Artistic prose in Croatian children's magazines 1864–1950], *Zaprešićki godišnjak* 9 (2000): 183–89; B. Pašagić, *Hrvatski dječji časopisi od 1864. do 1950. godine* [Croatian children's magazines 1864–1950] (Zagreb: Artist Association August Šenoa, 2003).

Another magazine issued by the Croatian Animal Protection Society, named *Živobran* (1894–1904),[11] is important for the beginnings of animal-themed Croatian children's and youth literature. The magazine was published for ten years and was designed primarily for adult readers. In some respects, however, *Živobran* can be considered a magazine for school students, since it had extensive cooperation with contemporary Croatian primary schools, with its publishers sending leaflets and booklets about animal protection to the schools and encouraging teachers to subscribe to the magazine. *Živobran*'s authors used the work of the contemporary children's authors as examples in their pedagogic discussions and articles about education. The beginnings of children's (animal-themed) literature in Croatia therefore can be traced to the pages of the Croatian Animal Protection Society's magazine as well.

Croatian Children's Magazines in the Second Period (1918–41)

Smilje remained the most influential Croatian children's magazine in the second period of development of Croatian children's literature. *Zoološki vrtić* (Little zoo, 1933–37), an illustrated animal-themed monthly for a general audience, was established in Zagreb in 1933. The Ministry of Education of the Kingdom of Yugoslavia approved the magazine as mandatory school reading, and it was included in school libraries' collections. Batinić argues that this was probably the first Croatian magazine specializing in zoology subjects.[12] Even though the magazine was not written primarily for children, we believe the magazine played an important role in the education of contemporary students, having been widely read by schoolchildren.

[11] The Croatian Animal Protection Society was established in Zagreb in 1884. Throughout the ten years of its publication, *Živobran* (1894–1904) mostly published administrative and business-themed texts (regulations, laws, ordinances, reports, statutes, criminal charges) related to animal protection.

[12] A. Batinić, *U carstvu životinja. Animalističko čitanje hrvatskih dječjih časopisa* [In the animal kingdom: Animal themes in Croatian children's magazines] (Zagreb: Hrvatska sveučilišna naklada, Perodica Croatica, 2013).

Croatian Children's Magazines in the Third Period (1941–91)

Between 1941 and 1991, Croatia's historical and cultural landscape was marked by communist rule. Croatia was one of the six socialist federal republics in Yugoslavia.[13] The most notable event in Croatian history at the beginning of World War II was the establishment of the Independent State of Croatia on April 10, 1941. *Smilje*, the Croatian children's magazine launched way back in 1873, continued to come out during World War II and in the Independent State of Croatia. Another children's magazine, titled *Ustaška uzdanica*, was launched in the Independent State of Croatia in 1941. This magazine was shut down in 1944, as expected, considering the political circumstances: the termination of the Independent State of Croatia and the establishment of the communist regime.

The rise to power of the communist regime after World War II brought about significant changes in Croatian children's literature. "Tradition was abandoned in favor of 'fighting for the world.' Old children's magazines, including *Smilje*, which came out without interruptions from 1873 to 1945, were terminated. All intellectual activities, including children's literature, were politicized."[14] The two dominant topics in children's literature of the period were the People's Liberation War and calls to participate in the development of the country. *Radost*, a significant magazine, highly popular among school students, was launched in 1951 in Yugoslavia by Grigor Vitez, one of the most important Croatian children's authors. *Radost* is still published. In a manner of speaking, *Radost* is an extension of *Smilje*, the longest-lived Croatian children's magazine. Two other important Croatian children's magazines were launched in Yugoslavia. In 1954, publishing company Školska knjiga launched the magazine *Modra lasta* for students aged eleven to fifteen, and this publication still exists. Another magazine launched in Yugoslavia, *Smib*, begun in 1970, made an important contribution to the development of Croatian children's literature.

[13] The Socialist Federal Republic of Yugoslavia consisted of six republics (Bosnia and Herzegovina, Montenegro, Croatia, Macedonia, Slovenia, and Serbia) and two autonomous regions (Kosovo and Vojvodina).

[14] Crnković and Težak, *Povijest hrvatske*, 397.

Croatian Children's Magazines in the Fourth Period (1991–Present)

The year 1991 marked the beginning of democratic changes in the Croatian historical and cultural landscape that would result in the official recognition of the Republic of Croatia as a sovereign country on January 15, 1992.[15] Three Croatian children's magazines launched in the previous period played an important role in the development of modern Croatian children's literature in this period as well—*Radost*, *Modra lasta*, and *Smib*—along with *Prvi izbor*, a children's magazine launched in 1993 in the democratic Republic of Croatia. They all still exist and are the most popular magazines among school students.

15.3 Research Methodology

The purpose of this study of animal-themed texts in Croatian children's magazines in chronological order, from the establishment of the first Croatian children's magazine *Bosiljak* in 1864 to the modern Croatian children's magazine *Prvi izbor*, which has been published continually since 1993, is to analyze the position of animals as subjects in Croatian children's literature and to identify any changes in the attitudes toward animals in literature.

We have used a comparative historical method to determine the differences in the approaches to animal-themed children's literature in the analyzed corpus.

The research corpus consisted of the following children's magazines: *Bosiljak* (1864–68), *Smilje* (1873–1945), *Bršljan* (1873–76, 1889–1903), *Ustaška uzdanica* (1941–44), *Radost* (1951–present), *Modra lasta* (1954–present), *Smib* (1970–present), and *Prvi izbor* (1993–present). The research corpus also included two magazines that were not primarily designed for children but that had a considerable influence on school students due to their successful cooperation with the contemporary school institutions. Their relevance for our research also stems from the fact that they are animal-themed magazines. They are *Živobran* (1894–1904), issued by the Croatian Animal Protection Society, and the illustrated monthly *Zoološki vrtić* (1933–37).

[15] Croatia freed itself of the communist regime that ruled it for half a century, ushering in the period of democracy.

Since the objective of the study is descriptive, the research is based on a representative sample in order to get the most objective idea of the development of animal-themed Croatian children's literature in chronological order. We collected information using the content analysis method (animal-themed texts in Croatian children's magazines).

We grouped animals into the following categories based on their relationship with people:

- *Domesticated animals*—(usually poultry and other farmed animals) are tamed animals who live in the proximity of people and are used as so-called "beasts of burden" or as food.
- *"Pets"*[16]—tamed animals with whom people have the closest relationship.
- *"Wildlife—game"* (animals who have not been domesticated but who are considered food and whom people hunt at certain times of the year) and "*beasts*" (also not domesticated but hunted because they are seen as a threat to domesticated animals; people usually have no control over their lives).
- *Insects*—animals perceived as "pests" because they feed on crops and destroy them. Their destruction and killing is considered justified. There are also certain insects who are considered useful, mostly because people exploit them for their own purposes (e.g., bees).
- *Birds*—they are free and live near people, who consider them useful because they destroy "pestilent insects."

15.4 Domesticated Animals in Croatian Children's Magazines

In the analyzed corpus, domesticated animals are portrayed as tamed and living in the proximity of people, who keep them as "beasts of burden" (livestock, oxen) or to be used as food (cows, calves, pigs, sheep, lambs, and poultry).

[16] We have retained the original language in order to show how it was used in children's magazines. We have placed language such as "pets," "pests," "beasts," and "game" in quotation marks to retain the language while indicating it is problematic from an animal ethics perspective.

In the historical overview of Croatian children's literature, domesticated animals are treated the worst. The slaughter of domesticated animals, in which both adults and children participate by way of ritual, is present as a theme in early Croatian children's literature. Builliet, as cited in Batinić, makes the following observation about the slaughter of domesticated animals: "Watching adults, who are their role models at this age, treat this as a normal everyday occurrence, in time children start to take such bloody scenes for granted."[17] The anonymous short story "Kolinje i klanje" (Pig slaughter) tells the tale of children's joy over taking part in the butcher's preparations for pig slaughter.[18] In the Croatian children's literary scene, pig slaughter not only is a normal activity for adults; the act of slaughter also transforms into an entertainment for children, a child's game, as described in Mijat Stojanović's short story "Posljedica nesmotrene igre" (The consequence of reckless play),[19] in which children's play imitates slaughter. Children move on all fours, mimicking "fat grunting pigs,"[20] and one plays the part of the butcher. For the children, slaughter is fun: the boys are described tumbling down "like a pig tumbles down when the butcher hits it in the forehead with his axe."[21] The short story "Neposlušno svinjče" (Defiant pig) by Jakov Majnarić bluntly discusses the fate of a pig on a farm, whom the family plans to slaughter around Christmas.[22] In the Croatian cultural landscape of the nineteenth and twentieth centuries, pigs were perceived as "very dirty, greedy and disgusting animals, but still of great use to people, which need to be fattened well for this purpose."[23] A description of how pigs were killed appears in the magazine *Živobran*: "A heavy blow to the head and a sure stab in the throat will kill the animal quickly," the writer says, remarking that children should not witness the act of slaughter because "this is not some sort of a festival that everyone should be invited to."[24]

[17] Richard W. Builliet, quoted in Batinić, *U carstvu životinja*, 20.

[18] "Kolinje i klanje" [Pig slaughter], *Smilje*, no. 3, 1878, 42–44.

[19] Mijat Stojanović, "Posljedica nesmotrene igre" [The consequence of reckless play], *Bosiljak*, no. 4, 1865, 62–63.

[20] Stojanović, "Posljedica nesmotrene igre," 62.

[21] Stojanović, "Posljedica nesmotrene igre," 63.

[22] Jakov Majnarić, "Neposlušno svinjče" [Defiant pig], *Bršljan*, no. 5, 1896, 155–56.

[23] *Živobran*, no. 6, 1896, 14.

[24] *Živobran*, no. 6, 1896, 14.

The topic of domesticated animal slaughter that both adults and children participate in by way of ritual was present in Croatian magazine production until the end of the 1980s. In the short story "Ljudožder" (Cannibal) the innkeeper holds a knife in his hand and tells the children this is the little piggies' "last hour" because he plans to slaughter them tomorrow.[25] Another morbid short story titled "Jezik i srdce" (Tongue and heart), published in *Ustaška uzdanica*,[26] discusses the slaughter of a sheep and the delivery of the animal's organs—tongue and heart—to the sheep's master. These organs are allegoric representations of the best (heart) and the worst (tongue) part of the human being. The magazine *Smib* perpetuated the image of the pig as a dirty, unclean animal. In one of its short stories, pigs are described as "living meat and fat factories that don't look nice and aren't clean, but are very intelligent, like sheepdogs; still, people have been unable to come up with any use for them other than as roasts on their plates."[27] Several children's compositions about pig slaughter were published in *Modra lasta* magazine in the 1970s and 1980s. Biserka Ribić, an eighth-grade student, penned the poem "Kolinje" (Pig slaughter) in which the pig, sensing that she is about to die, stays asleep for the slaughter, making the butcher's job easier.[28] Written from a child's perspective, the poem is completely devoid of any compassion for the animal who is killed. Instead, the girl is looking forward to roasted sausages. A poem by seventh grader Vlado Milec, also named "Kolinje," describes the slaughter with uncanny precision in disturbingly graphic detail—the poem was originally rhymed, but the rhyme is lost in the English translation:

> Let it be known / that slaughter has begun. / The fire is started, the water boils, / the pig is scalded, and then shaven. / Not a single hair must be allowed to remain, / it must be clean as a whistle. The pig is then hung, / its head chopped off, its entrails removed. / They then cut it into two halves / and take it down from the hook. / The pig is laid down on the table / and shoulders and hams are removed. / The bacon must be separated from the meat / and the ribs from the spine. / Stuffing is

[25] "Ljudožder" [Cannibal], *Ustaška uzdanica*, no. 1, 1941, 7–8.

[26] "Jezik i srdce" [Tongue and heart], *Ustaška uzdanica*, no. 4, 1943, 12.

[27] *Smib*, no. 4, 1979, 17.

[28] Biserka Ribić, "Kolinje" [Pig slaughter], *Modra lasta*, no. 13, 1972. Note: wherever children's magazines did not have numbered pages, no page number is specified.

prepared for the sausages, / fat is melted to make greaves. / This is pretty much what slaughter looks like: / that is what I wanted you to know.[29]

The poem is a testimony to its time: it was customary for children to participate in pig slaughter activities at family farms in the former Yugoslavia in the second half of the twentieth century, side by side with adults. Exposed to slaughter scenes in childhood, young people later started to perceive slaughter as a customary, normal, everyday activity. Mirela Bede, a seventh-grade student, also describes the joy of slaughtering pigs and eating pork in the text "Klanje bravaca" (Pig slaughter).[30]

The topic of animal slaughter in households continues to appear in modern Croatian children's literature, but from the perspective of the animal preordained for slaughter rather than from people's perspective. A girl named Marinela Rukavina published a poem penned from the perspective of a pig destined to be slaughtered (again, the poem originally rhymed, but the rhyme is lost in the English translation):

Two pigs are standing sadly / in the corner of the old sty. / Knowing that their lives are about to end, / they're counting their last minutes. / – What are you thinking about, / one pig asks the other. / – Are you feeling this great sadness? / All is quiet at the neighbor's, / they're preparing for slaughter too. / In a day or two, / this will all / be empty, / and we'll be turned into sausages. / A pig's life is short, / we only live until slaughtered.[31]

Another girl named Karla Česi published the poem "Gizdavo prase" (The gaudy pig), hinting at the fate of a pig destined for slaughter.[32] These poems, written by schoolgirls and published in children's magazines, indicate that children's perception of pigs has changed dramatically: from the former perception of pigs as dirty, unclean animals preordained for slaughter to empathy for the animals and a deep and sincere regret for the fate that awaits them.

[29] Vlado Milec, "Kolinje" [Pig slaughter], *Modra lasta*, no. 12, 1974.

[30] Mirela Bede, "Klanje bravaca" [Pig slaughter], *Modra lasta*, no. 12, 1988.

[31] Marinela Rukavina, poem in *Smib*, no. 6, 2010, 29.

[32] Karla Česi, "Gizdavo prase" [The gaudy pig], *Radost*, no. 6, 2019.

15.5 "Pets" in Croatian Children's Magazines

With respect to people's interactions with animals, "pets" are defined as animals tamed by people, with whom people have a very close emotional connection and share the same living space. In the Croatian cultural and linguistic landscape, dogs and cats are the most common "pets," even though many other animals can be included in the category.

The picture of "pet" dogs painted by the earliest Croatian children's literature and the first Croatian children's magazines is radically different from modern notions. From the first Croatian children's magazine (*Bosiljak*, 1864) to the 1970s, the motifs of chained dogs and various forms of dog neglect and abuse by adults and children alike often appeared in Croatian children's magazines. In the short story "Lisica i pas" (The fox and the dog), the dog accepts being kept on a chain as his fate, decreed by his master, directly suggesting to the youngest readers that a dog on a chain is a completely normal sight.[33] Even though the Croatian animal protection magazine *Živobran* argued at the turn of the twentieth century that it was unnecessary to keep dogs on chains, recommending to its readers that they stop doing it,[34] the fact is that the public did not heed this advice, and life on a chain would remain the reality of dogs in Croatia for the remainder of the twentieth century, along with many other forms of animal abuse, neglect, and killing, as confirmed by texts published in children's magazines. The short story "Lijepo piše gospodin Bog, ali..." (Our Lord's speaks well, but...) by the Croatian children's author Mato Lovrak describes a practice that was widespread at the time: getting rid of unwanted puppies by throwing them into the river.[35] In the story, the boy's mother proposes that they take away their dog's puppies, tie rocks around their necks, and toss them into the river. Animal cruelty and abuse does not appear in Croatian children's literature only from the perspective of adults: children are often portrayed as perpetrators of animal cruelty, abuse, and killing as well. Ivan Filipović's short story "Zloćudni Lovro" (Vicious Lovro) focuses on a boy who abuses, beats, and drags around dogs and cats and later moves on to abusing oxen and

[33] "Lisica i pas" [The fox and the dog], *Bršljan*, no. 3, 1899, 126.

[34] *Živobran*, no. 5, 1900.

[35] Mato Lovrak, "Lijepo piše gospodin Bog, ali..." [Our Lord's speaks well, but], *Smilje*, no. 6, 1940, 100–110).

other domestic animals.[36] In the second half of the twentieth century, Croatian children's magazines shifted toward the perception of dogs as humans' best friends. *Modra lasta* magazine published the informational article "Hrvatski ovčar," meaning "Croatian sheepdog," presenting this autochthonous Croatian dog breed known for loyalty to the dogs' people.[37] *Radost* magazine published an informational article depicting sheepdogs as "loyal guardians of sheep and people alike" and celebrating Yugoslavian shepherd dogs as "fiercely loyal to their owners and uncompromising fighters against wolves, bears and other beasts."[38] The article "Pasja vjernost" (A dog's loyalty) states that "dogs are sometimes so loyal that they will sacrifice their own lives for their owners, or die of grief for them."[39]

We have not found the motif of dog and cat abuse in the analyzed corpus of modern Croatian children's magazines, printed from 1990 onward, which confirms that Croatian society's view of dogs has changed. From being kept on a chain to protect people's homes, dogs transformed into an animal with whom people share their living space and form a very close emotional bond and companionship. Our analysis of the corpus confirmed that in the contemporary Croatian children's magazine *Smib*, "stories about dogs take on a happier note, and dogs appear in much different and more diverse contexts. The change in their role in people's lives—from a predominantly working animal in the past to a companion and friend today—also reflected on their portrayal in children's literature."[40] In truth, dogs are assigned a different role in modern times: "their main role is no longer to stand watch or hunt down rodents, but to make their owners happy and compensate for their increasing alienation from the natural world."[41] In addition to indoor companion animals, contemporary children's magazine *Prvi izbor* also discusses animals termed "outdoor pets,"[42] whom people can see while

[36] Ivan Filipović, "Zloćudni Lovro" [Vicious Lovro], *Bosiljak*, no. 1, 1864, 27–28.

[37] "Hrvatski ovčar" [Croatian sheepdog], *Modra lasta*, no. 16, 1974.

[38] "Psi ovčari—čuvari i ljubimci" [Sheepdogs—guardians and pets], *Radost*, no. 3, 1979.

[39] "Pasja vjernost" [A dog's loyalty], *Modra lasta*, no. 9, 1987.

[40] Batinić, *U carstvu životinja*, 161.

[41] Batinić, *U carstvu životinja*, 300.

[42] *Prvi izbor*, no. 6, 2017, 22–23.

they walk the streets or spend time outdoors and who do not share people's immediate living space but should still be shown empathy. In the second half of the twentieth century, "pets" also were discussed as possible carriers of infectious diseases in children's magazines, with advice to young readers to pay attention to their personal hygiene, as well as the hygiene of their "pets" and their living spaces.[43] Contemporary Croatian children's magazines also call on the youngest readers to treat their "pets" responsibly, because "dogs and cats are not toys,"[44] warning that "many want a "pet", but "pets" needs lots of time, attention, care and love.... Keep in mind that you might soon tire of your pet."[45] *Radost* magazine has a "pets" section, where it published the article "Svaki je jedinstven" (Each is unique), calling on readers to adopt mixed-breed dogs, especially those living at shelters, because these dogs deserve people's respect too.[46] *Prvi izbor* magazine has included articles about "pet" fish, turtles, and rabbits,[47] as well as guinea pigs, parrots, and hamsters.[48]

15.6 "Wildlife" in Croatian Children's Magazines

In terms of their relationship with people, "wildlife" can be divided into two categories. The category of animals considered "game" by many includes animals who are not domesticated but whom people treat with a certain awe and respect, hunting and killing them not necessarily for food, but for entertainment. The animals whom this chapter calls "beasts" are the "wildlife" over whom people have no control. Unlike so-called game, these animals are not eaten, but since people perceive them as a threat to their domesticated animals, people also hunt and kill these free-roaming animals.

In the historical Croatian social and cultural landscape, hunting was typically seen as a male activity. *Živobran* magazine reports that the

[43] *Radost*, no. 27, 1979.

[44] *Radost*, no. 1, 2016.

[45] *Radost*, no. 5, 2017, 5–7.

[46] "Svaki je jedinstven" [Each is unique], *Radost*, no. 10, 2017, 53.

[47] *Prvi izbor*, no. 9, 2012, 41.

[48] *Prvi izbor*, no. 6, 2017, 22–23.

purpose of hunting used to be "exercise, respectable and restrained entertainment, and promotion of chastity,"[49] but at the end of the nineteenth century, people's motives for hunting were "horrible slaughter and savage killing, an activity no longer practiced just by uneducated masses, but also by aristocrats, fine and illustrious gentlemen who care for blood-thirsty killing rather than hunting."[50] The themes of pheasant, duck, rabbit, deer, wolf, bear, and fox hunts appear in children's stories published in Croatian children's magazines by the end of the twentieth century. Different ways of killing "wildlife" are described. Animals are usually hunted with rifles or trapped and then killed or slaughtered in these stories; they are sometimes even poisoned.[51] Adults and children alike take part in hunts in the published works. In the poem "Lovac" (Hunter) by I. Lj. Varjačić, a boy voices his extreme excitement over his wish to participate in a hunt.[52] Another boy is the protagonist of the poem "U šumi" (In the forest), written in the form of dialogue between the boy, his mother, and a hunter, which glorifies hunting and killing.[53] His uncle's participation in a hunt inspires a feeling of unrestrained joy in the boy: "Dear Uncle, what are you carrying in your bag? / Look, something like a head is sticking out! / Did you kill anything in your hunt / or did you go in vain?"[54]

Throughout history, wolves were the most common victims of Croatian hunters, who maintained that wolves, unlike other "beasts," killed farmed animals without restraint: "they kill the entire herd, however many animals there may be in it, whereas other beasts usually only kill as many animals as they need to feed themselves."[55] Ivan Sertić's poem "Vuk" (Wolf) describes a man's killing of the animal in cold blood: "The forest is quiet / Only a wolf's howling is heard... / A hunter heard it / The wolf lies dead."[56] A change in attitude toward wolves is detected only

49 *Živobran*, no. 3, 1895, 9.

50 *Živobran*, no. 3, 1895, 9.

51 *Bršljan*, no. 4, 1893, 113–17; *Bršljan*, no. 3, 1899, 126; *Bosiljak*, no. 7, 1867, 105–7.

52 I. Lj. Varjačić, "Lovac" [Hunter], *Bršljan*, 1874.

53 "U šumi" [In the forest], *Bršljan*, no. 12, 1896, 385–87.

54 "U šumi" [In the forest].

55 *Modra lasta*, no. 7, 1974.

56 Ivan Sertić, "Vuk" [Wolf], *Radost*, no. 3, 1981.

in texts published by modern Croatian children's magazines. *Smib*, for instance, reports that the wolf has been a protected species in the Republic of Croatia since 1995.[57]

Croatian children's magazines have featured writings about the killing of many other animals, not just wolves. A text by V. Marik titled "Zec" (Rabbit) in the first Croatian children's magazine *Bosiljak* offers instructions on how to best kill a rabbit: "You can kill them easily by grabbing their hind legs and striking them behind the ears with the edge of your palm."[58] "Zečeva smrt" (Rabbit's death), a poem by Ivo Mlikota published in *Radost*, describes, without an ounce of sympathy or compassion, how a hunter killed a rabbit, merely commenting that the animal was unlucky and thus ended up dead.[59] In a short story by Drago Ivanišević, ironically named "Obična priča" (An ordinary story), a hunter kills a mother rabbit in the forest, and her children suffer.[60] As he walks home, with the bloody mother rabbit sticking out of his bag, the man is thinking about his sick little girl. In effect, the story invalidates the rabbit's parental feeling while portraying the hunter as a caring parent because he is concerned for his child's welfare. The magazine *Radost* published the short story "Lovački kalendar" (Hunting calendar) by Joža Horvat, written in dialogue form. As its title suggests, it contains a list of animals people kill throughout the calendar year:

> Boys asked a hunter: Mister, what do you hunt in January? Boars, he replied. In February? Foxes. In March? Pikes. In April? Grouse. In May? Roe deer. In June? Badgers. In July? Waterfowl. In August? Quail. In September? Deer. In October? Partridge. In November? Chamois. In December...? What I have here in my bag, the hunter said... and a rabbit jumped out of the bag.[61]

Modra lasta magazine published "U lovu" (On the hunt), an essay by a thirteen-year-old girl named Mia Vojnić about a deer hunt and a dog chase. Though initially excited to take part in the hunt, by the end of the

[57] *Smib*, no. 4, 2003, 26–27.

[58] V. Marik, cited in Batinić, *U carstvu životinja*, 255.

[59] Ivo Mlikota, "Zečeva smrt" [Rabbit's death], *Radost*, no. 4, 1954.

[60] Drago Ivanišević, "Obična priča" [An ordinary story], *Radost*, no. 4, 1979.

[61] Joža Horvat, "Lovački calendar" [Hunting calendar], *Radost*, no. 6, 1982.

story, the girl, horrified by the scene that has unfolded before her, sympathizes with the butchered animal: "I was troubled by guilt and left this sad place with a fierce resolve never to go hunting again, because I realized hunting was not for people who like animals."[62] Another student's essay, "Moj prvi lov" (My first hunt), relates the story of a boar hunt and a boy filled with a sense of "strange, savage satisfaction" as he stands over their kill.[63]

The motif of bear hunts is present in Croatian children's literature from its beginnings as well. The first Croatian children's magazine, *Bosiljak*, published the text "Sjeverni medvjed" (Polar bear), describing a brutal hunt of a she-bear who tries to defend her cub.[64] In 1933, *Zoološki vrtić*, an illustrated monthly magazine from the second period, published an environmentally conscious article clearly arguing for the protection of animal species, in this case the brown bear, who had been brought to the verge of extinction by hunting: "unless we protect it from hunting, it will soon disappear, like it already has in some other European regions."[65]

Some of the hunting-themed stories are told from the perspective of the hunted animal. In these stories, hunters are described as "people carrying rifles."[66] Humans are referred to as "the butcher of Mother Nature's gentle creatures."[67] "Hunters are usually hard of heart. And they're hard of hand too. They kill without asking: Does it hurt a lot?"[68] "Run, I tell you, this is a man, a cunning and dangerous animal. Fear his malice more than any weapon he might carry."[69] "These creatures are called men... and they are our only enemies."[70]

Croatian children's magazines also occasionally published anti-hunting stories, written to inspire compassion in the reader for the animal who gets wounded or killed. These stories usually end with the narrator abandoning the hunt, or children mocking adults as cowards for participating

[62] Mia Vojnić, "U lovu" [On the hunt], *Modra lasta*, no. 3, 1973.

[63] "Moj prvi lov" [My first hunt], *Modra lasta*, no. 6, 1989.

[64] "Sjeverni medvjed" [Polar bear], *Bosiljak*, no. 9, 1867, 138–40.

[65] *Zoološki vrtić*, no. 3, 1933, 53–56.

[66] *Bršljan*, no. 2, 1897, 63–64.

[67] *Bršljan*, no. 7, 1897, 222.

[68] *Bršljan*, no. 1, 1898.

[69] *Ustaška uzdanica*, no. 10, 1943, 11–13.

[70] *Prvi izbor*, no. 5, 2009, 37–39.

in a hunt. In the short story "Lovac i dijete" (The hunter and the child) by Ferdo Stražimir Kulundžić, the child mocks the hunter's "heroic act": "You know, Uncle Marko, killing a bear with a rifle is so easy even I could have done it. Why don't you try wrestling with one and clubbing him to death."[71] The magazine *Smilje* published an anti-hunting short story by Bogumil Toni named "Zečić" (The little rabbit), teaching readers not to take animals from their natural environment.[72] Bogumil Toni is the author of another anti-hunting story titled "Otac u lovu" (Father hunting), featuring a father, an excellent shot, who has his rifle trained on a doe and fawn, but is persuaded by his son not to shoot.[73] The same theme, a child persuading an adult to abandon the hunt, is found in the short story "Lovčevo obećanje" (The hunter's promise) by Drago Kozina.[74]

Unlike their predecessors, modern Croatian children's magazines rarely publish hunting or anti-hunting stories. The magazine *Radost* published "Djedov teški zadatak" (Grandpa's difficult task), a short story by Nada Iveljić in which a grandfather tries to explain to his grandson that "hunters are nature's friends and protectors, who set up feeders for the animals in the winter," to which the grandson replies: "So, you feed birds so someone can shoot them.... No, I can't understand that."[75]

Instead of hunting and anti-hunting stories, modern Croatian children's magazines started publishing articles about environmental protection and the need for coexistence between people and other living beings on the planet. Croatian children's magazines of the twenty-first century are mostly focused on endangered plant and animal species, who are in danger of extinction due to human activity, underlining that "the disappearance of one species often puts other species at risk."[76] Texts published in modern Croatian children's magazines "lay the blame for biodiversity loss not on natural processes, but on people and their irresponsible

[71] Ferdo Stražimir Kulundžić, "Lovac i dijete" [The hunter and the child], *Bršljan*, no. 11, 1899, 350.

[72] Bogumil Toni, "Zečić" [The little rabbit], *Smilje*, no. 5, 1941, 89–90.

[73] Bogumil Toni, "Otac u lovu" [Father hunting], *Smib*, no. 6, 1941, 103–4.

[74] Drago Kozina, "Lovčevo obećanje" [The hunter's promise], *Radost*, no. 6, 1982.

[75] Nada Iveljić, "Djedov teški zadatak" [Grandpa's difficult task], *Radost*, no. 4, 1998.

[76] *Radost*, no. 1, 2006.

treatment of nature."[77] *Prvi izbor* magazine points out that "big global corporations are farming for palm oil and destroying the only habitat available to orangutans."[78]

15.7 Insects in Croatian Children's Magazines

Some insects are viewed as "pests" because they feed on crops in the fields, vegetables in gardens, and fruits in vineyards and orchards. People claim the right to exterminate them. This role is assigned not only to adults but also to children, who are taught different ways of destroying "pests" in the pages of the first Croatian children's magazines (*Bosiljak*, *Smilje*, and *Bršljan*). The section "Sitnice" (Trivia) in the magazine *Smilje* thus featured an article in 1873 instructing children how to "have fun killing cockchafers: Children, you too should destroy this pestilent beetle, but do not torture the animals while you are at it. Rather, give them to the pigs to eat or drown them in hot water."[79] *Bršljan* magazine published the mocking poem "Prirodopis u pjesmicama" (Science in poems) describing how children kill cockchafers: "And the children hurriedly... / Trap it into their net, / Tie its legs, / And they call to it, buzz, buzz, buzz, / Let's see you fly now."[80] Our analysis of the magazine texts confirmed that very few texts took a different attitude toward insects perceived as "pests." The moral didactic story "Lov na hrušteve" (Cockchafer hunt), whose author is identified only by the initials Lj. T., portrays cockchafers as valuable beetles and states that "a good child treats all animals kindly, including cockchafers, and does not hunt or kill them, but lets them live in peace."[81] Caterpillars also were considered "pestilent" insects because they killed baby birds.[82]

Reflecting on the later magazine *Smib*, Batinić points out that flies, cockroaches, spiders, pigeons, sparrows, mice, and rats are viewed as "pests" and disease carriers in the publication, and he concludes: "Children will certainly not feel or learn empathy for these animals from the

[77] *Modra lasta*, no. 9, 2016.

[78] *Prvi izbor*, nos. 1–2, 2016, 48–50.

[79] "Sitnice," *Smilje*, no. 1, 1873, 14.

[80] "Prirodopis u pjesmicama" [Science in poems], *Bršljan*, no. 5, 1895, 156.

[81] Lj. T., "Lov na hrušteve" [Cockchafer hunt], *Bršljan*, no. 5, 1893, 138–41.

[82] *Bršljan*, no. 9, 1894, 287.

way they are portrayed, even though many people keep mice, rats and even spiders as pets nowadays."[83] There is no real reason for children's aggression against these tiny little animals in the magazines that depict these scenes. Sometimes the act of aggression is the result of boredom and the child's need for any kind of entertainment, even if it includes violence against another living being. In the 1898 poem "Kada motrim..." (While I watch...), a boy abuses bees out of sheer boredom, believing it to be an entertaining game.[84] In the article "Ne muči životinja!" (Don't torture animals!), boys think abusing a mouse is a form of entertainment too. "After they caught him, the boys tied a piece of string around his neck and immersed him in water to drown him."[85] In the poem "Janko," a boy abuses insects but is then haunted by guilt and dreams about a butterfly stabbing him like he stabbed butterflies.[86] The protagonist of the story "Usamljena glista" (Lonely earthworm) is a boy who cuts earthworms with a knife.[87]

Using nets to capture butterflies was a very popular children's game in the Croatian social and cultural landscape of the nineteenth and twentieth centuries. *Živobran*, the first Croatian animal protection magazine, harshly condemns this game, describing it as "a form of entertainment that may seem innocent at the first glance, but come to think of it carefully, we see the butterflies de-winged, thrown on the ground, pressed between the pages of a book, threaded and pinned, which makes good children evil and encourages them to commit many other evil acts."[88] Vjekoslav Majer's moving poem "Umirućem leptiru" (Ode to dying butterfly) describes a butterfly whose death was caused by children and their reckless games.[89] In the brutal children's short story "Strašan šumski razbojnik" (The dreadful forest brigand) by J. Peršić, a teacher and a ranger invite children to participate in the collective destruction of

[83] Batinić, *U carstvu životinja*, 224.

[84] "Kada motrim..." [While I watch], *Bršljan*, no. 3, 1898, 78.

[85] "Ne muči životinja!" [Don't torture animals!], *Bršljan*, no. 6, 1900, 185–88.

[86] "Janko," *Smilje*, no. 10, 1902, 148.

[87] "Usamljena glista" [Lonely earthworm], *Smib*, no. 6, 1993, 18.

[88] *Živobran*, no. 1, 1899.

[89] Vjekoslav Majer, "Umirućem leptiru" [Ode to dying butterfly], *Radost*, no. 1, 1954.

butterfly caterpillars in an oak forest.[90] The ranger hands out knives to the children to take the caterpillars off the tree bark and put them in tin containers. Later they burn them all in a clearing in the forest. The butterfly net game is hardly ever practiced anymore. Batinić believes this is due to dwindling butterfly numbers, a result of habitat destruction, but also to some level of awareness about the need to protect these insects.[91]

A 1980 *Modra lasta* article encourages children to catch (and ultimately kill) insects for a so-called home insectarium, a cage where they will keep the insects they catch and that soon will become their tomb: "We can observe butterflies, locusts, praying mantises, stag beetles and many other interesting representatives of the insect world in peace from up close if we make an insectarium."[92] When making the insectarium, the article instructs children to "be careful not to put insects who could eat each other in the same cage."[93] Children do not have to worry about providing food for the insects they trap "because they are going to die soon in the insectarium anyway."[94]

Changes in people's perception of insects as the tiniest of animals are very slow and barely noticeable in Croatian children's magazines. We will mention some examples found in the corpus we analyzed. In 1939 *Smilje* magazine published Šime Fučić's poem "Žalba na ljude" (Complaint against humans),[95] accompanied by "an illustration showing a fly and a flea with sad, human faces, offering a change in the perception of these tiny creatures—the poem calls for compassion for beings that people only ever see as a pest."[96] In the late 1980s, the children's magazine *Modra*

[90] J. Peršić, "Strašan šumski razbojnik" [The dreadful forest brigand], *Radost*, no. 1, 1954.

[91] Batinić, *U carstvu životinja*. Croatia included endangered butterfly species in the Red List of Endangered Plant and Animal Species in Croatia, available on the website of the Ministry for Environmental Protection and Green Transition. For more information, please visit https://www.haop.hr/hr/tematska-podrucja/prirodne-vrijednosti-stanje-i-ocuvanje/ugrozenost-vrsta-i-stanista/crveni-1.

[92] *Modra lasta*, no. 1, 1980.

[93] *Modra lasta*, no. 1, 1980.

[94] *Modra lasta*, no. 1, 1980.

[95] Šime Fučić, "Žalba na ljude" [Complaint against humans], *Smilje*, no. 9, 1939, 176.

[96] Batinić, *U carstvu životinja*, 73.

lasta published a poem by Nada Zidar Bogadi titled "Budućim sakupljačima leptira" (To future butterfly collectors), urging children not to capture butterflies on meadows, but to let them live.[97]

A significant change of attitude toward animals formerly perceived as "pests" is perceived in modern Croatian children's magazines only from the 1990s onward. Modern Croatian children's magazines no longer write about small animals—butterflies, mosquitoes, snails, earthworms, and insects—as animals who "destroy" the plants sown by people through their feeding on crops in the fields, fruits in vineyards and orchards, and plants in gardens. The small animals are given a certain anthropomorphized role in children's stories, poems, and plays: they are removed from their natural habitats and appear in situations typically associated with people rather than animals.[98] To bring the tiny animal world even closer to children and help little readers identify with these new literary heroes, authors regularly give them names: Fedor and Lord (mosquitoes), Hedviga (a fast and agile but not too bright fly), Žurkov and Nemirko (ants), Mirko (a scraggly ant), Bero (a snail), Mimi (a butterfly), Sara (a ladybug), Dorica and Zorica (a pair of earthworms), and so on.

Finally, bees and ants are held in the highest regard in Croatian children's literature among all insects. From their beginnings to modern times, Croatian children's magazines have portrayed bees as the most industrious insects. The very first Croatian children's magazines published informational articles about bees,[99] describing them as hardworking, industrious, and organized insects. An interesting short story in the magazine *Ustaška uzdanica* features a bee and a wasp arguing over which of the two is the more diligent and valuable insect.[100] The bee sees himself as busy and industrious, unlike the wasp, whom the bee describes as a straggler: "the bee builds, and the wasp steals."[101] Writers for modern Croatian children's magazines also write about bees as valuable, social insects whom people tamed a long time ago, building hives for them. In addition to being discussed as insects who play a clearly defined role in

[97] Nada Zidar Bogadi, "Budućim sakupljačima leptira" [To future butterfly collectors], *Modra lasta*, no. 1, 1989.

[98] *Radost*, no. 10, 2009; *Prvi izbor*, no. 10, 2009; no. 8, 2018; no. 8, 2019.

[99] *Bršljan*, no. 7, 1894, 217–18; no. 8, 1894, 250–53.

[100] "Pčela i osa" [The bee and the wasp], *Ustaška uzdanica*, no. 1, 1942, 6.

[101] "Pčela i osa," 6.

people's lives, bees are recognized for their importance for the preservation of biodiversity on Earth. *Radost* magazine thus quotes Albert Einstein: "If the bee disappeared off the surface of the globe, then man would only have four years of life left."[102]

Ants also are portrayed as very organized and industrious insects in Croatian children's literature. *Modra lasta* magazine in 1975 published a piece about ants as social creatures and as one of the most enduring and possibly strangest beings on our planet. The author of the informational article marvels at the competence and endurance of ants, "who can come back to life after having spent eight days under water. They are virtually indestructible. They can survive for as many as 20 days without a head and be constantly up and running."[103] Although ants are very peaceable in their interactions with their friends in their nests, the author of the article mentions that they are the only insects who have an organized army and wage offensive wars. There may be certain similarities between people and ants (wars, surprise attacks on their enemies, sieges), but the author concludes that "ants are never as cruel and bloodthirsty as people. They never abuse their power and they avoid conflict whenever they can."[104]

15.8 Birds in Croatian Children's Magazines

Birds have been the most celebrated group of animals in Croatian children's poetry from its beginnings in 1850 to the present. Poets voice their admiration for animals who have the ability to fly, along with impressive muscle power, flight speed, physical beauty, and a singing voice. Early Croatian children's poems published in children's magazines depict birds as God's creatures whom people must treat with care, kindness, and compassion. The poem "Prva lastavica" (The first swallow) by Marija Imova is an allegory of God's creation of all living things on Earth.[105] The poem "Da sam ševa..." (If I were a lark...) by Tugomil Janindolski suggests that birds celebrate God with their song.[106] The return of migratory birds, usually swallows, to their habitats in the spring is used as an

[102] *Radost*, no. 7, 2017, 37.

[103] *Modra lasta*, no. 17, 1975, 10.

[104] *Modra lasta*, no. 17, 1975, 10.

[105] Marija Imova, "Prva lastavica" [The first swallow], *Bršljan*, no. 3, 1894, 65–66.

[106] Tugomil Janindolski, "Da sam ševa..." [If I were a lark], *Bršljan*, no. 3, 1901.

allegory symbolizing the return of Croatian emigrants to their homeland—for example, by Stjepko Španić in his poem "Lastavicama" (To the swallows)[107] and by Vjekoslav Mancen in his poem "Lastama" (To the swallows).[108] In early Croatian children's poetry, birdsong also alludes to the glorious Croatian past and evokes the deepest patriotic feelings, as confirmed by several poems in the analyzed corpus, such as "Lastavice" (Swallows) by Mandaljena Martinac and "K dolasku lastavica" (Swallows coming home) by Josip Marijanin.[109] The hawk's nest in the poem "Sokol i sokolići" (The hawk and its chicks) by Ferdo Stražimir Kulundjić is also an allegoric representation of Croatia's glorious, heroic past.[110] During World War II, patriotic feelings and the pride in the independent Croatian state were symbolically extolled in the motif of an eagle's love for his mountain home in the Croatian children's magazine *Ustaška uzdanica*.[111]

Migratory birds are the most common animal theme in Croatian children's magazines. Children's poetry nostalgically dwells on the migration of birds to warmer climates and eagerly and impatiently expects their return as harbingers of spring.[112] The return of migratory birds is a motif encountered in Croatian children's poetry of the twentieth century too. Saying goodbye to migratory birds, usually swallows, and awaiting their return in the spring is the most common poetic motif in the magazine *Ustaška uzdanica*.[113] In addition to swallows, the modern Croatian children's magazine *Prvi izbor* often discusses the migrations of storks.[114]

Non-migratory birds, with whom the poet sympathizes, encouraging readers (both adults and children) to feed them in the winter, are

[107] Stjepko Španić, "Lastavicama" [To the swallows], *Bršljan*, no. 5, 1895.

[108] Vjekoslav Mancen, "Lastama" [To the swallows], *Bršljan*, no. 7, 1899.

[109] Mandaljena Martinac, "Lastavice" [Swallows], *Bršljan*, no. 3, 1896, 95–96; Josip Marijanin, "K dolasku lastavica" [Swallows coming home], *Bršljan*, no. 1, 1897.

[110] Ferdo Stražimir Kulundjić, "Sokol i sokolići" [The hawk and its chicks], *Bršljan*, no. 1, 1897.

[111] *Ustaška uzdanica*, nos. 6–7, 1941–42, 4.

[112] *Bršljan*, no. 3, 1893, 77; *Bršljan*, no. 11, 1893, 330; *Bršljan*, no. 3, 1894, 81; *Bršljan*, no. 10, 1895, 189–290.

[113] *Ustaška uzdanica*, no. 5, 1942, 10; *Ustaška uzdanica*, no. 7, 1942, 10; *Ustaška uzdanica*, no. 17, 1942, 19; *Ustaška uzdanica*, no. 2, 1943, 12.

[114] *Prvi izbor*, nos. 1–2, 2011, 11; *Prvi izbor*, nos. 1–2, 2017, 42–45; *Prvi izbor*, no. 7, 2019, 18–19; *Prvi izbor*, no. 9, 2019, 40–43.

also found in Croatian animal-themed children's poetry. In fact, Croatian children's literature, from its beginnings to the present, has shown more empathy for non-migratory birds than for any other animal group. Most poems encourage the reader to take action and protect the birds from hunger and frost, such as "Smilovanje" (Compassion) by Božić, "U zimi!..." (In the winter) and "Ptice u zimi" (Birds in the winter) by Slavoljub Magdić, "Tica u zimi" (Bird in the winter) by Vjekoslav Mancen, "Vrapčić i seljak" (The sparrow and the villager) by Jure Jurčić, and "Ptičice zimi" (Little birds in the winter) by Andrija Ivančan.[115]

In contrast to the approach to songbirds (both migratory and non-migratory), Croatian children's literature takes a diametrically opposite view of the group of birds who are perceived as "pests," and people claim the right to exterminate them. For instance, *Živobran*, the magazine issued by the Croatian Animal Protection Society at the turn of the twentieth century, called on its readers, including school students, to exterminate "pestilent" birds (eagles, hawks, falcons, crows, ravens, jays, magpies, sparrowhawks, herons, terns, and seagulls): "The destruction of these pestilent birds, their nests and their young is allowed always and everywhere."[116] The absurdity of the Croatian cultural and educational landscape at the turn of the twentieth century is reflected in the fact that *Živobran*, as the magazine issued by the Croatian Animal Protection Society, on the one hand called the Croatian public's attention to the widespread children's habit of shooting at birds with slingshots and destroying their nests, which the society condemned harshly and for which it called on the relevant institutions to sanction the perpetrators,[117] and, on the other, encouraged the citizens and school students to destroy "pestilent" birds.[118] The short story "Iznenadni pohod" (Surprise campaign) by J. Rastuški, published in *Ustaška uzdanica*, is a good illustration of the hypocrisy of then society with respect to people's

[115] Božić, "Smilovanje" [Compassion], *Smilje*, no. 4, 1877, 49–50; Slavoljub Magdić, "U zimi!..." [In the winter], *Bršljan*, no. 1, 1896, 30; Slavoljub Magdić, "Ptice u zimi" [Birds in the winter], *Smilje*, no. 6, 1896, 92; Vjekoslav Mancen, "Tica u zimi" [Bird in the winter], *Bršljan*, no. 11, 1899, 335; Jure Jurčić, "Vrapčić i seljak" [The sparrow and the villager], *Smilje*, no. 4, 1938, 67; and Andrija Ivančan, "Ptičice zimi" [Little birds in the winter], *Smilje*, no. 6, 1940, 107.

[116] *Živobran*, no. 3, 1895, 49.

[117] *Živobran*, no. 6, 1900.

[118] *Živobran*, no. 3, 1895.

perception of certain bird species as useful or "pestilent."[119] The story's protagonist is a boy who helps a wounded blackbird, aware of the moral obligation and duty to help anyone in trouble, even birds. Surprisingly, however, the adult protagonist praises the boy for "helping a valuable bird instead of a pest."[120]

Croatian children's magazines frequently published texts discussing the destruction of birds' nests, the theft of their young, and similar forms of bird abuse and killing. Occasionally, the stories' protagonists are adults, but usually they are children who think of bird killing as an entertaining game. In the short story "Niemi sinovi" (Mute sons), God curses a man who tortured birds by catching them and cutting out their tongues and gives him seven mute sons.[121] In Strižić's short story "Slavuljica" (Nightingale hen), the author condemns the theft of little birds from their nest, committed by "a sinful hand."[122] "Even though nightingales are usually used in poetry to invoke a lyrical atmosphere, this short story describes the suffering of the nightingale hen in an anthropomorphized fashion: she dies of grief for her stollen children."[123] Croatian children's magazines of the twentieth century discuss robbery and destruction of birds' nests too. In the Croatian children's magazine *Modra lasta*, we read about bird killing, a remnant of past times when children were encouraged to kill "pest" birds, but now this centuries-old children's entertainment has expanded to all other birds. *Modra lasta* published an informational article about a Croatian folk custom "that many seem unaware of, or, even more painfully, turn their heads away from. The custom in question is BIRD SLAUGHTER."[124] The perpetrators are children who visit their relatives in the country during the summer break, bringing air rifles and using them mercilessly to kill birds. According to the author J. Prica, "the noblest of birds are hurt the most by this custom:

[119] J. Rastuški, "Iznenadni pohod" [Surprise campaign], *Ustaška uzdanica*, no. 15, 1942, 3–6.

[120] Rastuški, "Iznenadni pohod," 3–6.

[121] "Niemi sinovi" [Mute sons], *Smilje*, no. 6, 1873, 90.

[122] Dor. Napoleon Špun Strižić, "Slavuljica" [Nightingale hen], *Bosiljak*, no. 1, 1864, 12.

[123] Batinić, *U carstvu životinja*, 63.

[124] *Modra lasta*, nos. 19–20, 1974.

tits, finches, goldfinches, blackbirds, starlings."[125] The text condemns parents who get air rifles for their children as a reward for good grades in school so that "children can have fun by killing birds."[126] The poem "Ubijena ptica" (Murdered bird), written by a schoolgirl by the name of Božena, describes a dying bird "killed by vicious male hand / killed by a bullet from an air rifle."[127]

The magazine *Radost* confirms that a peaceful coexistence of people and birds is indeed necessary and possible in texts published in the mid-twentieth century. In the short story "Priča o Lastavicama" (The story of swallows) by Drago Kozina, a grandma prevents a boy from destroying a swallows' nest, telling him that he must not kill swallows or destroy and tear down their nests: "Had you caught it, you hardly would have enjoyed seeing it helpless.... They are beautiful and dignified in their flight."[128] The magazine *Radost* published an article with the motivating title "Budimo prijatelji pticama" (Let's be friends with birds), describing the reality of the spring months, May and June, when many birds in Croatia nest and when irresponsible children look for birds' nests to rob."[129] The article concludes with a warning that "destruction of birds' nests and their young is punishable by law in Croatia."[130]

Caged birds have been an inevitable motif in Croatian children's literature from its very beginnings. The authors of Croatian children's stories and poems condemn the selfish act of caging birds and taking away their freedom, underlining that caged birds are denied the freedom to fly, one of their basic needs. The fate of caged birds is discussed by Mijat Stojanović in the short story "Mali Mato" (Little Mato), by an author writing under the pseudonym Young Patriot in the short story "Slavuj i slavujica" (Nightingale cock and hen), and by Rikard Katalinić Jeretov in the short story "Zarobljeni vrabac" (Trapped sparrow).[131] Croatian

[125] J. Prica, in *Modra lasta*, nos. 19–20, 1974.

[126] J. Prica, in *Modra lasta*, nos. 19–20, 1974.

[127] Božena, "Ubijena ptica" [Murdered bird], *Modra lasta*, no. 20, 1966.

[128] Drago Kozina, "Priča o Lastavicama" [The story of swallows], *Radost*, no. 7, 1970.

[129] "Budimo prijatelji pticama" [Let's be friends with birds], *Radost*, no. 9, 1982.

[130] "Budimo prijatelji pticama," *Radost*, no. 9, 1982.

[131] Mijat Stojanović, "Mali Mato" [Little Mato], *Smilje*, no. 4, 1879, 56–58; Young Patriot (pseud.), "Slavuj i slavujica" [Nightingale cock and hen], *Bršljan*, no. 7, 1897,

children's magazine writers of the contemporary period also write about caged birds, wishing that their cages would forever remain empty.[132]

Bird-related folklore made its way to Croatian children's literature as well. According to an old Croatian folk belief, swallows bring blessings to the house where they decide to make their nest, which is discussed in the short story "Dvije lastavice" (Two swallows) by Jakov Majnarić and "Lastino gnijezdo" (The swallow's nest) by Emil Kišebur.[133] An author writes in *Modra lasta*, "Ancient Slavs loved and respected swallows. They believed them to be holy birds that held the good spirits of their deceased ancestors.... Swallows were considered divine birds, harbingers of spring, and guardians of people's homes."[134] According to another folk belief, whoever dreams of a swallow in flight will fall in love, and whoever dreams of a swallow's nest will receive a gift. In all of these beliefs, swallows are regarded as bringers of love, well-being, and happiness to people. Croatian people also believe that the hooting of an owl hidden in a tree is the harbinger of death, a theme present in the short stories "Sova, smrt i čovjek" (Owl, death and man) by Ferdo Stražimir Kulundjić and "Sova" (Owl) by Ljudevit Krajačić.[135] "Birds are most often cast in the role of harbingers of death, and the sounds of night birds such as the owl or little owl are perceived as melancholic hooting and interpreted as a herald of death."[136] Another folk belief is that storks bring babies to new parents. According to the old belief, storks are bringers of happiness and well-being, and there is no greater happiness than a newborn child crying.[137] In the Middle Ages, couples who were unable to have children drank tea prepared from a plant popularly known as stork's bill. If this tea helped them conceive a child, "one could argue that a stork indeed

221–22; Rikard Katalinić Jeretov, "Zarobljeni vrabac" [Trapped sparrow], *Radost*, no. 1, 1954.

[132] *Radost*, no. 6, 1999.

[133] Jakov Majnarić, "Dvije lastavice" [Two swallows], *Bršljan*, no. 5, 1895, 144–47; and Emil Kišebur, "Lastino gnijezdo" [The swallow's nest], *Bršljan*, no. 1, 1897.

[134] *Modra lasta*, no. 2, 1989.

[135] Ferdo Stražimir Kulundjić, "Sova, smrt i čovjek" [Owl, death and man], *Bršljan*, no. 3, 1899; Ljudevit Krajačić, "Sova" [Owl], *Smilje*, no. 2, 1924, 33–37.

[136] Batinić, *U carstvu životinja*, 249.

[137] *Radost*, no. 7, 2016.

brought the child into this world."[138] There is another explanation for why storks were credited with bringing babies: "storks return from the south in spring, when the number of newborn children is allegedly the highest."[139] The red crossbill is a bird very similar to a parrot, recognizable by the male's long, shiny red tail. Lj. T. relates a biblical story about two birds, the red crossbill and the robin, removing thorns from Christ's body while he was nailed to the cross.[140] This is the reason the crossbill has a crooked bill, and the robin has red feathers (because they became soaked with blood). Ever since then, the two birds have been recognizable among all other bird species by their bills and feathers, respectively. The magazine *Ustaška uzdanica* also told the legend about the crossbill.[141]

15.9 Conclusion

Since its beginnings—the publication of the first Croatian children's book *Mali tobolac* in 1850—Croatian children's literature has gone through different developmental stages in its portrayal of animals and people's attitudes toward them. For almost a century and a half, from the mid-nineteenth century to the 1990s, animal killing, abuse, and neglect were frequent themes in Croatian children's magazines. Among all animal groups, it seems domesticated animals were treated the worst in Croatian children's literature. Domesticated animals are tame, live in the proximity of people, and are bred in people's households, usually as "beasts of burden" (livestock, oxen) or for use as food (cows, calves, pigs, sheep, lambs, and poultry).

Hunting of free-roaming animals, by both adults and children, also was a frequent motif in Croatian children's literature from its beginnings to the end of the twentieth century.

All the way to the 1990s, animal cruelty and abuse (toward cats, dogs, insects, rodents, and birds) was present in Croatian children's magazines. Children thought animal cruelty was fun and often killed animals for no other reason than sheer boredom.

[138] *Radost*, no. 7, 2016.

[139] *Radost*, no. 7, 2016.

[140] *Bršljan*, no. 6, 1891, 13–15.

[141] *Ustaška uzdanica*, nos. 16–17, 1943, 20.

Violence against animals, a frequent theme in the past, is no longer present in modern Croatian children's magazines. Any form of animal neglect, cruelty, or abuse is harshly condemned. Dogs are written about as the most coveted companion animals and no longer as animals kept on chains. Modern Croatian children's magazines are increasingly focused on environmental topics, calling attention to the need for environmental protection, conservation, and preservation of biodiversity on Earth.

Our analysis of animal-themed texts in Croatian children's magazines from the mid-nineteenth century to the present shows how attitudes toward animals in Croatian children's literature have taken shape, transformed, and evolved, reflecting the evolution of attitudes in Croatian society in general.

Bibliography

Batinić, Ana. *U carstvu životinja. Animalističko čitanje hrvatskih dječjih časopisa* [In the animal kingdom: Animal themes in Croatian children's magazines]. Zagreb: Hrvatska sveučilišna naklada, Perodica Croatica, 2013.

Crnković, Milan. *Hrvatska dječja književnost* [Croatian children's literature]. Zagreb: Školska knjiga, 1978.

Crnković, Milan, and Dubravka Težak. *Povijest hrvatske dječje književnosti od početaka do 1955. godine* [The history of Croatian children's literature from the beginning to 1955]. Zagreb: Znanje, 2002.

Filipović, Ivan. *Mali tobolac raznog cvetja za dobru i pomnjivu mladež* [A little pouch of assorted flowers for good and attentive youngsters]. Zagreb: Tiskom i troškom Franje Župana, 1850.

Ogrizović, Milan. *Ivan Filipović, učitelj učitelja* [Ivan Filipović, teachers' teacher]. Zagreb: HPKZ, 1974.

Pašagić, Blanka. *Hrvatski dječji časopisi od 1864. do 1950. godine* [Croatian children's magazines 1864–1950]. Zagreb: Artist Association August Šenoa, 2003.

Pašagić, Blanka. "Hrvatska dječja književnost od 1964. do 1950. godine" [Croatian children's literature 1964–1950]. *Zaprešićki godišnjak* 9 (1999): 167–82.

Pašagić, Blanka. "Umjetnička proza u hrvatskim dječjim časopisima od 1864. do 1945. g." [Artistic prose in Croatian children's magazines 1864–1950]. Zaprešićki godišnjak, 9 (2000): 183–89.

Magazine Sources

Bosiljak, no. 1, 1864; no. 4, 1865; no. 7, 1867; nos. 7, 9, 1867.

Bršljan, no. 6, 1891; nos. 3, 4, 5, 11, 1893; nos. 3, 7, 8, 9, 1894; nos. 5, 10, 1895; nos. 1, 3, 4, 5, 12, 1896; nos. 1, 2, 7, 1897; nos. 1, 3, 1898; nos. 3, 7, 11, 1899; no. 6, 1900; no. 3, 1901.
Modra lasta, no. 13, 1972; no. 3, 1973; nos. 7, 12, 16, 19–20, 1974; no. 17, 1975; no. 1, 1980; no. 9, 1987; no. 12, 1988; nos. 1, 2, 6, 1989; no. 9, 2016.
Prvi izbor, nos. 5, 10, 2009; nos. 1–2, 2011; no. 9, 2012; nos. 1–2, 2016; nos. 1–2, 2017; no. 8, 2018; nos. 7, 8, 9, 2019.
Radost, nos. 1, 4, 1954; no. 7, 1970; nos. 3, 4, 27, 1979; no. 3, 1981; nos. 6, 9, 1982; no. 4, 1998; no. 1, 2006; no. 6, 1999; no. 10, 2009; nos. 1, 7, 2016; nos. 5, 7, 10, 2017; no. 6, 2019.
Smib, no. 6, 1941; no. 4, 1979; no. 6, 1993; no. 4, 2003; no. 6, 2010.
Smilje, nos. 1, 6, 1873; no. 4, 1877; no. 3, 1878; no. 4, 1879; no. 6, 1896; no. 10, 1902; no. 2, 1924; no. 4, 1938; no. 9, 1939; no. 5, 1941; no. 6, 1940.
Ustaška uzdanica, nos. 1, 6–7, 1941; no. 15, 1942; nos. 1, 5, 7, 15, 17, 1942; nos. 2, 4, 10, 16–17, 1943.
Živobran, no. 3, 1895; no. 6, 1896; no. 1, 1899; nos. 5, 6, 1900.
Zoološki vrtić, no. 3, 1933.

CHAPTER 16

Stories from an Animal Rescue: Changing Behavior Through Indirect Education

Tarah Daly

Contents

16.1 Introduction

Involvement in animal rescue has become more commonplace in recent years, owing to an increase in rescues and sanctuaries within communities. This chapter sets out to explore how involvement in animal rescue

T. Daly (✉)
My Lovely Horse Rescue, Kildare, Ireland
e-mail: tarahdalyai@gmail.com

C. Linzey and A. Linzey (eds.), *The Palgrave Companion to Humane Education*, The Palgrave Macmillan Animal Ethics Series,
https://doi.org/10.1007/978-3-032-19795-5_16

can function as a form of indirect humane education. The participants involved in this study were not enrolled in any planned or structured humane education course but instead gained an indirect educational experience through their work with an animal rescue.

Humane Education

Milburn defined humane education as "an attempt to develop altruism and a sense of compassion in a world where all other pressures are in opposition to it."[1] This definition is quite appropriate in this context given that animal rescues and sanctuaries are seen as places of immense compassion in a sometimes unforgiving society where they must work hard to gain any improvements for animals or funding to continue their valuable work. Most humane education programs aim to create knowledgeable individuals who develop attitudes and behavior that are kind and compassionate toward both human and nonhuman animals. Encouragingly, for the most part, studies of the efficacy of humane education programs find a change of attitudes in the participants, in that they hold more humane attitudes toward animals after partaking in the course than before. It has been noted, however, that this has not necessarily led to changes in behavior that would be consistent with this change in attitude.[2] It has been suggested by Hein that more intensive instruction is necessary to bring about a change in behavior in participants.[3] Hornik also has suggested that in practice it is difficult to achieve lasting results without repeatedly exposing participants to humane messages and moral reasoning over a sustained period of time.[4] This position is supported by marketing science research, which finds consistently that unless any

[1] C. Milburn, "Introducing Animal Welfare into the Education System," in *The Status of Animals: Ethics, Education and Welfare*, ed. David Paterson and Mary Palmer (Wallinmgford: C.A.B. International, 1989), 73–78.

[2] G. Hein, *Massachusetts Society for the Prevention of Cruelty to Animals Outreach Program Evaluation: Final Report* (Boston: Boston SPCA, 1987).

[3] Hein, *Final Report*.

[4] Robert C. Hornik, "Exposure: Theory and Evidence about All the Ways It Matters," *Social Marketing Quarterly* 8, no. 3 (2002): 31–37.

program achieves high levels of exposure, it is unlikely to achieve a positive outcome.[5]

Most studies that have looked at the efficacy of humane education programs have involved children as participants, since children are the focus of most humane education programs. This study is novel in that it involves a sample of adult participants. This study also differs from most other studies of humane educational programs because the studied experience involves more contact with animals, especially farmed animals, than any classroom-based humane education course could. Finally, unlike most educational programs, the indirect education that was observed in this study was completely unstructured and unplanned. Through analyses of the personal accounts of volunteers from one animal rescue, this chapter explores how exposure to the animals and those who care for them contributed to a change in participants' own moral baseline—and thus their behavior.

Participants were asked to write their own personal account of what they believed were the factors that led to their shift toward vegetarianism or veganism. The accounts were then analyzed and common themes extracted. The factors that most influenced a change in a volunteer's behavior were found to be (1) exposure to farmed animals and their individual stories, (2) conversations with fellow volunteers, and (3) the normalization of vegetarianism and veganism within the organization. This chapter explores in some detail the participants' experiences of volunteering and the profound effect those experiences had on their lives and then briefly discusses why these factors may have had such an effect on the participants.

16.2 The Setting

My Lovely Horse Rescue

The rescue that provided the location and participants, not to mention the nonhuman animal teachers, was My Lovely Horse Rescue (hereafter MLHR). MLHR is an Irish animal rescue founded in 2011 by two Dublin sisters and their friend. The rescue is now caring for more than three hundred horses, donkeys, mules, goats, sheep, pigs, dogs, cats, turkeys

[5] Frank R. Ascione, "Humane Education Research: Evaluating Efforts to Encourage Children's Kindness and Caring toward Animals," *Genetic Social and General Psychology Monographs* 123, no. 1 (1997): 57–78.

and hens. Housed in three farms and in many foster homes throughout Ireland. MLHR not only cares for hundreds of animals but also works alongside law enforcement and other nonprofits to reduce the suffering of animals reported to them. The rescue saves many animals destined for euthanasia each year. The goal after a nonhuman animal has been rescued is rehabilitation and ultimate rehoming if possible. The rescue also works with authorities by providing education on the existing animal welfare legislation in Ireland and lobbying for greater enforcement of this legislation. The participants interviewed for this study are or were volunteers at the rescue's original and largest farm, which at the time of writing housed more than 150 animals of many different species. Care for the many inhabitants is provided by a small number of dedicated volunteers, making "the farm," as it is affectionately known, very much a multispecies community.

Participants

The six participants interviewed for this study were members of a group of twenty volunteers who helped at the farm at least twice per week at the time of the study. Of the twenty volunteers, eleven were already vegetarian or vegan before becoming involved with the rescue, and three were not vegetarian or vegan; however, the remaining six volunteers who turned vegetarian or vegan after becoming involved with the rescue are the participants of most interest in this study.

Volunteering duties listed by volunteers at the rescue include general cleaning, feeding, medical treatments and administration of medication, sick nursing, and assisting veterinarians and farriers, as well as very important time spent comforting and building trust with the residents. The six participants ranged in age from eighteen to fifty years, and each had spent different lengths of time at the rescue before making lifestyle changes. One participant turned vegan in the space of two weeks after joining the rescue; another spent three to four months at the rescue before turning vegetarian; another made the change after eight months; and the remaining three spent between three and four years working with the rescue before making a lifestyle change toward either vegetarianism or veganism.

16.3 Exposure to Farmed Animals

The most influencing factor that led to attitudinal and behavior change as reported by the participants was their exposure to farmed animals. The impact of spending time with the farmed animals at the rescue was mentioned in detail in each participant's personal story. In particular, caring for the rescued pigs at the farm seemed to have a large impact on the participants' attitudes toward eating animals. One participant noted of her first time meeting Wilbur, one of the (very large) pig residents,

> We met the pigs, and we met Wilbur. He was so big and so dirty and just so not my idea of a pretty pig. [A fellow volunteer] loved him so much, and I just couldn't understand why. She scratched him under the chin and chatted to him and made sure he got the best food, and something in me changed then without me even realizing it had.

Another resident who impacted one of the early volunteers at the farm was a pig called Mrs. Wobbles, who arrived at the rescue in a coma and in a terrible body condition from being neglected on the farm where she had lived: "Bottom line, I felt like such a hypocrite nursing Mrs. Wobbles back to health, but still going home to eat a bacon sandwich. She just opened my eyes to what I was actually putting into my mouth, and nearly three years later I'm still veggie."

The farm resident who had the most impact on the participants, in their own opinion, was Tina Sunshine. Tina is a 250-kilogram pig of the type who are classically killed for meat in Ireland. Following are two of the testimonials relating to Tina's impact:

> Playing with Tina last year and feeding her from my hands really changed my perspective, and that's when I seriously considered going vegetarian.

> The main deciding factor was when Tina the pig arrived to our rescue. She came to the farm, [and] we see her every day. She comes up to you, she has a sniff, she wants to have a play. If you give her a rub, she rolls over on her back, looking for belly rubs and just enjoying life. I remember sitting in the farm one evening, and we had pizza, and I had pepperoni and ham on the pizza, and I just looked at it and said, "I can't eat it, I can't do it, it's not right, it's not good, we shouldn't be doing it."

Deconstructing Carnism

In the attempt to understand how this exposure to farmed animals had such a profound effect on the volunteers, Melanie Joy's concept of carnism seems apt to potentially explain much of what is described in the testimonials.[6] Carnism is an invisible belief system or ideology that conditions people to eat certain animals. Joy proposes that carnism is deeply woven into our societies and is facilitated by mainstream media and the industries that profit from animal exploitation.[7] This is done through the justification of animal use as normal, natural, and necessary, denying the damage that is being done by rendering invisible the animals who are eaten. The truth about their lives and deaths and about the animal agriculture system as a whole is kept behind closed doors and high slaughterhouse walls. Over 150 million (a number difficult to quantify) land animals are killed for food every day around the world, but they are not represented in our media; there is no awareness among the general public of their existence, and their deaths are the absent referent at every meal.[8] The final piece of the carnism puzzle is the cognitive distortions that enable our societies to do the things we do to nonhuman animals. We break animals down into two categories: edible and inedible.[9] Adams suggests that through our commodification of nonhuman animals, we reduce them to objects and abstractions (e.g., "rashers," "sausages," "pork") as opposed to living sentient beings (e.g., Wilbur, Tina, and Mrs. Wobbles).[10] A distance is created between consumers and the lives they consume—physically by keeping slaughterhouses at a distance from populated areas and shielding from view what goes on there and also through the language that we use to talk about the animals who are consumed, which will be discussed in detail later in the chapter. It is suggested that three central processes occur in the minds of people who eat meat. These processes are objectification, de-individualization, and dichotomization.[11]

[6] Melanie Joy, *Why We Love Dogs, Eat Pigs, and Wear Cows: An Introduction to Carnism* (Newburyport, MA: Conari Press, 2011).

[7] Joy, *Why We Love Dogs*.

[8] Carol J. Adams, *The Sexual Politics of Meat: A Feminist-Vegetarian Critical Theory* (Bloomsbury, 2015).

[9] Joy, *Why We Love Dogs*.

[10] Adams, *Sexual Politics of Meat*.

[11] Joy, *Why We Love Dogs*.

During their time spent at the rescue among the animals, participants in this study were exposed to information that may have disrupted these all too often unconscious processes.

Objectification, De-Individualization, and Dichotomization

"Objectification" in this context refers to viewing a living being, usually a farmed animal, as an inanimate object. This is often brought about through the language that is used when referring to the beings in question and is further reinforced by their legal status as property. This process leads to our having less moral concern for these individuals and attributing less consciousness to them.

"De-individualization" in this context involves viewing individuals only in terms of their group identity. Individuals are often viewed as abstractions, especially through our use of language.[12] Cows when dead are referred to as "beef," and pigs as "pork" or "bacon." We further denigrate farmed animals by viewing their actions as instinctual (shared) rather than intellectual (individual). This process explains the seemingly contradictory views of people who eat cows and pigs but are disgusted by the thought of eating dogs, since dogs are thought to be intelligent individuals by most.[13] Dogs are perhaps viewed as individuals because the majority of people, at least in Western societies, have had a relationship with, or at least some contact with, an individual dog. There are fewer people, however, who have ever had contact or built a relationship with an individual pig, sheep, or cow. This study looks at a unique environment where participants meet, become familiar with, and often build relationships with the same nonhuman animals our media and society would have us think of as only food, animals supposedly lacking intelligence and emotionality and acting only on basic instinct and herd mentality.

It is a little difficult to objectify and deny the individuality and intelligence of pigs when Tina Sunshine, a 250-kilogram rescue pig, is lying in the kitchen or opening every gate on the yard, letting horses out of their stables and causing general mayhem, something none of the dogs at

[12] Iveta Šedová and Tereza Vandrovcová, "The Psychology of Meat Consumption," in *Handbook of Research on Social Marketing and Its Influence on Animal Origin Food Product Consumption*, ed. Diana Bogueva, Dora Marinova, and Talia Raphaely (Hershey: IGI Global, 2018), 1–16.

[13] Joy, *Why We Love Dogs*.

MLHR and only one of the horses has ever managed to achieve. From the participants' descriptions of their reasons for changing their views on nonhuman animals, it seems that the attribution of consciousness and intelligence came as a result of spending time with the nonhuman animals at the rescue, in particular the farmed animals. One participant said about Tina,

> You get to see the intelligence of her—she's roaming around opening gates, opening gates to the stables and letting all the horses out. And although it's a bit of work to get the horses back in, it's fantastic to see, and it emphasizes that these animals are just as intelligent as any other species... why should we have the right to kill them and eat them when they are of such super intelligence?

At MLHR, volunteers meet conscious individuals whose actions can clearly be seen as intellectual. Taking care of these nonhuman individuals, experiencing their strong personalities, and often fighting to keep them alive may allow volunteers to re-individualize farmed animals. Furthermore, the moral concern that the study participants felt toward the individual farmed animals they encountered seemed to then be generalized to the individuals' entire species. The farmed animals themselves seem to have become ambassadors of their species in the eyes of the participants. One participant commented on how their perspective had changed, saying that upon seeing a truck headed to the slaughterhouse with pigs on board, "instead of simply not noticing or acknowledging a truck full of pigs, I now see a truck full of Tinas."

The third process involved in enabling carnism to thrive is dichotomization, which in this context refers to the rigid categories we are encouraged to place animals into, in our minds and, as a result, in our hearts. The concept of edible animals and inedible animals is a categorization that dictates which animals we eat and which are worthy of our love, care, and companionship. The edible animal category usually includes all animals who are farmed in society—for example, cows, pigs, sheep, and chickens—and inedible animals usually include the animals most commonly kept as pets, such as dogs, cats, and horses.[14] Perhaps spending time with farmed animals, especially in a caring capacity as the participants of this study have done, leads to a breakdown of these

[14] Joy, *Why We Love Dogs.*

rigid categories and removes the barrier that carnism creates between the consumable and the cuddly. One participant remembers this encounter with Tina Sunshine: "I was sitting at the kitchen table, and she stood next to me. At that moment I realized she was not that different to my dog, Kobi, and I'd never eat him." Sitting at the kitchen table and looking down to see a pig where often a dog would stand led this volunteer to rethink the edible/inedible dichotomy. These occurrences seem to be common at the rescue and appear in many of the participants' testimonials. At MLHR it is very common to see the dogs sniffing around with Tina or sleeping in the kitchen beside Elvis, the young potbellied pig. If the carnistic idea of edibleness relies on reducing those who are eaten to objects and denying that they have any individuality, intelligence, or the ability to form meaningful relationships with humans, then the observations of this study are perhaps not surprising. These volunteers who come face-to-face with the victims of carnism while caring for their needs in a rescue environment changed not only their belief system but also their behavior.

16.4 Conversations with Fellow Volunteers

Another contributing factor that participants reported greatly influenced their change in lifestyle was the conversations and interactions they had with their fellow volunteers at the rescue. One participant said, "The first time we went to the farm for a visit was, looking back on it, quite intense. We sat in the kitchen having tea, and one of the founders spoke pretty much nonstop about just about everything, the horses, the farm, the struggle, the authorities, and I know now that it was her passion that began to seep into me right from the start without me even knowing it."

Participants also referred to the importance of the open and nonjudgmental nature of the other volunteers at the rescue, how they were made to feel welcome and that they did not have to be vegetarian or vegan in order to volunteer there. Looking back on her initial time at the rescue, one volunteer said, "You will never make anyone go vegan by making them feel guilty. It's not how I became vegan, and if the first vegan I met on the farm had done that to me, I might not have gone back. I might have felt that I didn't belong there. Yet there is exactly where I did belong. I just didn't know it at the time." Another volunteer had been at the rescue for four years, but one conversation was the tipping point for her to make a change in her life: "I thought I would never survive if

I gave up meat. Then I had a conversation with [a fellow volunteer], just casual, not a lecture! It was the passion in her voice that just made me stop in my tracks and give up meat right there. It was the best decision I ever made." Another volunteer spoke of the humor used at the rescue in order to discuss topics relating to animal use: "[a fellow volunteer] makes me a mug of tea and asks do I want soy milk or breast milk. That was the last time I drank dairy milk. I asked what's wrong with dairy, and she explained the dark side to something I had not even considered."

16.5 Normalizing Veganism

People are more likely to engage in a behavior if they feel that most other people are doing it[15]; this has been proven to be true time and again.[16] Unfortunately, in everyday life for most people, veganism is not viewed as a social norm, as most communities do not have a majority of vegan people. There has been huge growth in the plant-based sector,[17] but most individuals do not experience a concentration of vegan people in their workplace, sports club, or home. MLHR provides a veggie micro-community, a skewed society if you will, where the majority of volunteers and founders alike are either vegetarian or vegan, and therefore different norms pervade. Within this community, veganism is a social norm, and therefore, while being exposed to this community, volunteers are more likely to engage in similar behavior.

There are some limitations to this study, including the author's own work with the rescue in question, which may lead to bias, and also the use of a self-report measure, which is very much a subjective measurement of behavior and attitude. Although the parameters of this study are limited, with the sample size being small and limited to one organization, it is an example of indirect and unstructured humane education that is having an

[15] Robert B. Cialdini, "Crafting Normative Messages to Protect the Environment," *Current Directions in Psychological Science* 12, no. 4 (2003): 105–9.

[16] Muzafer Sherif, *The Psychology of Social Norms* (New York: Harper, 1936); Robert B. Cialdini and Melanie R. Trost, "Social Influence: Social Norms, Conformity and Compliance" (1998); Ernst Fehr and Urs Fischbacher, "Social Norms and Human Cooperation," *Trends in Cognitive Sciences* 8, no. 4 (2004): 185–90.

[17] Deloitte, *Plant-Based Alternatives: Driving Industry M&A* (London: Deloitte, 2019), https://www2.deloitte.com/content/dam/Deloitte/uk/Documents/consumer-business/deloitte-uk-plant-based-alternatives.pdf.

impact on a small scale and may be worthy of further investigation and development.

16.6 Conclusion

This study started out as simple observation of behavior change that was evident in volunteers at an animal rescue. After hearing from the participants why they believe their moral baseline shifted as a result of their time spent volunteering, the author then set out to examine what might have influenced this shift. The three factors that were discussed in this chapter are those the participants believe made the biggest impact on them: their exposure to farmed animals, their interactions with their fellow volunteers, and the normalizing of veganism and vegetarianism at the rescue. After a closer examination of these factors, this study postulates that the behavioral shift came partly as a result of the volunteers' change in attitudes toward farmed animals, which has been explained here through the ideology of carnism and how this was deconstructed in the minds of volunteers. Suggestions as to why interaction with fellow volunteers might have been influential include the open and nonjudgmental atmosphere at the farm and the passion of the participants' fellow volunteers. These findings show how important a supportive peer group is while someone is being exposed to and processing new information and, as a result, is going through attitudinal and behavioral shifts.

The presence of veganism and vegetarianism as social norms within the small community of volunteers at the rescue may have had an impact on participants also. It was found that a ready-made culture of empathy and caring was present at the rescue, one with which volunteers became involved, in which they became members. The findings of this study suggest that the norms present in this group may have stimulated the moral development of these individuals, leading them to make more compassionate, humane choices, and thus provide some insight into the power of invoking social norms as compelling reasoning for behavioral change. Although the sample in this study is small and perhaps only representative of the specific population involved, it provides an example of a situation where an unstructured humane education had an unplanned impact on a small scale. Further investigation into how to build on the most influential factors in this study may be beneficial in the development of effective structured humane education programs. This study provides indications of some of the key aspects needed for behavioral change as opposed to just a shift in participants' attitudes, which is often the outcome of educational programs.

Bibliography

Adams, Carol J. *The Sexual Politics of Meat: A Feminist-Vegetarian Critical Theory*. New York. Bloomsbury, 2015.

Ascione, Frank R. "Humane Education Research: Evaluating Efforts to Encourage Children's Kindness and Caring toward Animals." *Genetic Social and General Psychology Monographs* 123, no. 1 (1997): 57–78.

Cialdini, Robert B. "Crafting Normative Messages to Protect the Environment." *Current Directions in Psychological Science* 12, no. 4 (2003): 105–9.

Cialdini, R. B., & Trost, M. R. "Social influence: Social norms, conformity and compliance." In The handbook of social psychology, edited by D. T. Gilbert, S. T. Fiske, & G. Lindzey, pp. 151–192. Oxford: Oxford University Press, 1998.

Deloitte. *Plant-Based Alternatives: Driving Industry M&A*. London: Deloitte, 2019. https://www2.deloitte.com/content/dam/Deloitte/uk/Documents/consumer-business/deloitte-uk-plant-based-alternatives.pdf.

Fehr, Ernst, and Urs Fischbacher. "Social Norms and Human Cooperation." *Trends in Cognitive Sciences* 8, no. 4 (2004): 185–90.

Hein, G. *Massachusetts Society for the Prevention of Cruelty to Animals Outreach Program Evaluation: Final Report*. Boston: Boston SPCA, 1987.

Hornik, Robert C. "Exposure: Theory and Evidence about All the Ways It Matters." *Social Marketing Quarterly* 8, no. 3 (2002): 31–37.

Joy, Melanie. *Why We Love Dogs, Eat Pigs, and Wear Cows: An Introduction to Carnism*. Newburyport, MA: Conari Press, 2011.

Milburn, C. "Introducing Animal Welfare into the Education System." In *The Status of Animals: Ethics, Education and Welfare*, edited by David Paterson and Mary Palmer, 73–78. Wallingford: C.A.B. International, 1989.

Šedová, Iveta, and Tereza Vandrovcová. "The Psychology of Meat Consumption." In *Handbook of Research on Social Marketing and Its Influence on Animal Origin Food Product Consumption*, edited by Diana Bogueva, Dora Marinova, and Talia Raphaely, 1–16. Hershey: IGI Global, 2018.

Sherif, Muzafer. *The Psychology of Social Norms*. New York: Harper, 1936.

CHAPTER 17

The Importance of Seminary Education on "Creation Care" and "the Integrity of Creation"

Christina Nellist

17.1 Introduction

While the thrust of this chapter is the potential role of Eastern Orthodox Christian institutions in effecting real changes in the hearts and minds of Orthodox Christians, it has relevance for the wider Christian seminary and academic communities. For decades, there have been calls from senior Orthodox theologians for "creation care" to be incorporated into seminary and academic courses, yet few institutions have responded to those calls. One cannot say why this situation exists, but one possibility might be an intrinsic conservatism and belief that Orthodox tradition must be preserved because "the church" is opposed to change. I posit that our tradition is dynamic rather than static and that in educating our priests

C. Nellist (✉)
Independant Scholar and Fellow, Oxford Centre for Animal Ethics, England, United Kingdom
e-mail: panorthodoxconcernforanimals@gmail.com

C. Linzey and A. Linzey (eds.), *The Palgrave Companion to Humane Education*, The Palgrave Macmillan Animal Ethics Series,
https://doi.org/10.1007/978-3-032-19795-5_17

on creation care, which must include the recognition of animal suffering, and the need for compassionate care of animals, we will exercise our role as "priests of creation," create positive changes in our communities, and play our part in helping to reduce the damaging effects of climate change.

17.2 Religion's Authority to Engage with Contemporary Social Issues

At the 1987 World Council of Churches consultations, Limouris[1] and other Orthodox representatives called for more theological education on contemporary social and ethical disciplines in all of our seminaries and theological schools, together with guidance and encouragement from bishops, priests, and laity, in order to teach and inspire the youth of the church on environmental issues. In 1989, Stylios offered a parallel but equally positive route for effecting change: "This in practice means that Christians will be leaders in every ecological movement, which seeks to maintain and protect the natural environment."[2]

His All-Holiness Bartholomew has not been indifferent to such calls. Since Halki I in 1994, he has frequently called for the leaders of the church to develop programs of practical application and advises the clergy and others in parish ministry to encourage and promote love for nature, the planting of forests, cleaning lakes, and so on. In *Encountering the Mystery*, he gave such teachings theological context and justification:

> We cannot but be convinced environmentalists and firm believers in the sanctity of the material world... It is a pledge that we make to God that we shall embrace all of creation. It is what Orthodox theologians call "inaugurated eschatology," or the final state already established and being realized in the present.[3]

[1] Gennadios Limouris, ed., *Justice, Peace and the Integrity of Creation: Insights from Orthodoxy* (Geneva: WCC, 1990).

[2] Euthimios K. Stylios, *Anthrōpos kai physiko perivallon: Historikē, philosophikē, kai theologikē theōrēsē tou oikologikou provlēmatos* (Athens, Gregoris Publications, 1989).

[3] H.A.H. Bartholomew, *Encountering the Mystery: Understanding Orthodox Christianity Today* (New York: Doubleday, 2008), 107.

In 2013, H.A.H. Bartholomew acknowledged these calls when giving authority for Eastern Orthodox academics to involve themselves in contemporary issues:

> The scholarly community are, perhaps in a unique and even unprecedented manner, able to provide invaluable insights and influential incentives in the wider community with regard to the limitless treasures of the philosophical, patristic and prophetic tradition of our Church.[4]

In light of such statements and other material presented in my 2018 book, *Eastern Orthodox Christianity and Animal Suffering: Ancient Voices in Modern Theology*,[5] such teachings open theological space for Christian theologians and academics to engage with the subject of animal suffering. The question arising here is how many Eastern Orthodox (and other) Christian theologians and ethicists are knowledgeable on the various subthemes of animal suffering? If the answer is too few, this brings us to the need for urgent education on this subject for seminarians, academics, and existing priests who, with God's grace, will discuss the subject with their students and parishioners.[6]

Perhaps there is no need. Perhaps this is already being addressed in the wider Orthodox and Christian world. I have repeatedly asked this question. Initially, I contacted Fr. Chryssavgis in 2014 and asked if there was a seminary project or module for priests on the ethical treatment of animals and the environment. As far as he knew, no such program existed, but he believed the ecumenical patriarch would support such programs. In our 2014 interviews I asked Met. Kallistos and Bishop Isaias, and they gave a similar response. According to Met. John Zizioulas, this was still the case in 2015: "In the traditional manuals of theology, there is hardly

[4] H.A.H. Bartholomew, "Prefatory Letter," in *Toward an Ecology of Transfiguration: Orthodox Christian Perspectives on Environment, Nature and Creation*, ed. John Chryssavgis and Bruce Foltz (New York: Fordham University Press, 2013), xii.

[5] Christina Nellist, *Eastern Orthodox Christianity and Animal Suffering: Ancient Voices in Modern Theology* (Newcastle upon Tyne: Cambridge Scholars Publishing, 2018). See also Stanley Harakas, "Ecological Reflections on Contemporary Orthodox Thought in Greece," *Epiphany Journal* 10, no. 3 (1990): 46–61; H.A.H. Bartholomew, *Cosmic Grace and Humble Prayer: The Ecological Vision of the Green Patriarch Bartholomew*, rev. ed. (Grand Rapids, MI: Eerdmans, 2009), 336.

[6] This echoes calls from animal protectionists in my Cyprus case study for the church to educate its priests. Nellist, *Eastern Orthodox Christianity*, 2018, Chap. 5.

any place for ecology and the same is true for the academic curricula of the theological schools, Catholic, Orthodox and Protestant."[7] I asked a similar question to Professor James Skedros in 2017 and received a similar response.[8] In my 2018 book I posited that if this remains the case, would this not indicate a lack of commitment on behalf of the leaders of the Eastern Orthodox Church to ensure such teachings are available for the next generation of priests?

Other denominations have asked similar questions, as exemplified by two important studies. The first, a 2007 study titled *The Place of Environmental Theology: A Course Guide for Seminaries, Colleges and Universities*, has numerous areas of relevance that cannot be covered here; however, one of its concluding remarks is particularly relevant to this discussion: "In almost all cases environmental theology is still not really integrated into the curriculum of theological colleges."[9]

The second study, "Eco-Theology, Climate Justice and Food Security," is a global survey of seminary institutions conducted in 2014–15. One recommendation reiterates the conclusions of the 2007 study and my own views:

> The introduction of compulsory courses on ETCF at institutions of education should be seen as one part of a possible solution. This will affect students and future generation of teachers and pastors, who can pass on

[7] Met. John Zizioulas, "A Comment on Pope Francis' Encyclical Laudato Si" (Ecumenical Patriarchate, 2015), https://www.patriarchate.org/-/a-comment-on-pope-francis-encyclical-laudato-si-.

[8] James Skedros was at the time professor of Byzantine studies and early Christianity at Holy Cross Greek Orthodox School of Theology, Brookline, Massachusetts.

[9] John Weaver and Margot R. Hodson, eds. *The Place of Environmental Theology: A Course Guide for Seminaries, Colleges and Universities* (Whitley Trust, Oxford; International Baptist Theological Seminary, Prague: 2007), 120, available at (biblicalstudies.org.uk). See also Dietrich. Werner, David Esterline, Namsoon Kang, and Joshva Raja, eds., *Handbook of Theological Education in World Christianity: Theological Perspectives—Regional Surveys—Ecumenical Trends* (Oxford: Oxford Centre for Mission Studies, Regnum Books International, 2010), for an international debate on theological education, regional surveys, and selected websites for networks, associations, and resources for theological education. Available at (ocms.ac.uk); and Marc Wisnosky, *Contemporary Orthodox Christian Theological Education in the United States of America*, Submitted to the Graduate Faculty of The School of Education in partial fulfillment of the requirements for the degree of Doctor of Philosophy University of Pittsburgh 2015. Available at (99+) Contemporary Orthodox Christian Theological Education in the United States of America (Open Source) | Marc Wisnosky—Academia.edu.

> their knowledge in their daily work environment and therefore serve as multipliers.[10]

Sadly, this report also confirms the lack of creation care in Orthodox seminary and academic institutions:

> Despite the comparatively important place occupied by the ecologically related course in the university curriculum and the development of Orthodox networks and programs on ecological topics... eco-theology is almost absent from theological education.[11]

As previously noted, H.A.H. Bartholomew has been at the forefront in asking educators to involve themselves in environmental issues. His Religion, Science and Environment symposia and the Halki Summits on various aspects of the contemporary environmental crisis are examples of his continuing efforts to give our Orthodox educational institutions the opportunities and authority to unite theory and praxis on contemporary issues.

17.3 Halki III Summit: Kairos—The Final Opportunity for Progress?

Calls from senior theologians for curriculum development continue to today:

> Aware of the urgency of environmental education, the Ecumenical Patriarchate devoted the Third in its series of international Halki Summits to the subject of "Theological Education and Ecological Awareness" (Istanbul, May 31st to June 4th, 2019) with a view to incorporating ecology and environmental awareness into programs and curricula of theological schools

[10] Dietrich Werner and Elisabeth Jeglitzka eds. *Eco-Theology, Climate Justice and Food Security: Theological Education and Christian Leadership Development*, (Geneva: Globethics.net, 2016), Sect. 17.4, p. 87.

[11] Dietrich Werner and Elisabeth Jeglitzka eds. *Eco-Theology, Climate Justice and Food Security: Theological Education and Christian Leadership Development*, (Geneva: Globethics.net, 2016), Sect. 13, pp. 232–33.

> and seminaries. The solution to the great challenges of our world is unattainable without spiritual orientation.[12]

My argument has always been that the aforementioned conferences and seminars should include the important subject of animal suffering. This appears to have been acknowledged, for I was invited to speak at the 2019 Halki III Summit on behalf of the trillions of animals who are generally ignored in our academic work and daily lives. My presentation outlined how I could facilitate the inclusion of the important subject of animal suffering into Orthodox academic and seminary education programs by offering an outline of a master's dissertation/seminary module and a copy of my 2018 book for each participating institution.

I posited that ignorance has produced and will continue to produce bad philosophy and theology. We cannot shy away from knowing the types or scale of animal suffering any more than we should shy away from knowing about deforestation or the extent of pollution in our oceans and the air we breathe. I understand why we would wish to shy away from this knowledge because it is indeed harrowing, but it is absolutely necessary if we are to fully understand the subject itself and provide informed opinions on difficult issues. I argued that in educating our priests on the subject of animal suffering, we would help ensure a broader appreciation of the suffering of God's creation in the laity.

I acknowledged that while some seminary and faculty heads might argue that topics such as human dietary habits and recreational activities that delight in the killing of innocent creatures are outside of the remit of the church, this is not the case. I explained that my book includes teachings from the earliest periods of the church on the need for kindness, mercy, and compassion in our dealings with animals. I argued that if, as the early and contemporary church fathers teach, we are to embrace all of creation,[13] and if we are to follow Christ's teachings and act immediately to prevent animal suffering,[14] we ought to be willing to hear *and*

[12] Encyclical of Ecumenical Patriarch Bartholomew for the Feast of Indiction and the First Day of the Ecclesiastical Year and the Day of Environmental Protection—Greek Orthodox Archdiocese of America—Greek Orthodox Archdiocese of America (goarch.org) (2019).

[13] Saint Irenaeus, *Against Heresies* (Whitefish, MT: Kessinger Publishing's Rare Reprints, 2004), 2.2:5.

[14] Luke 14:5 in original Greek.

respond to the very real cries of the suffering animals in contemporary societies. This is reinforced in H.A.H. Bartholomew's September 1, 2019, encyclical: "And we proclaim once more the truth that there can be no genuine progress, when the 'very good' creation and the human person made in the image and likeness of God suffer."

In the Halki III summary session, the head of one of the seminaries asked what he should drop from the curriculum if he were to incorporate care for the environment or animals into his seminary's courses, and this is a valid point. This led to surprisingly bold discussions, including suggestions that the seminaries should be innovative and, yes, give some thought to dropping or reducing aspects of their present courses in order to incorporate creation care; alternatively, they could include creation care across the entire curriculum. This leader also made the point that changes to such courses could take a long time. In light of the previously mentioned theological surveys, one could at this point be forgiven for thinking that some Orthodox institutions might continue to resist calls for curriculum change. It is therefore important to offer two encouraging comments. During the summit, several seminary administrators from different countries informed me that they would be using my book with their students. In August 2019, I met the aforementioned faculty head, and he informed me that his seminary was beginning discussions on curriculum change.

At the time of this writing, it is too early to state whether these intentions have turned into concrete proposals for change, but a questionnaire was to be circulated to all participating institutions, which should provide details of any proposed developments in curriculum and, consequently, the fusion of Orthodox theory and praxis.[15] In addition, a steering committee has been established to keep this momentum moving forward, and it took the author's "Creation Care: Christian Responsibility Course"

[15] I can add that the Hellenic College Holy Cross Greek Orthodox School of Theology's Huffington Ecumenical Institute, will offer in Spring 2025, a course on "The Soul and the Soil: Orthodox Christianity and the Natural Environment." This course provides an opportunity for students to become acquainted with the past, present, and future of the Orthodox Church's commitment to creation care in an ecumenical context and is a direct result of the Halki 111 Summit. See https://huffingtoninstitute.hchc.edu/courses/. It will be taught by Dr Nikolaos Asproulis of the Volos Academy for Theological Studies in Greece, and the Chair of Pan Orthodox Concern for Animals Charity, https://panorthodoxconcernforanimals.org/.

under review.[16] In Met. John of Pergamon's summation of the Halki 111 Summit, he took the opportunity to call for education at all levels, so that as children grow, creation care becomes a natural part of their character.

Finally, touching on a different but related point, Limouris identified another important reason that the leaders of the church need to participate in teachings on reducing all forms of suffering. The Christian church is not the only authority in today's society, and some of these ideologies are antithetical to the Christian message. He raises the challenge for the leaders of our church and institutions to provide a vision "which the world desperately needs to hear, adopt and realize in practice."[17] I agree entirely with his views and would simply add that we must include the nonhuman animal creatures who suffer and will continue to suffer, until we hear that spiritual voice expressed on their behalf in our seminary and academic courses, towns, and villages.

17.4 Concluding Remarks

We have assuredly reached a kairos—with calls from across the globe for seminaries and academies to respond to the suffering of God's creation. How Orthodox institutions will respond is as yet unknown. Certainly, for some, there is trepidation about making changes, yet as I have intimated, Orthodox theology is traditionally dynamic. Our theological institutions should therefore respond courageously and engage with that pneumatologically inspired dynamism by incorporating important contemporary issues such as creation care and the integrity of creation into their curricula.

For far too long, the teachings of a few ancient Greek philosophers have determined our definitions and applications of justice and influenced certain aspects of our theology. A significant part of our creation has suffered every bit as much as a result of the classification of animals as irrational, non-eternal creatures as they have suffered from the misinterpretation of dominion as domination. What should guide us in determining who receives justice, mercy, and compassion should not be scholasticism with its separationist interpretations of language and

[16] "Creation Care: Christian Responsibility Course," Pan-Orthodox Concern for Animals, http://panorthodoxconcernforanimals.org/creation-care-christian-responsibility-course.

[17] Limouris, Gennadios *Justice, Peace and the Integrity of Creation*, 22.

rationality, but rather the guiding spirit of love, as exemplified in an all-loving God who offers love, justice, mercy, and compassion to all created beings. It is time to cast such teachings aside, just as we have done with ancient philosophical teachings on slavery and women as property and replace them with teachings from contemporary church fathers and patristic teachings, which offer a unification of theological and ethical thought, where justice, peace, respect, and compassion, coupled with the responsible use of our freedom, extend in practice to all creature beings.

I fully expect that in educating new and existing priests and consequently the laity on the sin of animal abuse and, for example, the detrimental environmental impacts of consuming animal food products, our professors and administrators will help to achieve the long-called-for reformation in the hearts and minds of our communities. In so doing, they will also assist in the reduction of animal suffering and some of the impacts of global warming. The overarching problem to all of this, however, is that we are quickly running out of time for them to act.

Bibliography

Bartholomew. His All Holiness. *Cosmic Grace and Humble Prayer: The Ecological Vision of the Green Patriarch Bartholomew*. Rev. ed. Grand Rapids, MI: Eerdmans, 2009.

Bartholomew. *Encountering the Mystery: Understanding Orthodox Christianity Today*. New York: Doubleday, 2008.

Bartholomew. Encyclical for the Ecclesial New Year. 2019. http://www.ec-patr.org.

Bartholomew. "Prefatory Letter." In *Toward an Ecology of Transfiguration: Orthodox Christian Perspectives on Environment, Nature and Creation*, edited by John Chryssavgis and Bruce Foltz. New York: Fordham University Press, 2013.

Harakas, Stanley. 1990. Ecological Reflections on Contemporary Orthodox Thought in Greece. *Epiphany Journal* 10 (3): 46–61.

Irenaeus, Saint. *Against Heresies*. 5 vols. Whitefish, MT: Kessinger Publishing's Rare Reprints, 2004.

Limouris, Gennadios., ed. *Justice, Peace and the Integrity of Creation: Insights from Orthodoxy*. Geneva: WCC, 1990.

Nellist, Christina. 2018. *Eastern Orthodox Christianity and Animal Suffering: Ancient Voices in Modern Theology*. Newcastle upon Tyne: Cambridge Scholars Publishing.

Stylios, Euthimios. K. *Anthrōpos kai physiko perivallon: Historikē, philosophikē, kai theologikē theōrēsē tou oikologikou provlēmatos.* Athens, Gregoris Publications 1989.

Weaver, John and Margot R. Hodson, eds. *The Place of Environmental Theology: A Course Guide for Seminaries, Colleges and Universities.* Oxford, The Whitley Trust/

International Baptist Theological Seminary, Prague: 2007. Available at (biblicals tudies.org.uk).

Werner, Dietrich., David Esterline, Namsoon Kang, and Joshva Raja, eds. *Handbook of Theological Education in World Christianity: Theological Perspectives—Regional Surveys—Ecumenical Trends.* Oxford: Oxford Centre for Mission Studies, Regnum Books International, 2010.

Werner, Dietrich and Elisabeth Jeglitzka eds. *Eco-Theology, Climate Justice and Food Security: Theological Education and Christian Leadership Development*, ETCF study project. Geneva: Globethics.net, 2016. www.globethics.net.

Wisnosky, Mark. *Contemporary Orthodox Christian Theological Education in the United States of America.* University of Pittsburgh, 2015. Available at (99+) Contemporary Orthodox Christian Theological Education in the United States of America (Open Source) | Marc Wisnosky—Academia.edu

Zizioulas, Met. John. "A Comment on Pope Francis' Encyclical Laudato Si." Ecumenical Patriarchate, 2015. https://www.patriarchate.org/-/a-comment-on-pope-francis-encyclical-laudato-si-.

PART III

Developing Practice

CHAPTER 18

Building a Bright Future: Humane Education from Theory to Practice

Magdalena Smrdelj, Amanda McKibbon, Elizabeth Davis, and Alyse Patterson

Generally, humans tend to regard themselves as separate from the rest of the world in many ways. But the truth is, we are part of an integrated system that is constantly being affected by and responding to the movements and actions of humans. When decisions are to be made, it is critical that as members of this vibrant system, we consider the well-being of all

M. Smrdelj (✉) · A. McKibbon
OneKind, Medicine Hat, AB, Canada
e-mail: magdalenas33@gmail.com

A. McKibbon
e-mail: Almtkibb@gmail.com

E. Davis · A. Patterson
York Region District School Board, Aurora, ON, Canada
e-mail: Elizabeth.davis@yrdsb.ca

A. Patterson
e-mail: Alyse.patterson@yrdsb.ca

C. Linzey and A. Linzey (eds.), *The Palgrave Companion to Humane Education*, The Palgrave Macmillan Animal Ethics Series,
https://doi.org/10.1007/978-3-032-19795-5_18

living beings. But how is such an idea translated into action—in other words, how do we get from theory to practice? This chapter begins with an exploration of the OneKind concept, actualized by examples that illustrate the interconnectedness of the living world. Humane education is proposed as a vehicle through which to disseminate the ideas inherent in the concept, and a learning resource called AnimalSmart is presented as a powerful tool to support these understandings. The development and impact of a school-based program are presented as an inspiring example of how the concept of OneKind can be brought to life through humane education, to positively impact our future leaders.

18.1 OneKind

When we contemplate why humane education matters, it is essential that we recognize that compromising or harming animals or the environment ultimately affects our well-being, as well as that of the rest of humanity.

OneKind is a concept that easily justifies the critical need for humane education. It enables us to elevate our thinking to a holistic vantage point from which we can consider the interconnections and intersections within the areas of animal well-being, human well-being, and environmental well-being and conservation. We can then apply these considerations to various areas of focus or programs.

The OneKind framework is a conceptual framework that describes the interrelationships between animal well-being, human well-being, and the earth that supports us.

Earlier representations of how these three areas interact exhibited very small areas of intersection, as seen on the left of Fig. 18.1; a more contemporary appreciation of the breadth and depth of interconnections and interrelationships is represented on the right in the OneKind philosophy and highlights just how interwoven the needs, interests, and well-being of these three areas are.

Within this framework, there are many components, with overlapping influences that impact the well-being of humans, animals, and the environment, as represented in Fig. 18.2. To illustrate the nature of these interconnections as represented in Fig. 18.2, a few examples follow.

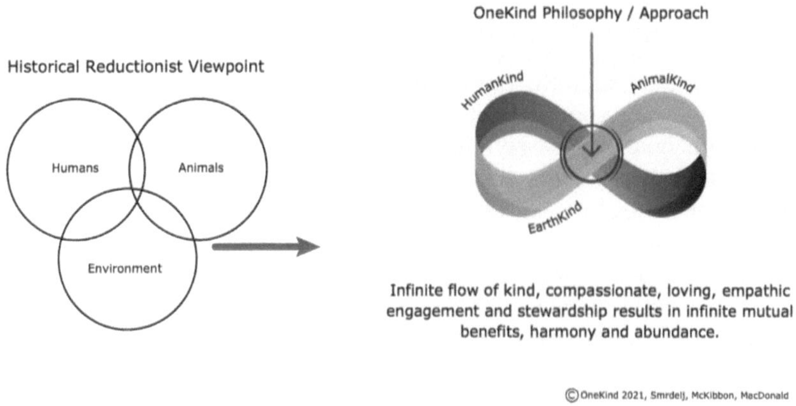

Fig. 18.1 Evolving towards OneKind philosophy (Image by Kate MacDonald and authors Magdalena Smrdelj and Amanda McKibbon, 2021)

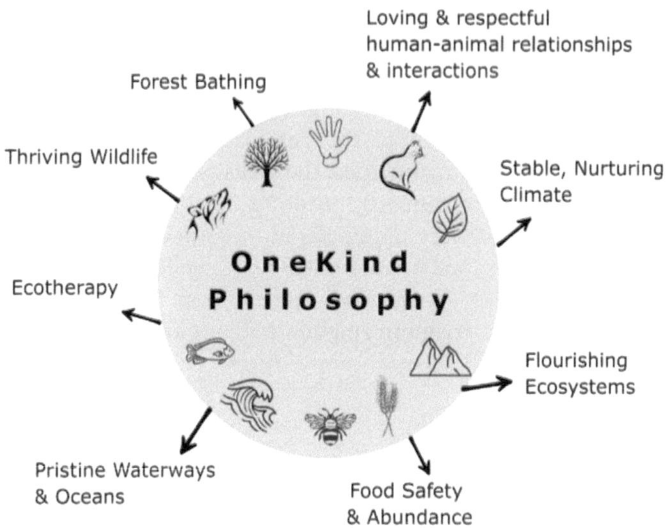

Fig. 18.2 OneKind philosophy (Image by Kate MacDonald and authors Magdalena Smrdelj and Amanda McKibbon, 2021)

18.2 Animal Health and Wellness, Human Well-Being, Food Safety, and Sustainability

Multiple sources state that approximately 75 percent of emerging infectious diseases are zoonotic (meaning that they originate in animals and spread to people),[1] that the rate of emergence of zoonotic diseases is on the rise, and that the majority of emerging diseases are driven by human activities. Bats are increasingly recognized as an important reservoir of zoonotic viruses; however, there are many other reservoirs in free-ranging animal populations that are of importance to the local human populations. These infectious agents exist within free-ranging populations and can cause disease or exist benignly. Either way, they can be passed on to other species such as humans or domesticated animals and cause disease as a result of encountering a naive immune system. They also can mutate and adapt to the new host species and be spread from person to person or from animal to animal. Human activities, such as urbanization, agricultural expansion, deforestation, hunting, "wildlife" trade, and global travel, are the major drivers of emerging zoonoses.[2]

Nipah virus emerged in Malaysia in 1998, as a result of increased and intensified pig farming practices. Fruit bats, a Nipah virus reservoir species, made their homes in overhanging fruit trees in nearby orchards where pigs could be exposed to the bat urine and fruit debris. These dense populations of pigs maintained the virus population and facilitated the spread of the virus to humans, resulting in serious illness and death. Almost one million pigs were depopulated to contain the virus.[3]

Many other viruses, such as Ebola virus, yellow fever virus, and Kyasanur forest disease virus, have spilled over to humans due to encroachment onto free-roaming animals' habitats and agricultural expansion.

[1] R. M. Voreau, V. G. Papavassiliou and S. Tsiodras, "Emerging Zoonoses and Vector-Borne Infections Affecting Humans in Europe," *Epidemiology and Infection* 135, no. 8 (2007): 1231–47, https://www.ncbi.nlm.nih.gov/pmc/articles/PMC2870710/.

[2] Jonathan H. Epstein and Hume E. Field, "Anthropogenic Epidemics: The Ecology of Bat-Borne Viruses and Our Role in Their Emergence," in *Bats and Viruses: A New Frontier in Emerging Infectious Diseases*, ed. Lin-Fa Wang and Christopher Cowled (Hoboken: John Wiley & Sons, 2015), 249–51.

[3] Epstein and Hume E. Field, "Anthropogenic Epidemics," 252.

The process of increased interaction between humans and free animals and the subsequent spillover of disease is easily recognized in the case of yellow fever in Brazil. Clear-cutting, degradation of natural habitats, and encroachment on forested areas due to the expansion of agricultural lands have brought about increased encounters with monkeys, the reservoir of yellow fever virus, which in turn has caused increases in yellow fever incidence in the human population. Tragically, many people have not recognized that the virus is carried by the mosquito vector and have killed monkeys out of fear. Monkey populations have subsequently been severely compromised by both yellow fever and human acts of violence.[4]

18.3 Connections Between Biodiversity, the Environment, Wildlife, Animal Well-Being, and Human Well-Being

Due to the nature of intensive beef and dairy practices, large populations of cattle are held in dense housing systems that emit a preponderance of CO_2 greenhouse gas, which is well accepted as a contributor to climate change.[5] Furthermore, the massive volumes of manure require a nutrient management program for the manure to be utilized in a way that maintains a balance of soil nutrients for the respective crops being grown. If nutrient management is not carried out effectively and efficiently—for example, in the cases of spreading manure before the land has thawed or spreading excess amounts—runoff and eutrophication of the water can lead to harmful algal blooms in fresh and ocean water.[6]

Serious or even tragic outcomes can result from harmful algal blooms. These blooms can produce domoic acid, which shellfish and small fish ingest and concentrate, thus delivering dangerous or lethal doses to

[4] Cristina Possas et al., "Yellow Fever Outbreak in Brazil: The Puzzle of Rapid Viral Spread and Challenges for Immunisation," *Memorias do Instituto Oswaldo Cruz* 113, no. 10 (2018): 1–12, doi:10.1590/0074-02760180278.

[5] Pierre Gerber et al., *Tackling Climate Change through Livestock: A Global Assessment of Emissions and Mitigation Opportunities* (Rome: Food and Agriculture Organization of the United Nations, 2013), accessed November 11, 2019, http://www.fao.org/3/a-i3437e.pdf.

[6] Sydney Jewell, "Manure and Algal Blooms," Penn State, accessed November 11, 2019, https://sites.psu.edu/skj5242civicissues/2018/02/21/manure-and-algal-blooms/.

their predator, the sea lion. Domoic acid causes toxicity and neurological disease such as disorientation, seizures, head bobbing, beaching, and even death.[7] Many documented cases exist in which dogs swimming in freshwater have become ill and even died from exposure to harmful algal blooms.[8]

18.4 Therapeutic Interactions Involving Animals, Humans, and the Environment

Although the examples of interconnections previously detailed involve negative outcomes, there are also many ways to bring about positive effects. Research has illustrated that many interventions utilizing the human–animal bond can have a profound impact on the health of humans. These effects include lowering blood pressure, increasing physical activity, reducing stress, reducing aggression and agitation in people with Alzheimer's disease and dementia, promoting social behavior and interaction both in people with Alzheimer's disease and dementia and in those with autism spectrum disorder, assisting individuals with PTSD to manage their symptoms, and a myriad of other positive outcomes.[9]

Many factors in modern society, including digital technology, have caused us to feel disconnected from our world. Stress, depression and anxiety, diminished cognitive performance, decrease in empathy, fatigue, and poor sleep from overstimulation are all symptoms of daily immersion in technology.[10] An essential counterbalance is a green prescription, or time spent in nature, including interventions as simple as forest bathing or walking in nature, or more prescriptive approaches such as ecotherapy.[11]

The remainder of this chapter will focus on humane education as an assisted intervention to foster empathy, well-being, leadership, and a

[7] "Domoic Acid Poisoning," Oceans of Fun, accessed November 11, 2019, https://www.oceansoffun.org/domoic_acid_poisoning.

[8] Jordan Davidson, "Dogs Are Dying from Toxic Algae in Lakes and Ponds," *EcoWatch*, August 12, 2019, https://www.ecowatch.com/dogs-lakes-ponds-toxic-algae-2639783608.html?rebelltitem=3#rebelltitem3.

[9] Human–Animal Bond Research Institute, "Research: Understanding the Human–Animal Bond," HABRI, accessed November 11, 2019, https://habri.org/research/.

[10] Eva M. Selhub and Alan C. Logan, *Your Brain on Nature* (Toronto: Collins, 2012), 38–44.

[11] Selhub and Logan, *Your Brain on Nature*, 3–4, 170–72, 216.

connection with the natural world. If we are to recognize these interconnections between human, animal, and environmental well-being, it is essential that humane education be implemented to facilitate the embodiment of feeling connected to animals and the environment.

18.5 Humane Education

Concepts such as OneKind are critical in that they act as a compass to guide us in our work. Concepts and theories offer a foundation from which to make decisions as well as inspire change and new directions.[12] However, for a concept such as OneKind to have an impact, it is critical to connect the concept to practice. Humane education aligns closely with the ideas put forth by OneKind and presents a powerful channel through which to make that connection.

18.6 Why Humane Education?

Broadly defined, humane education is focused around nurturing compassion and respect for living beings and things.[13] With a focus on respect and empathy, humane education takes learning beyond the basics of classic instruction. It adds context and values and imparts an understanding of the interconnected nature of the world that children share with other people, animals, and the environment. Humane education moves away from narrowly focused answers to broader, more sustainable solutions that not only consider multiple stakeholders and their priorities but also, in line with the tenets of the OneKind framework, take into account the causes and effects of each domain on the other.

Ultimately, humane education aims to shape attitudes and build knowledge and skills that will enable children to make decisions to create a better world for humans, animals, and the environment. In a world facing major social injustice and issues related to animal and environmental

[12] Kathleen Stassen Berger, "Theory and Practice: Teaching in the Real World," *The NEA Higher Education Journal* 17, no. 2 (2001–02): 107, http://www.nea.org/assets/img/PubThoughtAndAction/TAA_01Win_12.pdf.

[13] Humane Education Coalition, "What Is Humane Education?," accessed November 11, 2019, https://www.hecoalition.org/what-is-humane-education.html; R. Arbour, Tania Signal, and Nik Taylor, "Teaching Kindness: The Promise of Humane Education," *Society & Animals* 16, no. 2 (2009): 136, https://doi.org/10.1163/156853009X418073.

ethics, children must be provided with an opportunity to develop the tools needed to think critically about these matters. As children grow into positions of leadership, we want them to arrive with a sense of connection to the world and feel empowered to take action that considers the priorities of all living things. To that end, humane education emphasizes the value of applying both the heart and mind to the process of integrating compassion, kindness, and integrity with fact-finding and critical thinking, to arrive at thoughtful and creative solutions. The Institute for Humane Education refers to this as "solutionary" thinking.[14] According to Zoe Weil, cofounder of the institute, a solutionary is "a person who can identify inhumane and unsustainable systems and then develop solutions that are healthy and just for people, animals, and the environment."[15] To prepare and empower children to become solutionaries, Weil proposes that educators integrate the following four "Essential Elements" for humane education:

1. "**Acquire knowledge**: by preparing them to be enthusiastic and effective researchers who can obtain accurate information about interconnected global challenges and discern fact from opinion and conjecture.
2. **Think deeply**: by developing their critical, creative, strategic, and systems-thinking skills.
3. **Make compassionate and responsible choices**: by fostering wonder and appreciation for the natural world; empathy for people and animals; and a commitment to doing the most good and least harm.
4. **Focus on solutions**: by offering opportunities to collaboratively engage in problem-solving, implement ideas, and assess and improve upon them."[16]

[14] Institute for Humane Education, *The Solutionary Guidebook*, accessed November 11, 2019, https://humaneeducation.org/wp-content/uploads/2019/11/The-Solutionary-Guidebook-AN-FINAL2.pdf.

[15] Zoe Weil, "Becoming a Solutionary: What It Means and Why It Matters," *Psychology Today*, March 14, 2019, https://www.psychologytoday.com/ca/blog/becoming-solutionary/201903/becoming-solutionary.

[16] Zoe Weil, "Essential Elements of Humane Education" (blog post), September 8, 2016, https://humaneeducation.org/essential-elements-humane-education/.

Humane education is well-suited for integration into programs that rise to meet the needs of the child, including those that seek to build capacities such as well-being, resilience, and leadership. It can be delivered in a variety of ways, including formal or informal education, media, or literature and through firsthand or shared experiences. While all learning is valuable and contributes to raising awareness and creating change, humane education must become integrated into formal programming and education to achieve large-scale impact. Such a grand initiative presents opportunities for collaboration and new possibilities. It is a chance for experts in the well-being of the living world to take part in the solution through developing resources and programming that educators can utilize in their teaching. In keeping with the multidisciplinary nature of humane education, the solution to widespread humane education delivery involves multiple stakeholders coming together, to invest in the hearts and minds of our future leaders.

18.7 The AnimalSmart Humane Education Series

As the largest animal well-being charity in Ontario, Canada, the Ontario SPCA and Humane Society is a major stakeholder in animal well-being. In 2018, under the leadership and guidance of Kate MacDonald, Chief Executive Officer, the Ontario SPCA launched the first edition of a humane education series called AnimalSmart.[17] This learning resource consists of a series of lesson plans developed to foster empathy, promote a sense of responsibility, and inspire connection with animals and the environment. The lessons also promote mutual respect among learners, encourage learners' collaboration with their classmates and family, and promote participation in their local and global community.

In the development of content for the first and future editions in the series, two main factors have been identified as integral to the success and reach of the program: (1) incorporating the needs of the audience into content development and (2) carefully considering factors related to the accessibility of the resource. Other elements are also considered, but the preceding are of particular importance to ensure the program meets the needs of and is utilized by educators.

[17] Ontario SPCA and Humane Society, *AnimalSmart* (Stouffville: Ontario SPCA, 2018), https://www.animalsmart.ca.

18.8 Considering Audience Needs and Access

When creating a resource, it is essential to understand the needs and goals of the intended audience and to use those insights to guide the creation of the tool. This is critical to ensuring the resource is valuable in both content and design. For content development, the primary audience identified for the AnimalSmart program is teachers, administrators, and librarians. While the learning goals of the program are informed by the animal-centric mission of the society, the lessons themselves are developed by educators. This approach facilitates the integration of key elements identified as important to teachers. In this respect, AnimalSmart is a learning resource made by educators, for educators. Each lesson includes all information and resources required to deliver the lesson, including reproducible worksheets that complement the teaching, and provides links to the curriculum expectations of the Ontario Ministry of Education. For educators outside of Ontario, these links can still serve to highlight where this lesson might align with their requirements. All lessons can be adapted to suit individual needs and aptitudes and to integrate with other programming to augment learning in all contexts.

Once the lesson plans are finalized, a group of teachers pilot the lesson plans in their classrooms and provide feedback for consideration. The feedback is reviewed and integrated into the final version before the resource is released for public access.

Accessibility of a resource should be carefully considered: even the most comprehensively developed material can gather dust if access is challenging. To meet the goal of maximizing the reach of the AnimalSmart program, a free-access approach was taken: the manual is accessible online, free of charge to anyone. In addition, printed copies are available as an on-site resource—for example, at municipal libraries and educator resource centers, for distribution at events, and to send to remote communities with restricted online access. Furthermore, each AnimalSmart edition is translated into both French and Spanish to make it available to educators and students teaching and learning in those languages.

With the aforementioned barriers addressed, AnimalSmart is accessible to a wide range of educators as a powerful tool that integrates humane education to engage and empower the next generation of community and global leaders, focused on the well-being of people, animals, and the environment.

The final section of this chapter will discuss the development of humane education programming for students at risk of not graduating. The programs take a holistic approach to build student well-being and couple the learning with connection to animals and the natural world to build resilience and leadership skills.

> There just comes a point where we need to stop just pulling people out of the river. We need to go upstream and find out why they are falling in.
>
> —Desmond Tutu

18.9 Humane Education in Practice

Well-being is that positive sense of self that is felt when our cognitive, emotional, social, and physical needs are being met. By supporting the well-being of children in school, at home, and in the community, we are meeting the conditions needed to help them learn, grow, and develop a positive understanding of themselves.[18]

As discussed earlier, the relationship between human well-being and that of our animal friends and the environment is well documented.[19] Humane education presents a powerful opportunity to build this learning and connection in children.

Through strategic partnerships, such as with the Trails Youth Initiatives and the Ontario SPCA, our staff involved in this work utilized humane education to build resilience and leadership skills in our most at-risk students, as well as foster a sense of connection to other people, animals, and the environment. As will be explored in further detail, our partnership with the Ontario SPCA has resulted in the development of a humane education program focused on developing the well-being of at-risk students through challenges, experiences, and opportunities. This program uses components of the OneKind framework to connect the well-being of these students with that of the living world.

[18] Ministry of Education, *Ontario's Well-Being Strategy for Education* (Queen's Printer for Ontario, 2016), http://www.ontla.on.ca/library/repository/mon/30005/334837.pdf.

[19] Pinillos, *One Welfare: A Framework to Improve Animal well-being and Human Well-Being*. Wallingford: CAB International, 2018.

18.10 Aligning and Implementing Well-Being

The York Region District School Board (YRDSB), in Ontario, has applied a tiered model to each area of the well-being framework. This framework considers socio-emotional, cognitive, and physical and behavioral well-being to create a strong sense of mattering and belonging for all students. Called the Aligned and Integrated Model, or AIM, it is a powerful tool for building foundations of well-being by working to prevent risk, offer support, and reinforce protective factors, while also recognizing students who may require more intensive supports.[20] In 2014, the York Region District School Board embarked on a study with York University to explore the factors that improve resilience in youth.[21] The goal of this research project was to increase resilience and well-being by working collaboratively with community partners to narrow the gap for children with learning challenges and also to build capacity and sustainable solutions. The research demonstrated that there were two primary groups of students who struggled with the skills necessary to graduate:

1. Invisible middle—students who struggled without displaying overt signs of risk in terms of academic behaviors.
2. Double jeopardy—students who struggled as a result of academic and behavioral issues and were confronted by many challenges, including destructive peer groups.

As a result of this initial project, schools in YRDSB implemented a brain-based resource that focused on teaching students about neuroscience, to help them better understand their brains, and provided mindfulness training to support social-emotional learning and well-being. While this was effective for some students, it became apparent that this resource was not helping a certain group of students who were at risk of not graduating due to social and academic factors, as well as the availability of

[20] School Mental Health ASSIST, *Leading Mentally Healthy Schools: A Resource for School Administrators* (Toronto: SMH ASSIST, 2013), https://smh-assist.ca/wp-content/uploads/LMHS-Flipchart-EN-Web.pdf; School Mental Health ASSIST, *Leading Mentally Healthy Schools: A Vision for Student Mental Health and Well-Being in Ontario Schools* (Toronto: SMH ASSIST, 2013), https://smh-assist.ca/wp-content/uploads/LeadingMentallyHealthSchools.pdf.

[21] Gordon Flett and Susan Ball, *Resilient Students, Resilient Schools, Resilient Communities: What Everybody Needs* (Ontario: York Region District School Board, 2014).

support beyond the school. This prompted school leaders to explore other programming that might better suit the needs of these students.

18.11 Working in Partnership

A partnership with an organization called Trails Youth Initiatives (Trails) provided a promising approach to meeting the needs of these students. Trails is a preventive program that provides programming for students from vulnerable communities through an intensive program, as well as through outreach. Its focus is mainly on understanding the self, developing interpersonal skills, and integrating the outdoors.

Our partnership with Trails followed a three-year program, where five students from three different schools visited Trails once a month with the same teachers, thereby creating a sense of community. The students also met at their schools with the teacher linked with the Trails program. This approach helped build a relationship with a caring adult in the student's life, which is a protective factor for student resilience. This program began with three schools and has since grown to include seven schools and more than sixty students who either have completed or are completing the three-year program. Data for students who have already completed the program show that throughout the program, students' self-esteem, well-being, and self-awareness improved drastically, as did their behavior at school and the quality of their relationships with others. In addition, a rise in school attendance was coupled with a reduction in school suspensions.

The success of the preventive partnership with Trails prompted school leaders to consider other community affiliations, leading to the partnership with the Ontario SPCA. This new affiliation presented an opportunity to create a program that incorporated humane education that was geared specifically toward building capacity in students identified as "the invisible middle." These are students who tend to be highly underserved; they often are underperforming, may not participate in school, and may exhibit poor self-esteem. These elements place the students at risk of falling through the cracks because they are unlikely to exhibit behaviors that might draw attention to their struggle. This new program sought to do the following:

- Focus on assets
- Build skills in confidence, competence, and emotional regulation

- Improve communication, conflict resolution, and problem-solving skills
- Integrate work with animals to build trust and empathy as well as leadership skills and service
- Build an understanding of the importance of stewardship for the environment

In addition, this program was developed around three protective factors determined to have the most impact on resilience: high expectations with support, caring relationships with healthy adults, and opportunities to contribute and participate.[22]

18.12 Developing the Humane Education Program

Working from this preventive framework, the partnership with the Ontario SPCA followed a similar framework to Trails, using the three years and once-a-month strategy. The program used OneKind as a theoretical framework and AnimalSmart as a tool to enhance well-being and foster compassion for the living world. The format of the program applied an outdoor learning approach combined with humane education lessons and integrated a core element of fostering connections with animals. The ultimate goals of the program were to build emotional, interpersonal, and academic resilience through developing skills such as self-compassion and advocacy, mindful awareness, self-regulation, problem-solving, a growth mindset, mattering, and leadership, rather than focusing on deficits.

Throughout the program, facilitators (i.e., teachers or school staff) met biweekly with students in groups of five to six to discuss new skills and ideas, as well as complete exercises and explore their feelings and thinking about their self and their connections to others, nature, and the animals from the Ontario SPCA. Outcomes were collected over the first year of the humane education program, specifically at the beginning and at the end. Before the program started, the selected students were interviewed about their thoughts, hopes, and fears about the program. Common themes that arose were feelings of fear, doubt, and being unworthy. For

[22] Sara Truebridge, *Resilience Begins with Beliefs: Building on Student Strengths for Success in School* (New York: Teachers College Press, 2013).

example, when one student was told that multiple teachers had chosen them for the leadership program, their response was "Someone chose me? I've never been chosen for anything." Another student said, "I don't think I will be able to do it." When asked to elaborate further, the student shared, "I'm not really good in school, and I'm definitely not a leader."

The selected students were also asked to write a short application letter explaining why they would be a good fit. The first prompt for the letter was "Part of the role is working with people and other students—why would you be a good leader?" Most students stated that they could not do the program, they were not good with people, it was too hard, they did not possess any good qualities, or they would not be good for the position. These responses were discussed as a group, followed by a conversation that highlighted the good qualities of others in the group. Hearing staff and peers say positive things about them made students smile or laugh; two students cried. When asked about it, one student shared, "I just feel like it's a lot to hear. I don't always see myself that way."

The next prompt for the letter was "Part of the role is standing up for and working with animals—do you think you could be a kind leader for animals?" All participants said yes, and all participants were able to write why they would be great leaders working with animals. While students reportedly did not consider themselves "good leaders" in the context of connecting to people, they easily considered themselves "good leaders" in the context of connecting to animals. Drawing on the tenets of the OneKind concept, the following hypothesis for the program was posed: Students perceived themselves as confident, worthy individuals when connecting with animals and the outdoors. Could building capacity through the students' preexisting connections to animals and nature lead to students transferring this positive sense of self and capacity to connect with others in their everyday lives?

When visiting the Ontario SPCA site once a month, the students participated in outdoor activities such as team building, nature observations, and environmental stewardship. The students also worked with AnimalSmart to develop empathy and understanding of animals. With this knowledge, they created projects and ideas to support the animals in care. All projects were student-created and included such compassion-based ideas as a blanket drive for animals and the creation of enrichment activities and toys for both dogs and cats. Students also had the opportunity to interact with and learn how to identify the behaviors and needs of animals. They were able to then use this learning to understand and explain their own needs, triggers, and reactions.

At the end of the year, the students facilitated two full-day sessions on “Connecting to Animals, Nature and Others” to grade 3 classes. Through these sessions, the participants created their lesson plans using the AnimalSmart program and their knowledge of the Ontario SPCA facility.

After students had spent a year in the program, discussions with them revealed that they either saw or had begun to see themselves as leaders, felt connected to a peer group, felt they could better understand their own emotions and needs, and had a caring adult they could approach at school. Most of the students said they were doing better in their classes, participated more frequently, and had started to take on more leadership roles in their classroom, in their school community, and at home.

The highs and lows of developing this program revealed the importance of acknowledging certain truths: (1) some students are underserved, (2) these students are not achieving because of the systematic barriers in education and the community, and (3) school boards need to take measures to increase accessibility for all students. Identifying students as “underperforming” places the responsibility on the shoulders of those students and their families. Educators can change that dynamic by taking ownership of barriers experienced by underserved groups and making changes to meet their needs.

The program that arose from the partnership between the Ontario SPCA and the York Region District School Board was born not from a single idea, but rather from a flowing together of multiple ideologies and pedagogies. The partnership created an intensive program to support students who are underserved and underperforming, by improving resilience with protective factors, coupled with rich learning experiences connected to people, animals, and the environment.

18.13 Conclusion

The OneKind framework is a powerful concept to guide the decisions we make every day. It can help us understand how the work we do affects the rest of the living world. Humane education seeks to nurture respect and compassion for all living things and equips children with the perspective, skills, and knowledge they will need to make decisions that consider the needs of all living beings and things. Most children have a natural affinity for nature and animals, and learning acquired in the context of this connection can build resilience and a strong sense of self and develop leadership skills.

The desired outcome is empathy for our fellow humans, regardless of their circumstances; for our fellow creatures; and for the earth itself, which supports us all and is not an inexhaustible resource. Ultimately, at the center of all these concepts and frameworks, there is only one humanity, only one species, that can truly make a difference in acknowledging, nurturing, and curating OneKind across our planet. Through humane education, the interconnected nature of living things can become incorporated into the way children come to think about global society and their place in it. Humane education invites children to connect with the rest of the world and to develop a sense of empowerment and responsibility in creating a better place for people, animals, and the environment.

Bibliography

Arbour, R., Tania Signal, and Nik Taylor. "Teaching Kindness: The Promise of Humane Education." *Society & Animals* 16, no. 2 (2009): 136–48. https://doi.org/10.1163/156853009X418073.

Davidson, Jordan. "Dogs Are Dying from Toxic Algae in Lakes and Ponds." *EcoWatch*, August 12, 2019. https://www.ecowatch.com/dogs-lakes-ponds-toxic-algae-2639783608.html?rebelltitem=3#rebelltitem3.

Epstein, Jonathan H., and Hume E. Field. "Anthropogenic Epidemics: The Ecology of Bat-Borne Viruses and Our Role in Their Emergence." In *Bats and Viruses: A New Frontier in Emerging Infectious Diseases*, edited by Lin-Fa Wang and Christopher Cowled, 249–79. Hoboken: John Wiley & Sons, 2015.

Flett, Gordon, and Susan Ball. *Resilient Students, Resilient Schools, Resilient Communities: What Everybody Needs.* Ontario: York Region District School Board, 2014.

Gerber, Pierre, Henning Steinfeld, Benjamin Henderson, Anne Mottet, Carolyn Opio, Jeroen Dijkman, Alessandra Falcucci, and Giuseppe Tempio. *Tackling Climate Change through Livestock: A Global Assessment of Emissions and Mitigation Opportunities*. Rome: Food and Agriculture Organization of the United Nations, 2013. Accessed November 11, 2019. http://www.fao.org/3/a-i3437e.pdf.

Human–Animal Bond Research Institute. "Research: Understanding the Human–Animal Bond." HABRI. Accessed November 11, 2019. https://habri.org/research/.

Humane Education Coalition. "What Is Humane Education?" Accessed November 11, 2019. https://www.hecoalition.org/what-is-humane-education.html.

Institute for Humane Education. *The Solutionary Guidebook.* Accessed November 11, 2019. https://humaneeducation.org/wp-content/uploads/2019/11/The-Solutionary-Guidebook-AN-FINAL2.pdf.

Jewell, Sydney. "Manure and Algal Blooms." Penn State. Accessed November 11, 2019. https://sites.psu.edu/skj5242civicissues/2018/02/21/manure-and-algal-blooms/.

Ministry of Education. *Ontario's Well-Being Strategy for Education.* Queen's Printer for Ontario, 2016. http://www.ontla.on.ca/library/repository/mon/30005/334837.pdf.

Oceans of Fun. "Domoic Acid Poisoning." Oceans of Fun. Accessed November 11, 2019. https://www.oceansoffun.org/domoic_acid_poisoning.

Ontario SPCA and Humane Society. *AnimalSmart.* Stouffville: Ontario SPCA, 2018. https://www.animalsmart.ca.

Pinillos, Rebeca Garcia. *One Welfare: A Framework to Improve Animal Well-Being and Human Well-Being.* Wallingford: CAB International, 2018.

Possas, Cristina, Ricardo Lourenço-de-Oliveira, Pedro Luis Tauil, Francisco de Paula Pinheiro, Alcides Pissinatti, Rivaldo Venâncio da Cunha, Marcos Freire, Reinaldo Menezes Martins, and Akira Homma. "Yellow Fever Outbreak in Brazil: The Puzzle of Rapid Viral Spread and Challenges for Immunisation." *Memorias do Instituto Oswaldo Cruz* 113, no. 10 (2018): 1–12. https://doi.org/10.1590/0074-02760180278.

School Mental Health ASSIST. *Leading Mentally Healthy Schools: A Resource for School Administrators.* Toronto: SMH ASSIST, 2013. https://smh-assist.ca/wp-content/uploads/LMHS-Flipchart-EN-Web.pdf.

School Mental Health ASSIST. *Leading Mentally Healthy Schools: A Vision for Student Mental Health and Well-Being in Ontario Schools.* Toronto: SMH ASSIST, 2013. https://smh-assist.ca/wp-content/uploads/LeadingMentallyHealthSchools.pdf.

Selhub, Eva M., and Alan C. Logan. *Your Brain on Nature.* Toronto: Collins, 2012.

Soriano, Vanessa Souza, Arendy Ribeiro VillaBranco, Maria Alice Schnaider, and Carla Molento. "Compassionate Conservation: Concept and Applications." *Archives of Veterinary Science* 22, no. 4 (2017): 116–30. https://www.researchgate.net/figure/Representation-of-one-well-being-concept-in-which-the-compassionate-conversation-approach_fig1_322213317.

Stassen Berger, Kathleen. "Theory and Practice: Teaching in the Real World." *The NEA Higher Education Journal* 17, no. 2 (2001–02): 107–16. http://www.nea.org/assets/img/PubThoughtAndAction/TAA_01Win_12.pdf.

Truebridge, Sara. *Resilience Begins with Beliefs: Building on Student Strengths for Success in School.* New York: Teachers College Press, 2013.

Voreau, R. M., V. G. Papavassiliou, and S. Tsiodras. "Emerging Zoonoses and Vector-Borne Infections Affecting Humans in Europe." *Epidemiology and Infection* 135, no. 8 (2007): 1231–47. https://doi.org/10.1017/S0950268807008527.

Weil, Zoe. "Becoming a Solutionary: What It Means and Why It Matters." *Psychology Today*, March 14, 2019. https://www.psychologytoday.com/ca/blog/becoming-solutionary/201903/becoming-solutionary.

Weil, Zoe. "Essential Elements of Humane Education." Blog post. September 8, 2016. https://humaneeducation.org/essential-elements-humane-education/.

CHAPTER 19

"Cool to Be Kind": An Evaluation of a Classroom-Based Humane Education Intervention

Brynn M. Wiessner, Julia E. Senecal, Sarah M. Bexell, and Caroline A. Griffin

19.1 Introduction

High rates of school-based violence and bullying are alarming issues impacting youth and communities across the United States today. During the 2015–2016 school year, 79 percent of public schools recorded that one or more incidents of violence, theft, or other crimes had taken place on school grounds, amounting to a rate of twenty-eight crimes per

B. M. Wiessner (✉)
Aurora Public School District, Aurora, CO, USA
e-mail: brynnwiessner@gmail.com

J. E. Senecal
Center for Sustainability, University of Denver, Denver, CO, USA
e-mail: Julia.Senecal@du.edu

C. Linzey and A. Linzey (eds.), *The Palgrave Companion to Humane Education*, The Palgrave Macmillan Animal Ethics Series,
https://doi.org/10.1007/978-3-032-19795-5_19

one thousand students enrolled.[1] Furthermore, between July 1, 2014, and June 30, 2015, a total of forty-seven student, staff, and other non-student school-associated violent deaths in the United States were recorded, including twenty-eight homicides, seventeen suicides, and two "legal intervention" deaths.[2] Colorado in particular, where suicide is a leading cause of death among ten- to twenty-four-year-olds, has seen an increase in suicide attempts in high school–aged youth, rising from 6 percent in 2011 to 8 percent in 2015.[3] Because suicide-related behavior and bullying behavior are closely related, creating a safe school climate is essential to preventing violence and suicide.[4] Humane education, broadly defined as the use of educational settings to foster compassion and reverence for living beings and the environment, has been theoretically connected to promoting positive change in schools through reducing school-based violence and bullying.[5]

S. M. Bexell
University of Denver, Denver, CO, USA
e-mail: Sarah.Bexell@du.edu

C. A. Griffin
Show Your Soft Side, Baltimore, MD, USA
e-mail: cag@carolineagriffin.com

[1] Lauren Musu-Gillette et al., *Indicators of School Crime and Safety: 2017* (Washington, DC: National Center for Education Statistics and Bureau of Justice Statistics, 2018), https://nces.ed.gov/pubs2018/2018036.pdf).

[2] Musu-Gillette et al., *Indicators of School Crime and Safety: 2017*.

[3] Colorado Department of Public Health and Environment (CDPHE), *Mental Health among Youth in Colorado, Healthy Kids Colorado Survey 2015* (CDPHE, 2015), https://www.colorado.gov/pacific/sites/default/files/PF_Youth_HKCS_MH-Infographic-Digital.pdf.

[4] National Center for Injury Prevention and Control, *The Relationship between Bullying and Suicide: What We Know and What It Means for Schools* (Chamblee, GA: National Center for Injury Prevention and Control, Centers for Disease Control and Prevention, 2014), https://www.cdc.gov/violenceprevention/pdf/bullying-suicide-translation-final-a.pdf.

[5] Catherine A. Faver, "School-Based Humane Education as a Strategy to Prevent Violence: Review and Recommendations," *Children and Youth Services Review* 32, no. 3 (2010): 365–70, https://doi.org/10.1016/j.childyouth.2009.10.006.

Broader issues of interpersonal violence and animal abuse concerns also may be addressed through the implementation of humane education interventions. The link between animal abuse, neglect, and cruelty and other forms of family violence has been widely researched and documented, with recent research indicating a strong correlation between interpersonal violence and animal abuse.[6] In fact, a study addressing concurrent engagement in animal abuse and bullying behavior among adolescents found that witnessing animal abuse was a common predictive factor for engaging in both animal abuse and bullying behavior.[7] Humane education may be part of the solution for preventing animal abuse, thereby disrupting this relationship between interpersonal violence and animal cruelty.

Finally, the research outlined in this chapter sought to answer the following question: Would the implementation of the humane education lesson "Cool to Be Kind" in a classroom setting have the capacity to increase children's critical thinking about how companion animals are treated? Additionally, would the lesson impact children's views about how societal values impact the treatment of not only animals but other humans as well?

19.2 Literature Review

Early research on humane education programs indicates that humane education has the potential to positively influence the development of prosocial behaviors and children's empathy toward animals and humans.[8] Validated measurement tools, such as the Primary Attitude Scale, the Intermediate Attitude Scale, and the Bryant Index of Empathy for Children and Adolescents, have been utilized to demonstrate increased empathy in elementary-aged students after participation in humane

[6] Colorado LINK Project, "Interpersonal Violence," accessed November 15, 2018. http://coloradolinkproject.com/.

[7] Eleonora Gullone and Nerida Robertson, "The Relationship between Bullying and Animal Abuse Behaviors in Adolescents: The Importance of Witnessing Animal Abuse," *Journal of Applied Developmental Psychology* 29 (2008): 371–79.

[8] Frank R. Ascione, "Humane Education Research: Evaluating Efforts to Encourage Children's Kindness and Caring toward Animals," *Genetic, Social, and General Psychology Monographs* 123, no. 1 (1997): 57–77.

education programming.[9] Results also have indicated that both elevated measures of empathy and positive attitudes toward animals may be sustained for at least one year after a person receives humane education programming.[10] Though these findings are encouraging, research in this area is still limited, underscoring the need for additional studies to examine the long-term impacts of humane education programming.[11]

Additionally, humane education may be more effective in developing prosocial skills than other types of interventions and activities without an animal component. Research has illustrated that elementary-aged children who receive humane education programs show a greater increase in prosocial behaviors than their peers who participate in alternative programming, such as a chess club.[12] Students exposed to humane education also self-report increased awareness of—and concern for—humane issues, such as animal welfare, the wellness of other humans, and the wellness of the environment.[13] By supporting prosocial behavior, empathy, and kindness in children, the implementation of humane education curricula also has the capacity to decrease aggression toward humans and animals. These positive impacts of humane education not only benefit individual students by promoting kindness to others in the classroom but also undoubtedly carry over into other areas of the students' lives outside of school.

Humane education programming also may serve as a violence-prevention tool, increasing the likelihood of detecting and intervening

[9] Frank R. Ascione, "Enhancing Children's Attitudes about the Humane Treatment of Animals: Generalization to Human-Directed Empathy," *Anthrozoos: A Multidisciplinary Journal of the Interactions of People & Animals* 5 (1992): 176–91; B. K. Bryant, "An Index of Empathy for Children and Adolescents," *Child Development* 53 (1982): 413–25.

[10] Frank R. Ascione and Claudia V. Weber, "Children's Attitudes about the Humane Treatment of Animals and Empathy: One-Year Follow Up of a School-Based Intervention," *Anthrozoos: A Multidisciplinary Journal of the Interactions of People & Animals* 9, no. 4 (1996): 188–95, https://doi.org/10.2752/089279396787001455.

[11] Ascione, "Humane Education Research."

[12] William E. Samuels, Lieve L. Meers, and Simona Normando, "Improving Upper Elementary Students' Humane Attitudes and Prosocial Behaviors through an In-Class Humane Education Program," *Anthrozoös* 29, no. 4 (2016): 597–610, https://doi.org/10.1080/08927936.2016.1228751.

[13] Samuels, Meers, and Normando, "Improving Upper Elementary Students' Humane Attitudes."

in violence that is already occurring in children's home environments.[14] When children participate in humane education lessons, they often compare their own experiences to what they are learning about the proper care of animals, as well as humans and the environment, allowing teachers and other school officials to stay attuned to the circumstances and needs of their students.[15] Indeed, humane education programs are more likely than character education, violence prevention, or anti-bullying programs to maintain students' attention, contributing to the efficacy of such programs in the detection and prevention of both interpersonal and animal-directed violence.[16]

Interpersonal and interspecies violence impacting children and animals are complex social problems demanding creative and proactive solutions that promote critical thinking and support prosocial behaviors both in and outside of school. Humane education has the capacity to address this need, with its focus on development of compassion and altruism. Indeed, humane education may serve as a protective factor against interpersonal and animal-directed violence both in school and at home.[17]

19.3 Methods

The humane education lesson "Cool to Be Kind," designed for use with grades 3–6, was developed by graduate students at the University of Denver with the support of the director of humane education at the Institute for Human–Animal Connection and in conjunction with the anti–animal cruelty campaign "Show Your Soft Side" (SYSS). The lesson accompanies a SYSS poster campaign that features photographs of famous athletes, musicians, actors, and other celebrities displaying care and affection toward their companion animals. The lesson was developed to challenge perceptions of "cool," by exploring how internal traits, such as empathy and kindness, might be considered as cool, as opposed to external traits and materialistic values. The intervention also aimed to

[14] Faver, "School-Based Humane Education."

[15] Faver, "School-Based Humane Education."

[16] Kelly Thompson and Eleonora Gullone, "Promotion of Empathy and Prosocial Behaviour in Children Through Humane Education," *Australian Psychologist* 38 (2003): 175–82.

[17] Mary R. Jalongo, ed., *Teaching Compassion: Humane Education in Early Childhood* (New York: Springer, 2014).

ensure that students could not only define empathy but also identify why compassion and empathy are critical attributes for the happiness and health of individuals and communities.

"Cool to Be Kind" was piloted in two classrooms at Academy 360, a Denver Public Schools charter school in the Montbello neighborhood that utilizes a health and wellness approach to serving approximately 230 students. The local community is diverse, with approximately 59.7 percent of residents identifying as Hispanic; 24.2 percent as African American; 11.5 percent as non-Latino White; 1.8 percent as Asian; 2.5 percent as Mixed; and 0.4% as Other. In addition, approximately 35.6 percent of individuals in the Montbello community over the age of twenty-five have less than a twelfth-grade education, and 49.9 percent have only a high school education.[18] Moreover, approximately a quarter of Montbello families live below the poverty line, higher than metro Denver's overall poverty rate.[19] Academy 360's diverse student population is largely representative of school demographics in the greater United States; thus, the humane education intervention serves as a valid indicator of the efficacy of such programs nationwide.

The forty-five-minute lesson was presented in a multi-grade classroom of third and fourth graders and in a fifth-grade classroom during a separate class period. In total, fifty-five students were present for the intervention and the "Cool to Be Kind Post-Assessment" (Appendix A). After accounting for student absences during the lesson and on pre- and post-assessment days, the total sample size was forty-one students. Prior to students' participation in the intervention, parents and guardians were given an opportunity to have their child(ren) opt out of the study. In addition, students were given an opportunity to voice concerns or ask clarifying questions before being read an assent script and providing their assent via signature.

In order to assess the efficacy of the "Cool to Be Kind" lesson, both quantitative and qualitative data were collected through the use of two questionnaires administered to the students and through teacher and

[18] "Household Income in Montbello, Denver, Colorado." Statistical Atlas, September 14, 2018. https://statisticalatlas.com/neighborhood/Colorado/Denver/Montbello/Household-Income.

[19] Siebrase, Jamie. "In Montbello, A Future of Promise and Challenge." Confluence Denver, January 25, 2017. https://www.confluence-denver.com/features/montbello_012517.aspx.

student interviews. The validated assessment tool, the Children's Treatment of Animals Questionnaire (CTAQ) (Thompson & Gullone, 2003), was administered to the students one week prior to the lesson and then two weeks after the lesson. In addition, the students were given a short, knowledge-based questionnaire, "Cool to Be Kind Post-Assessment" (see Appendix A), immediately after participation in the lesson. This assessment tool provided information about the students' understanding of the presented material and whether they met each of the three identified lesson objectives. All questionnaires were read out loud to the students.

Three to four weeks after participating in the intervention, the classroom teacher and sixteen randomly selected students participated in an interview to share their perspectives on the lesson and what they may have learned (see Appendix B). Results of the pre- and post-CTAQ assessment were analyzed using a t-test statistical approach to identify whether students' attitudes regarding the treatment of animals had changed after their participation in the intervention. The research team also used the results of the knowledge assessment and the qualitative interviews to evaluate the efficacy of the lesson in meeting its objectives.

19.4 Data Analysis

A dependent-samples t-test was conducted to compare children's scores on the CTAQ before and after the humane education lesson intervention to help the research team understand whether significant change had occurred in the way the participants treated their companion animals (or *would* treat their companion animals if they did not live with a companion animal in their house). Descriptive statistics were used to analyze the results of the knowledge-based post-assessment through examination of themes relating to the students, such as (1) their understanding of ways in which being a kind and caring citizen can be "cool," (2) whether they perceived value in being cool and kind, (3) their ability to identify empathetic behavior, and (4) their ability to identify healthy ways to cope with a frustrating or upsetting situation.

Finally, the qualitative interviews were reviewed for common themes across the responses to the four questions asked in each of the sixteen student interviews, as well as the teacher interview. Common themes that emerged included a deeper understanding of concepts such as kindness and respect and an interest in sharing information from the lesson with family members, friends, or community members.

19.5 Results

In the dependent-samples t-test, no significant difference was found in the CTAQ scores before and after the lesson (M = -0.01, SD = 0.246; t = -0.308, p = 0.7600). These results indicate that there was no significant change in the way children treated their companion animals before and after receiving the humane education lesson. For the six-item "Cool to Be Kind Post-Assessment," results were as follows: (1) 76 percent of students selected at least three (of four) actions that were kind from a list of eight choices; (2) 100 percent of students indicated that they valued showing kindness to other humans and nonhuman animals; (3) 81 percent of students demonstrated that they believed someone could be both "cool" and "kind" at the same time; (4) 71 percent of students were able to identify the correct definition of empathy; (5) 100 percent of students identified at least one positive way to deal with feelings of frustration or sadness, while 78 percent identified two (of two) healthy ways to cope with emotions; and (6) 84 percent of students included "help and listen to others" as a way they could identify that someone was "cool" and "kind." These results suggest that humane education may be an effective way to build critical thinking skills, social-emotional skills, and understanding of and capacity for empathy.

In examination of the results of the qualitative assessment, two prominent themes were identified: a deepened understanding of concepts related to the lesson and a sharing of knowledge related to these concepts.

19.6 Qualitative Themes

Deepened Understanding

Students demonstrated both a change in their beliefs as to what constitutes "kindness" and an increased ability to think critically about societal values, by acknowledging that a person can be both "cool" and kind at the same time. Multiple students verbalized that due to the intervention, they now believed being "cool" included showing kindness to other people and nonhuman animals. One student mentioned that being cool means that you should be nice to other kids, while another student mentioned that being cool means showing kindness to both people and animals. "Cool isn't what everyone thinks it is," another student stated, and "cool can be that you're being nice to animals and friendly." One participant also demonstrated his ability to challenge societal viewpoints about what it means to be cool: "Some people think that you can't be kind and cool [at the same time], but the lesson taught us you can be both." Students were able to engage critical thinking skills through participation in the lesson, and this resulted in a change in viewpoint for some. For example, one student acknowledged their change in perspective, stating, "Now I know the kinder you are, the cooler you are, and it saves more animals." Although most of the students demonstrated a deepened understanding of what kindness means to them, a few students reported that this had not changed much for them because they were "already nice to animals and people." Last, one student reported they did not remember anything from the lesson and were unable to describe what they had learned, underlining the challenges of a one-time intervention.

The results of the teacher interview also aligned with the theme of a deepened understanding of kindness in the participants. She noted that prior to the lesson, some students had verbally shared that they were unkind to dogs in their home. By bringing different voices into the classroom to talk about kindness, the lesson really "opened students' eyes," she felt, and allowed them to think more critically about how society defines "cool." It was especially impactful, she stated, for those who had told her about their unkind actions, as she noted that after the intervention, she heard students say on several occasions in the classroom, "That's not cool, don't do that anymore," in regard to unkind behavior toward animals.

Sharing Knowledge

The data analysis also revealed that many participants shared their intervention experience with others, largely through verbal communication and by physically sharing the poster with friends, family members, or community members. Ten of sixteen students reported that they had shared what they had learned during the lesson, while one additional student indicated that he would still like to share it with a family member after the interview. Three participants shared pieces of the lesson related to the care and treatment of animals, including "how you take care of animals at home and anywhere you are" and how everyone "should always treat our pets with respect" and be "kind and caring." One student shared broader attitudinal takeaways related to "coolness" and "kindness," including that "you don't have to be mean to be cool" and that "you can be [both] nice and cool." Some participants made connections between the lesson and broader concepts of empathy by communicating to others to "[not] judge a book by its cover and don't judge somebody once you meet them because you don't really know what's going on with their lives." Similarly, one participant reported that immediately following the intervention, he had shared the poster with a security guard at the school, to whom he explained "what it means to be cool." The participant reported that during this discussion, the guard then engaged him in a discussion about the Golden Rule (i.e., treating others as you wish to be treated). Also of note, one participant showed his family members the poster and shared that "all these people can affect you and affect others to do the right thing," exhibiting an understanding of the impact that celebrities and cultural leaders may have on societal attitudes toward animals.

19.7 Discussion and Conclusions

While results from the CTAQ suggest that a one-time humane education lesson may not impact how children treat the animals in their lives, results from the "Cool to Be Kind Post-Assessment," as well as from the qualitative interviews, indicate that the intervention was effective in meeting its objectives, by developing students' critical thinking skills concerning how societal values impact the treatment of not only animals but also other humans.

Because the lesson was not designed to teach children how to treat animals with kindness, the researchers acknowledge that the CTAQ might not have been the most effective tool for measuring change for this intervention. Due to limited research in the field of humane education, however, there is a notable lack of validated instruments designed to assess this type of intervention. Therefore, the researchers also included a proprietary knowledge-assessment tool designed to evaluate the effectiveness of the lesson in meeting its objectives. Although not a validated tool, this evaluation also provided useful information regarding the effectiveness of the lesson and the students' understanding of the material. The researchers acknowledge that results would have been more demonstrative had the evaluation been administered in a pre/post format, but they opted to give this questionnaire as a post-survey to limit assessment fatigue in the children.

Results from the CTAQ also may indicate a need for longer-term humane education programming in order to make a difference in not only attitudes but also the actions of youth participating in such lessons. Though only one student indicated that he could not recall the lesson at all, this raises concern regarding the efficacy of one-time interventions. Overall, more research surrounding comprehensive and long-term humane education programming is needed to help educators develop effective curricula that can be integrated into daily school programming.

Additionally, results from the "Cool to Be Kind" post-assessment indicated that students were able to think critically about ways that being a caring and empathetic individual can be "cool." Students demonstrated their understanding of what empathy means, as well as how they valued showing kindness to humans and animals. This understanding was demonstrated through the results of the qualitative interviews as well, given that common themes of deepened understanding and sharing of knowledge emerged from the teacher and student interviews. Students were provided an opportunity to challenge individual viewpoints as well as societal viewpoints on the definition of "coolness" and to incorporate new information into their own world perspective. These results indicate the potential for humane education to be linked with social-emotional learning goals and competencies required in many schools, as these interventions have the potential to be an engaging medium for students to build social-emotional skills while also promoting positive school culture. Although the lesson demonstrates potential for building students' skills in critical thinking and capacity for empathy, it did not show a significant change in actions concerning treatment of animals quantitatively. Perhaps a longer-term intervention with more opportunities for practicing skills would yield more significant results in change of student actions toward companion animals.

The qualitative component of this study also underscored the potential for humane education as a means to promote student safety and provide a comfortable space for students to share their own personal challenges. During one interview, a participant revealed that his family is "mean" to his dog, that his parents and sister regularly "yell at her," and that his father has kicked the dog. This disclosure presents tremendous implications for the detection and prevention of in-home violence and suggests that the incorporation of humane education in educational and mental health settings may ultimately promote the safety of children and adolescents.

In considering the strengths of the present study, it is important to note that the humane education lesson was presented to a student population that is diverse in ethnicity, race, socioeconomic status, and age. Because much of the existing research on humane education has been with children from middle-class white neighborhoods, this study provides valuable information regarding the efficacy of a humane education curriculum within a diverse population representative of many urban schools in the United States. Research conducted in such diverse populations advances the field of humane education by ensuring that programming is accessible to and inclusive of all individuals and communities. Similarly, the qualitative component of the study ensured that the diverse student population was heard and enabled the researchers to gain a deeper understanding of the impact the intervention had across the sample population.

Since this was a pilot study of a new curriculum, the researchers acknowledge that there may have been small inconsistencies in the way the lesson was taught in the two classrooms. The researchers also acknowledge that there may have been social desirability bias in students' responses to the CTAQ survey; that is, students may have overreported "good" behavior so as to appear more favorable to the researchers who both taught the lesson and administered the questionnaires (due to limitations in personnel resources for the project). The research team worked to limit this bias by informing students that their responses would remain anonymous and would not be evaluated for a class grade. Furthermore, if time and resources had permitted, the study might have been strengthened by repeated administration of the lesson in multiple schools.

Because humane education is a developing field, continued research is necessary to develop effective curricula that serve the interests of individual students, classrooms, schools, and the broader community. Additional research will also aid humane educators in developing effective practice guidelines. Studies to date confirm that humane education is effective for increasing empathy and helping youth develop socially and emotionally, thereby decreasing incidents of bullying and violence toward human and nonhuman animals. In light of this research, it is crucial that we continue to assess the efficacy of humane education interventions with the ultimate goal of creating a safer and kinder future for all species.

Appendix A: Cool to Be Kind Post-Assessment Questionnaire

Cool to Be Kind Post-Assessment Questions

1. Which of the following actions are "cool"? Circle as many as you think are correct.

    a. Showing kindness by saying hello to someone with a smile
    b. Teasing a classmate because they got an answer wrong
    c. Inviting someone who looks lonely to sit with you at lunch
    d. Getting into a fight with someone at recess
    e. Talking in class when the teacher has asked the class to be quiet
    f. Having the newest shoes that are the most popular
    g. Feeling happy with your friend who is happy
    h. Saying you are sorry when you know you have made a mistake

2. Is it important to show kindness to animals and other people? Circle: Yes or No
3. Someone who is "cool" and "kind" at the same time is

    a. A nerd
    b. It is not possible
    c. Amazing
    d. Not someone I would be friends with

4. Which of the following demonstrates someone showing empathy (or caring behavior?)

    a. A student interrupts the teacher to tell a funny joke in class
    b. A friend got new shoes over the weekend and shows them to you at school
    c. A classmate is really sad because their pet dog is sick and their friend can feel their sadness and also feels sad with them
    d. A student is frustrated during class and calls the person next to them a bad name
    e. All of the above

5. When feeling upset or angry, it is a good idea to....

    a. Yell at someone else because then you will feel better
    b. Try to show kindness anyway and talk it over with a friend after class

c. Get angry at the teacher too because it is probably their fault
d. Take some space and explain to a trusted adult or friend how you are feeling.
e. Both b and d.

6. I can tell someone else is cool and kind because …

a. They have amazing clothes
b. They help and listen to other people
c. They always know what new music is the best
d. They tell the best jokes
e. They win Fortnite every time

Appendix B: Post-Lesson Qualitative Questions

Students:

1. What did you learn from the lesson we taught a few weeks ago on being cool and how to treat animals?
2. How did this lesson change the way you think about kindness/being caring?
3. What message do you think the posters tell people?
4. Have you shared anything that you learned in that lesson, or from the posters with anyone you know?

Teachers:

1. How have you seen the lesson and posters impact your students?
2. How do you think this lesson assisted you in teaching your students about kindness and empathy?
3. How did this lesson and the posters impact you personally and professionally?
4. How would you improve the implementation of this lesson to best support the needs of your school?

Bibliography

Ascione, Frank R. "Enhancing Children's Attitudes about the Humane Treatment of Animals: Generalization to Human-Directed Empathy." *Anthrozoos: A Multidisciplinary Journal of the Interactions of People & Animals* 5 (1992): 176–91.

Ascione, Frank R. "Humane Education Research: Evaluating Efforts to Encourage Children's Kindness and Caring toward Animals." *Genetic, Social, and General Psychology Monographs* 123, no. 1 (1997): 57–77.

Ascione, Frank R., and Claudia V. Weber. "Children's Attitudes about the Humane Treatment of Animals and Empathy: One-Year Follow Up of a School-Based Intervention." *Anthrozoos: A Multidisciplinary Journal of the Interactions of People & Animals* 9, no. 4 (1996): 188–95. https://doi.org/10.2752/089279396787001455.

Bryant, Brenda K. "An Index of Empathy for Children and Adolescents." *Child Development* 53 (1982): 413–25.

Colorado Department of Public Health and Environment (CDPHE). *Mental Health Among Youth in Colorado, Healthy Kids Colorado Survey 2015.* CDPHE, 2015. https://www.colorado.gov/pacific/sites/default/files/PF_Youth_HKCS_MH-Infographic-Digital.pdf.

Colorado LINK Project. "Interpersonal Violence." (accessed November 15, 2018. http://coloradolinkproject.com/).

Faver, Catherine A. "School-Based Humane Education as a Strategy to Prevent Violence: Review and Recommendations." *Children and Youth Services Review* 32, no. 3 (2010): 365–70. https://doi.org/10.1016/j.childyouth.2009.10.006.

Gullone, Eleonora, and Robertson, Nerida. "The Relationship between Bullying and Animal Abuse Behaviors in Adolescents: The Importance of Witnessing Animal Abuse." *Journal of Applied Developmental Psychology* 29 (2008): 371–79.

"Household Income in Montbello, Denver, Colorado." Statistical Atlas, September 14, 2018. https://statisticalatlas.com/neighborhood/Colorado/Denver/Montbello/Household-Income

Jalongo, Mary R., ed. *Teaching Compassion: Humane Education in Early Childhood.* New York: Springer, 2014.

Musu-Gillette, Lauren, Anlan Zhang, Ke Wang, Jizhi Zhang, Jana Kemp, Melissa Diliberti, and Barbara A. Oudekerk. *Indicators of School Crime and Safety: 2017.* Washington, DC: National Center for Education Statistics and Bureau of Justice Statistics, 2018. https://nces.ed.gov/pubs2018/2018036.pdf.

National Center for Injury Prevention and Control. *The Relationship between Bullying and Suicide: What We Know and What It Means for Schools.* Chamblee, GA: National Center for Injury Prevention and Control, Centers for

Disease Control and Prevention, 2014. https://www.cdc.gov/violenceprevention/pdf/bullying-suicide-translation-final-a.pdf.
Samuels, William E., Lieve L. Meers, and Simona Normando. "Improving Upper Elementary Students' Humane Attitudes and Prosocial Behaviors through an In-Class Humane Education Program." *Anthrozoös* 29, no. 4 (2016): 597–610. https://doi.org/10.1080/08927936.2016.1228751.
Siebrase, Jamie. "In Montbello, A Future of Promise and Challenge." Confluence Denver, January 25, 2017. https://www.confluence-denver.com/features/montbello_012517.aspx.
Thompson, Kelly & Eleonora Gullone. "The Children's Treatment of Animals Questionnaire (CTAQ): A psychometric investigation." *Journal of Applied Psychology*, 88(1) (2003): 158–163.
Thompson, Kelly, and Eleonora Gullone. "Promotion of Empathy and Prosocial Behaviour in Children Through Humane Education." *Australian Psychologist* 38 (2003): 175–82.

CHAPTER 20

Building Empathy through Picture Books: Grades 2 and 5 Teacher and Student Perspectives on an Animal-Themed Book Program

Donna Crawford and Melissa Logan

20.1 Introduction

Why Use Picture Books?

Using literature to promote social-emotional learning, including empathy development in young people, is well documented as an effective approach.[1] As Hall notes, "stories engage our sentiments and make us feel

[1] Gina M. Almerico, "Building Character through Literacy with Children's Literature," *Research in Higher Education* 26 (October 2014).

D. Crawford
Calgary, AB, Canada

M. Logan (✉)
Alberta SPCA, Edmonton, AB, Canada
e-mail: mlogan@albertaspca.org

C. Linzey and A. Linzey (eds.), *The Palgrave Companion to Humane Education*, The Palgrave Macmillan Animal Ethics Series,
https://doi.org/10.1007/978-3-032-19795-5_20

deeply about people. Use of literature makes perfect sense to help children foster the ability to imaginatively reach beyond self."[2] Literature also can be a powerful tool to help students develop and shape their moral code. Stories expose students to a variety of issues or situations and provide them with the opportunity to discuss ideas, feelings, and behaviors in a safe environment.[3]

Picture books specifically have many advantages for the fostering of empathy in children of all ages.[4] They are concise, essentially providing a short "case study" for students to evaluate.[5] Lessons presented through picture books are not restricted to younger students; older students can process the same content and gain insight at higher levels of maturity.[6]

Books featuring animals are an intrinsic hook for students, as children have a natural affinity for living beings.[7] Humane educator Phil Arkow highlights that animals engage the minds of children in a way that subject matter focused on people does not: "Teaching abstract concepts like character and compassion can be easier, more engaging and more fun when animals are the springboard for discussion."[8] Caine also notes, "Although anyone at any age can learn more humane ways of living, young children are far more flexible in their habits, attitudes and behaviors."[9] Picture books are a great vehicle to explore our beliefs, attitudes, and actions toward others, including animals.

[2] Susan Hall, *Using Storybooks to Teach Character Education* (Phoenix: Oryx Press, 2000), vii.

[3] Alberta Education, "Instructional Strategies That Support Character and Citizenship Education," in *The Heart of the Matter: Character and Citizenship Education in Alberta Schools* (Edmonton: Alberta Education, 2005), 124–25.

[4] Maria Nikolajeva, "Picture Books and Emotional Literacy," *The Reading Teacher* 67, no. 4 (2013).

[5] Patricia Crawford, "Beyond Words: Using Language and Literature to Teach Compassion for Others," in *Teaching Compassion: Humane Education in Early Childhood*, ed. Mary Renck Jalongo (New York: Springer, 2014), 168.

[6] Alberta Education, "Instructional Strategies," 126.

[7] Gail Melson, *Why the Wild Things Are: Animals in the Lives of Children* (Cambridge: Harvard University Press, 2001), 19.

[8] Phil Arkow, "Animal-Assisted Interventions and Humane Education: Opportunities for a More Targeted Focus," in *Handbook on Animal-Assisted Therapy*, ed. Aubrey H. Fine (New York: Academic Press, 2010). 457–480

[9] Robert S. E. Caine, "Humane Education: A Foundation for Connecting with All of Earth's Inhabitants," *Green Teacher*, Summer 2009, 10.

Simply reading a book may not be the most effective way to change attitudes or behaviors.[10] To promote affective learning that has the potential to shift attitudes and behaviors, students should be provided with the opportunity to reflect upon and respond to the literature.[11] Activities such as discussion, debate, research, role-playing, and journaling help students to process new knowledge, create meaning, and apply it to their own lives.[12]

The Alberta SPCA AnimalTales Book Program

The Alberta Society for the Prevention of Cruelty to Animals (SPCA) provides humane education resources for teachers across the province of Alberta, in order to inspire compassion for animals, people, and the environment. The AnimalTales program consists of four animal-themed picture books, a teacher's guide, and accompanying materials. Books feature animals across many species, from cats and dogs to ducks and ants. Most books in the program do not employ an anthropomorphized depiction of the animals. Of the books that do, the anthropomorphization highlights realistic elements of the animals, such as their physical and psychological needs, and their intrinsic value, for the promotion of empathy. The AnimalTales teacher's guide includes pre-reading discussion questions to introduce the theme and concepts in the book, as well as post-reading activities that prompt critical thinking, perspective taking, and making connections to one's own life. The activities are tied to the provincial programs of study and address learner outcomes in a variety of subject areas, including language arts, social studies, science, health, and math. Alberta teachers request the program online, receive it in the mail for free, and have one month to use the materials as they wish.

[10] Shawna Brynildssen, "Character Education through Children's Literature," ERIC Digest (2002), ED 4699299, http://www.vtaide.com/png/ERIC/Char-Ed-Literature.htm

[11] Brynildssen, "Character Education"; Alberta Education "Instructional Strategies," 125–26.

[12] Hall, *Using Storybooks to Teach Character Education*, vii; Karen Bohlin, *Teaching Character Education through Literature* (New York: Routledge, 2005), 31.

20.2 Purpose of Study

The purpose of this study was to determine the range and nature of the effectiveness of the Alberta SPCA AnimalTales book-lending program in increasing empathy in elementary students. Here "empathy" is assumed to be demonstrated by students' attitudes as to how animals, people, and the environment are, and should be, treated.

20.3 Problem Statement

The overarching question guiding this study was this: How do elementary school students and teachers experience the Alberta SPCA AnimalTales book program, and what is the nature of the impact of this program on students?

20.4 Research Questions

Four major research questions were designed in order to investigate both the individual and the collective experiences with the AnimalTales program:

1. *Students' perspectives:* Does the AnimalTales program impact students' perceptions regarding how animals and people should be treated?
2. *Students' observable behavior*: Are there any indications that participation in the AnimalTales program influences student behavior relating to kind treatment of people and animals?
3. *Teachers' experiences with the program and the students' learning*: What are teachers' perspectives on using the AnimalTales program, in grades 2 and 5?
4. *Use of the program in different school system contexts:* How are teachers utilizing the program in the classroom?

20.5 Methods

Recruitment of teachers was done through opportunistic sampling. Teachers were already signed up to receive the book program, were recruited through the Alberta SPCA teacher e-newsletter, or showed interest at a teachers' convention and volunteered to participate in the study. Semi-structured teacher interviews were recorded during a school site visit. Once teacher participation was confirmed, principals were asked about their willingness to participate in an interview. Interviews with school staff took place at the earliest convenient time for the teacher, following the completion of the program in the classroom. Parents could participate via a survey sent home with their child.

During the site visit, the researcher read a grade-appropriate picture book with a humane education focus to all students, discussed the themes in the book, and spent time in the classroom speaking with students in order to increase comfort levels during individual student interviews. Students also had the option to share related work samples; only data from students with parental consent was included in the analysis.

Ethical Considerations

The study had ethical approval from the University of Calgary, as well as each school district that had teacher participation. Students had formal parental consent and a demonstrated willingness to participate. All data was anonymized using codes to represent type of participant, gender, location in Alberta, and relative size of surrounding area (rural, urban, and metro).

Context of the Research Study

The data for this study was collected from teachers, students, administrators, and parents in eighteen classrooms (fifteen schools) located throughout rural, urban, and metropolitan settings in northern, central, and southern Alberta. Teachers had from the current year to three years of experience using the AnimalTales program and from the current year to more than ten years of classroom experience.

Five Catholic schools (six teachers), including one Francophone and one Montessori setting, and ten public schools (twelve teachers), including one French immersion dual-track school and one Logos Christian setting, provided a variety of school types. Among the schools, a single school was described as generally middle-class. Eight sites were described as drawing from lower socioeconomic populations, and six as mixed. At eleven sites, students were described as having a variety of complex needs. Classrooms were almost always described as being culturally diverse, regardless of other demographics. All schools reported partnerships with the community, in support of their own students, in support of others with needs in the community, or both.

20.6 Results

1. *Students' perspectives: Does the AnimalTales program impact students' perceptions regarding how animals and people should be treated?*

In total, eighty-seven students were individually interviewed, representing each of the participating schools and classrooms: forty-five grade 2 and forty-two grade 5 students. Additional information regarding student perceptions was gathered during classroom discussions with the researcher present, as well as from student work samples provided by the teachers or students. At both grade levels, there was evidence of new knowledge and understandings regarding empathy for both animals and people. In all the classrooms, discussions about kindness were ongoing. At all schools with the exception of two, there was a schoolwide focus on empathy and character education–related topics. The amount of time between the completion of the AnimalTales program and the interviews varied from school to school. It was difficult for students to separate what they learned from any one specific initiative such as AnimalTales, or to discern where one initiative started and another concluded.

Students in both grades were easily able to speak about kindness in general, for people, for animals, and for the environment. They could clearly identify times they had been kind, when someone had been kind to them, when they had been unkind or had been treated poorly, and how kindness applied to animals. One example follows:

> Being kind is really important because if you are not kind, then there is a ripple. If you are kind to someone, they will pass it on to someone else, and it will just keep going, and it can spread kindness all over the world. (Grade 4/5 student, male, rural, north)

At the grade 2 level, seven ways to be kind to animals were identified by students. Not hurting animals, being nice to them, and treating animals the way you would want to be treated were the most frequent responses. At the grade 5 level, kindness to animals focused on treating animals as you would want to be treated and ensuring all their needs were taken care of. Students offered up many examples of personal actions related to kindness toward animals, people, and the environment:

> We talked about kindness, a little about people and animals. That people need to be treated the same as animals. Like, animals still need sort of the same things as humans or else they might die. They need food, shelter and water. (Grade 4/5 student, female, urban, central)

> I think the message was that people need to pay more attention to how they are changing the land, and the animals were here before us, so we need to respect their land, because it is not really ours, it is theirs. (Grade 5 student, male, rural, north)

Students reported learning many new things about the responsibilities that come with being a caregiver to a companion animal and appropriate ways to treat animals:

> The books tell you that you should actually take care of your pets. Feed them every day, give them a walk, water. You should treat them very nice. You shouldn't ride on your dog, pull their tails, or pull your cat by the feet, that is not a good pet owner. (Grade 2 student, female, rural, south)

Twelve responsibilities of companion animal caregiving were identified by students; providing food and water, providing exercise, and cleaning up after a companion were most commonly mentioned. Grade 5 students were more aware of cost items, such as veterinary care and buying items that companion animals need.

At both the grade 2 and grade 5 levels, very few students commented on classmates acting differently with each other following the lessons, either positively or negatively. They were unable to put a specific cause to any improvements they did see, other than being aware that the teacher had worked on caring and kindness all year:

> You have to care about every living thing, even plants. She [the teacher] does her point of view on the book after we read it, and we discuss it. If you are kind you are going to have a lot more friends and a lot more living things in your life like pets, plants, and people. (Grade 5 student, female, urban, north)

Students were unanimously clear that they enjoyed using the books as well as the accompanying materials, although they had different opinions regarding best or least favorite book and activity. They hoped the teacher would continue using the materials with her next class (or in some cases the next grade's materials with them):

> She should do it again next year, because we learned how important animals are to the environment and reading stories helps your brain, and for fun. (Grade 5 student, female, metro, north)

> I would bring the books back next year, to teach everybody else what a pet can actually do and needs and what people can do for pets. We had not talked about that before. (Grade 5 student, male, rural, north)

The majority felt that they had learned something new about animals and how to treat them or that their classmates had:

> I think most people learned a lot. I don't think there was anybody who didn't learn at least one thing. People should be more aware of the way animals are, they are kind of like people, they are just trying to look after themselves. (Grade 4/5 student, male, metro, north)

2. *Students' observable behavior: Are there any indications that participation in the AnimalTales program influences student behavior relating to kind treatment of people and animals?*

Teachers, students, parents, and administrators all responded to this question. Although no group could conclude that there was a direct causal relationship between the programming and student actions, all groups found using AnimalTales to be a positive experience, recommended its continuation, and viewed it as supportive of family and school goals and values with regard to character education in general and humane education in particular.

Transfer was reported by teachers to occur in two ways: at the individual social-behavioral level and at the action level. A number of classrooms undertook action or research projects regarding environmental or animal causes as a result of increased sensitivity to these topics.

The importance and value of both human and animal life was a theme common to many classrooms, especially those that were faith-based.

Eleven teachers commented specifically on the discussion value of materials. This included the following remarks:

> Our best class was when compared ants and people and came to the conclusion that the more we know about something or someone, the more we understand them and have feelings for them. We extended that to people we meet in our classes from other schools or countries. If we know them, we will be able to understand them and have empathy for them. (Urban, north)

> We shared a lot of our thoughts on responsible pet ownership, including debating whose fault it would be if a person got bit. We were able to discuss dog behavior/warning signs and body language. (Metro, north)

> I think it really started to sink in when they started discussing the proper treatment of people and animals outside of the set discussion time. They also started coming to school and telling me they had been teaching family members how to properly care for animals. (Rural, north)

At both the grade 2 and 5 levels, there were no direct comments from students about acting differently as a result of exposure to the content. Students were, however, very informed regarding how to act appropriately with animals and gave examples of doing so, or of family members and others being kind to animals. Older students noted paying for animal needs from allowances or gifts, taking in stray animals or taking them to a shelter, or reporting others' poor behavior with animals.

Parent surveys were sent home with students. The response rate was low from all schools. Of the forty-six parents who did respond, 36 percent (nineteen) noted that their children already had been empathetic and aware of the responsibilities of companion animal caregiving. They generally felt that the lessons had reinforced these concepts but had not introduced them or caused significant change in attitudes or behavior. Thirty-two percent (fifteen parents) reported new knowledge or levels of responsibility regarding companion animals from their children. Sixty-five percent (thirty parents) reported discussing responsible animal caretaking with their child. All parents who participated were in favor of teachers continuing to offer the program. The top reported reasons, other than generalized positive effects, were to build an understanding of appropriate treatment of animals, to foster empathy, and to teach responsibility, all key objectives of AnimalTales. One of the more impactful reports was the following:

> This became especially difficult for him when he realized the extent of the abuse one of our rescues had suffered through before coming to our home. The dog was already recovered and becoming well-adjusted before our oldest child was born but he had never heard the story. Being able to relate the animals' [in the stories] suffering to that of our own dog was hard for him, but ultimately helped him understand the benefit of rescue programs and responsible pet ownership, especially having witnessed firsthand what amazing animals they are. (Urban, north)

Administrators reported not feeling sufficiently informed to comment on this question, and like teachers, they were unable to differentiate the effects of this particular program from the ongoing focus on caring and compassion in the school.

3. *Teachers' experiences with the program and the students' learning: What are the teachers' perspectives on using the AnimalTales program, in grades 2 and 5?*

All teachers reported positive experiences, both as educators and for their students. They commented on the applicability of the materials for any type of student, the ease of using the materials, the fit with other character or faith programs, and their experience of getting better at using the materials each year they presented them.

Twelve teachers reported using AnimalTales because their class was difficult or in need of such programming, with emotional/behavioral concerns, English language learner backgrounds, or poverty being among those identified. Thirteen teachers reported the animal focus as a draw. Many teachers were animal lovers themselves, increasing the appeal.

> The students were very attentive during the reading. Which is surprising because they can get easily distracted at times, but I think the length of the book was good and the subject was interesting. Even if you don't have a pet, you have seen animals. We have students from different backgrounds,

> but everybody could say something. I enjoyed it, I would do it again, I liked the choice of books. (Teacher, metro, central)

Eleven teachers specifically chose to use the programming because of the focus on kindness, empathy and being a caring person:

> Just knowing your math, as good as it is, if you are not a kind person, it might not get you very far. (Urban, north)

> Since we are a Catholic school we are always talking about kindness and making a good choice, the best choice they can make. It reinforces it. We just transferred it to animals, what would you do as a good choice? (Metro, central)

> With each other, there are some students who are fairly selfish and entitled, so when there is a conflict, it is very "you did this to me." Every day I have to walk the kids through how the other person has felt. (Rural, central)

> A combination of need in the class and good practice. Earlier in the year there were some definite bullying issues, so it helped with that. (Urban, north)

4. *Use of the program in different school system contexts.*

There were no significant differences in program effectiveness for northern, southern, or centrally located districts; for rural, urban, or metro locations; or between male and female participants. On average, teachers reported spending one month on the program, with weekly instructional time varying between 1.5 and 10 hours per week. School leadership support and alignment with district objectives both were considered helpful precursors to an effective program. Principals were consistent in their support for the program and believed that it aligned well with district and school initiatives, as well as with their personal beliefs about what is generally important in a school. Three principals highlighted that books with content that supports social-emotional learning, such as those included in AnimalTales, are an important component of school libraries. One principal noted that library books will now be examined to ensure they depict an appropriate human–animal relationship, highlighting that the program can bring awareness of problematic depictions of animals or human–animal relationships in existing literature.

20.7 Summary

Students reported learning about compassion and kindness and about how others, including animals, are impacted by their actions, both through the use of the books and activities and through the efforts of the teacher to engage them in the work. With regard to kindness and animals specifically, students reported learning many new things about the responsibilities that come with being a caretaker to a companion animal and appropriate ways to treat animals:

Because students were exposed to a wide variety of social-emotional programming, the transfer to behavior from any specific set of materials was difficult to determine. This did not diminish the perceived value of AnimalTales: "Our world is changing, so it is the programs like this that get you talking about important stuff that matters" (teacher, urban, north). Again, all teachers reported positive experiences as educators and for their students. One teacher noted, "Having resources with animals as the focus helps the kids transfer the concept of kindness back and forth between animals and people. We can focus on treating all living things with kindness—and how you treat animals is how you treat others, treat others as you want to be treated" (teacher, metro, central).

Program efficacy was not impacted by geographic location or demographics, however school leadership support, and congruence with district objectives were identified as key factors for program usage and success.

As with most programs, the consensus from all groups reporting was that "it is what the teacher does with the content and conversation that will make the difference." The teacher who said this continued, "I love it, I love that it is free and that it is quality" (teacher, rural, central).

> Overall, the continued use of humane education materials such as AnimalTales in schools, in order to increase empathy for all living things, is well supported by the findings of this program evaluation. As one educational assistant commented, the teacher using the materials was just one more stepping stone to making them all-around citizens. We still talked about

empathy and how that relates not only to animals but to people as well. It's a whole classroom, whole-school emphasis. Definitely in this classroom it's a big emphasis. And if you can connect with animals, your next step is to understand people. You just have a better appreciation for any living thing, when you can connect your empathy to others. (EA, urban, north)

References

Education, Alberta. 2005. Instructional strategies that support character and citizenship Education. In *The heart of the matter: Character and citizenship Education in Alberta schools Alberta.* Edmonton: Alberta Education.

Almerico, Gina M. 2014. Building character through literacy with children's literature. *Research in Higher Education* 26: ERIC Digest EJ1055322.

Arkow, Phil. 2010. Animal-assisted interventions and humane Education: Opportunities for a more targeted focus. In *Handbook on animal-assisted therapy*, ed. Aubrey Fine, 457–480. New York: Academic Press.

Bohlin, Karen. 2005. *Teaching character Education through literature.* New York: Routledge.

Brynildssen, Shawna. 2002. Character Education through children's literature. *ERIC Digest.* ED 4699299. http://www.vtaide.com/png/ERIC/Char-Ed-Literature.htm.

Caine, Robert S. E. Summer 2009. *Humane Education: A Foundation for Connecting with all of earth's inhabitants*, 9–13. Green Teacher.

Costello, Bill, and Nancy J. Kolodziej. 2006. A middle school teacher's guide for selecting picture books. *Middle School Journal* 38 (1): 27–33. www.nmsa.org/publications/middleschooljournal/articles/september2006.

Crawford, Patricia. 2014. Beyond words: Using language and literature to teach compassion for others. In *Teaching compassion: Humane Education in early childhood*, ed. Mary Renck Jalongo, 161–173. New York: Springer.

Hall, Susan. 2000. *Using storybooks to teach character Education.* Phoenix: Oryx Press.

Melson, Gail. 2001. *Why the wild things are: Animals in the lives of children.* Cambridge: Harvard University Press.

Nikolajeva, Maria. 2013. Picture books and emotional literacy. *The Reading Teacher* 67, no. 4:ERIC 2013249-254.

Wolk, Steven. 2009. Reading for the world. *Educational Leadership* 66: http://www.ascd.org/publications/educational-leadership/jul09/vol66/num10/Reading-for-the-World.aspx.

CHAPTER 21

Can Humane Education Motivate Students to Learn? Teachers' Perceptions

Julie O'Connor

21.1 Introduction

Motivation and engagement are important in academic achievement.[1] When students suffer from motivational apathy, they have little reason to engage in classroom learning activities.[2] The significant academic problem

[1] Mitchell Ross, Helen Perkins, and Kelli Bodey, "Academic Motivation and Information Literacy Self-Efficacy: The Importance of a Simple Desire to Know," *Library & Information Science Research* 38, no. 1 (2016): 2, https://doi.org/10.1016/j.lisr.2016.01.002.

[2] Sung Hyeon Cheon and John Marshall Reeve, "A Classroom-Based Intervention to Help Teachers Decrease Students' Amotivation," *Contemporary Educational Psychology* 40 (2015): 99, https://doi.org/10.1016/j.cedpsych.2014.06.004.

J. O'Connor (✉)
School of Education, Touro College, New York, NY, USA
e-mail: julie.oconnor@aplnj.org

C. Linzey and A. Linzey (eds.), *The Palgrave Companion to Humane Education*, The Palgrave Macmillan Animal Ethics Series,
https://doi.org/10.1007/978-3-032-19795-5_21

of lack of student motivation impacts elementary, middle, and high school educators.[3]

Character education is the explicit instruction of positive values and has been found to improve student motivation and engagement.[4] Humane education is a type of character education integrating animal-related curricula to foster compassion in children's relationships with animals, people, and the environment. Because most children are naturally interested in animals, humane education lessons have the potential to stimulate motivation to learn and engagement in class activities more than conventional character education lessons.[5]

21.2 Humane Education

Humane education is a broad curriculum framework that has not been thoroughly explored by researchers. According to Weil,[6] it can be defined as an educational program promoting compassion and respect related to animal welfare, the environment, and social justice. The Academy of Prosocial Learning defines humane education as encouraging growth through the development of empathy as well as problem-solving skills related to people, animals, the planet, and the intersectionality among them.[7] For the purposes of this study, humane education involves only curricula related to animal well-being. Fostering compassion and empathy

[3] Suzanne Hidi and Judith M. Harackiewicz, "Motivating the Academically Unmotivated: A Critical Issue for the 21st Century," *Review of Educational Research* 70, no. 2 (2000): 151, https://doi.org/10.2307/1170660; Lisa Legault, Isabelle Green-Demers, and Luc Pelletier, "Why Do High School Students Lack Motivation in the Classroom? Toward an Understanding of Academic Amotivation and the Role of Social Support," *Journal of Educational Psychology* 98, no. 3 (2006): 567–82, https://doi.org/10.1037/0022-0663.98.3.567.

[4] Joseph A. Durlak et al., "The Impact of Enhancing Students' Social and Emotional Learning: A Meta-Analysis of School-Based Universal Interventions," *Child Development* 82, no. 1 (2011): 412–13, https://doi.org/10.1111/j.1467-8624.2010.01564.x.

[5] Catherine A. Faver, "School-Based Humane Education as a Strategy to Prevent Violence: Review and Recommendations," *Children and Youth Services Review* 32, no. 3 (2010): 365, https://doi.org/10.1016/j.childyouth.2009.10.006.

[6] Zoe Weil, *Power and Promise of Humane Education* (Gabriola Island, BC: New Society Publishers, 2013).

[7] Academy of Prosocial Learning, "Humane Education Training: Academy of Prosocial Learning," APSL, accessed August 5, 2019, https://www.prosocialacademy.org/.

for others, including nonhuman animals, is an overarching goal of humane education.[8]

Most teachers hold positive perceptions of humane education involving animal well-being.[9] Its encouragement of critical thinking is appealing to educators. Humane education is intended not to provide students with right or wrong answers, but to make information available so that students can make informed decisions.[10]

Many proponents of humane education promote student action and problem-solving as central to its mission. The Institute of Humane Education calls students who participate in humane education "solutionaries." A solutionary is "a person who is able to identify inhumane and unsustainable systems and then develop solutions that are healthy and just for people, animals, and the environment."[11]

Research exploring students' interactions with animals and the natural environment typically builds on children's natural affinity for animals and intrinsic motivation to learn about animals and nature.[12] Animals should be used as a pedagogical support and motivator for learning.[13]

[8] Faver, "School-Based Humane Education," 366; Jane Bone, "The Animal as Fourth Educator: A Literature Review of Animals and Young Children in Pedagogical Relationships," *Australasian Journal of Early Childhood* 38, no. 2 (2013): 62, https://doi.org/10.1177/183693911303800208.

[9] Beth Daly and Suzanne Suggs, "Teachers' Experiences with Humane Education and Animals in the Elementary Classroom: Implications for Empathy Development," *Journal of Moral Education* 39, no. 1 (2010): 101, https://doi.org/10.1080/03057240903528733.

[10] Laura Barraza, "Attitudes to Animal Dilemmas: An Exploratory Comparison between Mexican and English Children," *International Electronic Journal of Environmental Education* 5, no. 2 (2015): 41, https://doi.org/10.18497/iejee-green.64245; Piers Beirne and Meena Alagappan, "A Note on Pedagogy: Humane Education Making a Difference," *Journal for Critical Animal Studies*5, no. 2 (2007); Bone, "The Animal as Fourth Educator," 62.

[11] Institute for Humane Education, home page, March 11, 2019, accessed August 6, 2019, https://humaneeducation.org/.

[12] Jürgen Drissner, Hans-Martin Haase, and Katrin Hille, "Short-Term Environmental Education—Does It Work? An Evaluation of the 'Green Classroom,'" *Journal of Biological Education* 44, no. 4 (2010): 155, https://doi.org/10.1080/00219266.2010.9656215; Marj Francis, Kathy Paige, and David Lloyd, "Middle Years Students' Experiences in Nature: A Case Study on Nature-Play," *Teaching Science* 59, no. 2 (June/July 2013): 20–21, https://search.informit.com.au/documentSummary;dn=687559195442456;res=IELHSS.

[13] Bone, "The Animal as Fourth Educator," 59.

Though humane education does not require the presence of a live animal, human–animal interaction is sometimes a component of humane education–themed activities. People experience benefits during those interactions. Animals facilitate conversation, decrease anxiety, and dissolve communication barriers for children.[14] Children increase their ability to cope with stressors in their daily lives by interacting with the natural world, including animals.[15]

Humane education serves two important purposes. The first is its role as a beneficial learning pedagogy. As this research will present, it increases student motivation to learn. The lessons focused on kindness to animals serve as conduits to learning new information. Humane education's second vital justification is that society benefits from an increased awareness of the plight of animals and how people can help them. One could argue that the health of humankind is connected to the well-being of nonhuman animals.[16]

21.3 Talking to Teachers

Building on the foundation of children's natural interest in animals as well as the fundamentals of research in motivation, teachers were asked to explore whether students who received lessons in humane education, specifically animal well-being, demonstrated increased engagement in learning. The educators also were asked to elaborate on how such lessons could be most effectively delivered. Given humane education's diverse teaching methods and variety of topics, researchers have argued that the use of humane education programs involving animal well-being

[14] Anna Goddard and Mary Gilmer, "The Role and Impact of Animals with Pediatric Patients," *Pediatric Nursing*, 41, no. 2 (March/April 2015): 65, https://pdfs.semanticscholar.org/fd07/5d7ea6121852bebeeb2ffa8b0a56cbb72fa0.pdf?_ga=2.146038003.679191520.1565094097-521204368.1565094097.

[15] Nicole Migliarese, "Researching the Child–Nature Connection" (California State Parks' Children in Nature Campaign white paper, March 2008), https://www.parks.ca.gov/pages/24914/files/microsoft%20word%20-%20copy%20of%20white%20paper%20for%20kids%20and%20nature%20campaign%20ready%20to%20pub2%20single%20space.pdf.

[16] Aysha Akhtar and Carl Safina, *Our Symphony with Animals: On Health, Empathy, and Our Shared Destinies* (New York: Pegasus Books, 2019).

is outpacing research about them.[17] In general, curriculum research has focused on effects on children but has largely neglected the role teachers hold as practitioners and implementers. Teachers' beliefs should become the focus of research on educational effectiveness because teachers guide classroom activities.[18] Teachers sift through information, deciding how it will be conveyed in the classroom, and they are thereby illuminative in the study of any curriculum.[19] Curriculum designers more broadly and humane education designers in particular would benefit more from understanding teachers' views of these issues.

21.4 Methods

Sample and Design

The study discussed here was self-funded by the researcher. The study design and full results have been published elsewhere.[20] Ethical approval was granted by the relevant academic institution. A ten-question interview protocol was used on eight teachers of humane education, some of whom also provided physical evidence of classroom activities. Therefore, this qualitative phenomenological case study used individual interviews and artifact collection as two sources of evidence.[21] The interview method allowed the participating educators to provide their lived perspectives on

[17] Tania Signal, R. Arbour, and Nik Taylor. "Teaching Kindness: The Promise of Humane Education," *Society & Animals* 17, no. 2 (2009): 139, https://doi.org/10.1163/156853009x418073.

[18] Gary Fenstermacher, "A Philosophical Consideration of Recent Research on Teacher Effectiveness," *Review of Research in Education* 6 (January 1, 1978): 157–85, https://doi.org/10.2307/1167245.

[19] Frank Pajares, "Teachers' Beliefs and Educational Research: Cleaning Up a Messy Construct," *Review of Educational Research* 62, no. 3 (1992): 307–32, https://doi.org/10.2307/1170741.

[20] Julie O'Connor, "A Qualitative Case Study of Teacher Perceptions of the Motivation of Students in Humane Education" (PhD diss., Northcentral University, 2018), ProQuest.

[21] Robert K. Yin, *Case Study Research: Design and Methods* (London: Sage Publications, 2014).

their students' motivation and engagement.[22] The humane educator artifacts—that is, lesson plans and student work samples—were obtained to triangulate the data and bolster the credibility of the qualitative analysis.

Purposeful criterion-based sampling was used to select eight educators who had at least two years' experience teaching humane education curricula, specifically related to animal well-being. Educators were invited to participate via emails that humane education organizations sent to their memberships on the researcher's behalf. The sample of humane educators included a diverse population representative of the actual population. However, given the regional nature of some of the organizations that assisted with recruitment, seven out of the eight participants lived in just two states of the United States. Participants included four teachers from New Jersey (one of whom had recently retired), three instructors from New York (one was a school psychologist), and one educator from California who had diverse experiences, including work within the prison education system.

External validity was addressed through the use of a cross section of humane educators who had experience in diverse academic content as well as differing school environments. Respondents had both primary and secondary teaching experience in both public and private schools. The instrument for the semi-structured individual interviews consisted of an interview protocol of ten open-ended questions designed to elicit descriptive responses. These semi-structured interviews occurred in person and via telephone. Reliability was supported by the use of an identical typed interview protocol for each interview.

Recording open-ended responses, the interviewer asked participants to indicate how they defined humane education, how their students responded to humane education lessons, whether and how gender impacted student reaction, how different animal welfare issues elicited student responses, and what aspects of humane education enhanced student motivation and engagement. The educators' true lived experiences informed their responses.

Interviews were recorded, both face-to-face and over the telephone. Responses were transcribed verbatim, and manual data analysis was used. Teachers' classroom artifacts were analyzed. The classroom artifacts included classroom photos, student work samples, and teachers'

[22] Linda Dale Bloomberg and Marie Volpe, *Completing Your Qualitative Dissertation: A Road Map from Beginning to End*, 3rd ed. (Thousand Oaks: SAGE Publications, 2016).

lesson plans related to animal welfare lessons. The researcher conducted a line-by-line coding, identifying concepts among the interview transcripts and finding relations between them. Patterns and emergent themes were coded in homogenous categories. The in-depth analysis and corresponding findings were the result of an unbiased examination.

21.5 Results

Six themes related to teacher perception of humane education and its impact on student motivation and engagement became apparent through careful analysis of the collected interview data. These themes included teaching techniques; safe topics and species; food and animals on farms; student and teacher connection; age, gender, and culture of students; and administrator, colleague, and parent reaction. The degree to which students were motivated and engaged by humane education depended on the interaction of various factors within these themes, such as the curriculum aspects and the manner in which the curriculum was delivered. All participants perceived that humane education had a positive impact on student motivation and engagement. Participants perceived humane education's inclusion of animals as a key motivating and engaging aspect for students.

21.6 Discussion

Eight out of eight teachers defined humane education as instruction on the topic of animal welfare. All of the study's participants also noted enhanced student interest when animals were integrated into the lesson. One of the teachers explained how she recognized a motivated student and how humane education was particularly interesting for her class. "Body language, participation, raising their hands, and wanting to talk about it—that motivation is pretty easy to read as a teacher. You've got the kids who love science projects, and the kids who love math, or a particular book. But what I'm saying is, I knew if I did a humane education type of lesson, I would hook them all," she said. The implication is that the inclusion of animal-related topics motivates and engages students. If teachers were to introduce animals when conveying lessons in their curriculum, educators could observe students giving deeper attention and displaying more positive attitudes toward their learning. Many children experience positive feelings toward animals. Positive emotions enhance academic performance whereas negative emotions undermine it.[23]

Humane education could be considered a nontraditional curriculum and also could be delivered in different ways. In a previous study of what methods motivated unenthusiastic students, researchers presented the use of multiple teaching styles, extracurricular activities, and technology as some of the methods found to motivate children.[24] Participants in this case study agreed. According to the humane education teachers, their lessons should be delivered to effectively motivate and engage students in myriad ways. Data in this study showed teachers using current events, teaching lobbying techniques, using technology, conveying lessons with humor, and inviting guest speakers with animals. The message for teachers is to use diverse teaching methods to deliver humane education in a way appealing to them and their students.

The study indicated that teachers could engage and motivate children when animals were introduced throughout different areas of the curriculum. Teachers provided what they perceived as positive examples of focusing on animals in literacy, math, social studies, and science. The study indicated that teachers did not find it necessary to bring live animals to the classroom to motivate their students. Even if they integrated only discussion of humane education topics throughout their teaching, they found that this engaged their students more than if they did not include animal-related topics.

Within the curriculum of humane education, teachers found that real-world applications, also referred to as experiential learning, presented the opportunity to motivate and engage students to do challenging academic work. Students debated, wrote letters, conducted fundraisers, and met with politicians on animal-related issues, current events, and legislation, such as the captivity of "wild," undomesticated animals or humane treatment of dogs. Having students act on what they learned was credited as particularly motivating by some educators. The implication is that when

[23] Carolina Mega, Lucia Ronconi, and Rossana De Beni, "What Makes a Good Student? How Emotions, Self-Regulated Learning, and Motivation Contribute to Academic Achievement," *Journal of Educational Psychology* 106, no. 1 (2014): 121, https://doi.org/10.1037/a0033546.

[24] Alexandra Usher, "What Nontraditional Approaches Can Motivate Unenthusiastic Students?" (Washington, DC: Center on Education Policy, 2012).

students are asked to act upon the information, their academic skills, as well as their knowledge of the topic, will improve.[25]

In addition to experiential learning being evident in the teachers' examples of humane education, data in this study showed that the educators' approach included a constructivist framework. Constructivism is a learning theory predicated on the belief that children construct new knowledge on the basis of their preexisting knowledge. Learning is active, not passive.[26] In a constructivist classroom, students play more leading roles in their own learning than in a traditional classroom.[27] The implication is that the humane pedagogy the teachers discussed in this study promotes a constructivist framework.

The findings of the study can be applied to educational practice as to teachers' selection of topics to cover in class. The case study indicated that teachers could introduce "safe" or familiar animals such as common companion animals to initially motivate students. Teachers observed that learning about dogs initiated a high level of student engagement, and some educators believed that familiarity with other species of animals also elicited student motivation to learn lessons about them. Teachers noted that children in urban environments were familiar with different animals than children in other living situations. A first-grade teacher in an inner-city, urban environment explained about which animal her students wanted to learn. She presented to the students four animal welfare–themed comic books with four different animals as subjects: *A Chicken's Life*, *A Cow's Life*, *A Dog's Life*, and *A Rat's Life*. The comic books had been provided by the animal protection organization People for the Ethical Treatment of Animals. The students chose the dog comic book first. After the dog's story was read and discussed, it was the students' second choice that surprised the teacher: "The next one they chose was the rat, and I was very surprised. I said, 'Do you all know what rats are?'

[25] Jeffrey Coker and Desiree Porter, "Student Motivations and Perceptions across and within Five Forms of Experiential Learning," *The Journal of General Education* 65, no. 2 (2017): 138, https://doi.org/10.5325/jgeneeduc.65.2.0138.

[26] Rajesh Kumar Sharma, "Constructivism—An Approach to Enhance Participatory Teaching Learning," *GYANODAYA—The Journal of Progressive Education* 7, no. 2 (2014): 13, https://doi.org/10.5958/2229-4422.2014.00003.6.

[27] Nuket Gunduz and Cigdem Hursen, "Constructivism in Teaching and Learning; Content Analysis Evaluation," *Procedia—Social and Behavioral Sciences* 191 (2015): 526–33, https://doi.org/10.1016/j.sbspro.2015.04.640.

They said, 'Yes, like mice. They're cousins.'... I said, 'Boys and girls, why did you choose the rat?' Believe it or not, their response was because they see rats in their building! That is their nature. That's their wildlife—their rats." Whereas students in urban environments were more familiar with rodents, the students in suburban neighborhoods had more experience with companion animals kept at home and in a yard. The educators believed that students' familiarity led to their increased motivation to learn about and discuss those animals. The implication for teachers is that focusing on animals that students know and applying the lessons to the students' real-world experiences can be a point of entry in the use of humane education for engagement purposes. Teachers also believed that safe humane education topics could be presented as introductions to more controversial lessons regarding the treatment of animals.

Teachers reported that discussions of food and issues surrounding animals living on farms in particular evoked powerful responses from students. Animals used in the agriculture industry were mentioned by all the participants in the study, even though questions about farmed animals were not explicitly asked. Participants believed that students' daily interaction with food made it an important topic to address. Most of the teachers explored the food system and its impact on animals, people, and the planet. The topics of food and animals on farms were recognized as controversial. One of the teachers taught *Charlotte's Web* by E. B. White to her students and pointed out, "You think of it as a sweet story, but it's pretty heavy-handed. You see the father going out with the ax because he's [the main character Wilbur] the runt. I say, 'That's the way it is. That is what would happen.'" Even when they elicited a strong reaction, teachers believed that discussions of animals living on farms, as well as other controversial animal topics, could aid in motivating students.

Additionally, some teachers in this study shared details of their own vegetarian or vegan lifestyles with their students, and they reported observing interest. Those case study participants believed that providing students with non-meat alternative foods in class was engaging and helped students understand their teachers' choices. One of the implications for teachers is that food can be used to motivate students to live a more humane lifestyle.

Of note, some educators experienced student defensiveness when cruelty on farms was discussed. Additionally, discussions about food and animals on farms were the main sources of parents' and administrators' negative reactions. One of the middle school teachers in the study taught a humane education–themed social justice class in a private school, and she elaborated about some parents' reactions to her teaching about animal agriculture in the following conversation:

> We talked about companion animals, farmed animals, and wildlife, so you spend a lot of time on each topic. We talked a lot about animal agriculture. Of course, I do that *last* .... So companion animals like dogs and cats, *especially* dogs, are like the gateway animal. Then you could go to wildlife because there are a lot of injustices that most people, given the choice, wouldn't participate in, hopefully. And then, when you get to the farmed animals, that's something that's hard because most people eat meat, dairy, and eggs. That is the only part of the curriculum that I've ever gotten any kind of parent complaints about.

She believed that a part of those complaints was rooted in the inconvenience of children possibly changing their diet, but still living with the rest of their family. "Suddenly, their kids didn't want to eat meat or dairy anymore, and that affected their lives. I once had these two parents go to the principal of the school and say, 'I don't want a vegetarian kid.'" However, the controversy that this topic can sometimes evoke was not found to be a reason to avoid teaching it. The participants believed that topics pertaining to animals on farms evoked a strong reaction and a high level of student engagement. The implication is that under the umbrella of humane education topics, students may be more attentive to lessons that include information about animals used in the agriculture industry.

This study indicated that teachers could sense that some students had an enhanced feeling of closeness to them because of their application of humane education. When teachers shared their own interest in animals, they believed some of their students connected with them more deeply. The positive emotions evoked by humane education could translate into a more positive student–teacher relationship. As one teacher put this, "I'm not saying kids dislike me. I'm saying certain kids will get what I'm saying about animals, and I will notice their behavior in the way they approach me, and... the way they act toward me is really different than other kids. I'll notice those kids who didn't have an opinion of me will gravitate towards me." When humane educators discussed what animals they had at home, students shared about their own companion animals. Additionally, students attended pet adoption locations where their teachers volunteered, after school hours. The study indicated that teachers could share about their own interest in animals to allow students to connect with them on a deeper, more personal level. The implication is that teachers will be rewarded with enhanced student motivation and engagement.

Age, gender, and culture of students arose as factors that teachers perceived as important when implementing humane education themes to motivate and engage learners. One of the teachers noticed a difference in the number of boys who were interested in her animal rights lunchtime club as they got older. "In second grade, the boys and girls responded exactly the same. It did not matter. With the older kids in fifth and sixth grade, it became mostly girls who wanted to be involved. As they got older, the boys just stepped away." However, data from this study indicated that the perceived drop in boys' interest in humane education could be addressed in a deliberate and effective manner. In short, it is important to ascertain what is interesting to students within humane education. Boys and girls should be allowed, where possible, to choose their animal topic to read and research. Some of the teachers believed boys expressed interest in different animals than girls did. One teacher made the case for student options within humane education as related to gender when he observed, "Girls are generally more caring and more interested in animals than the boys. But then you get the boys who are into wildlife and things like that. So if you bring in the picture of a hawk, they are interested in that. You talk about a cat, and you can get a lot of the girls who are interested." Humane education that includes wildlife issues may be more motivating for boys than teaching that covers only domesticated animal concerns. However, teachers perceived that both binary genders of students were motivated equally when learning about lessons including dogs.

Other educators believed that the activities students were asked to do (e.g., poster making or letter writing) determined the level of engagement students showed toward the curriculum. These teachers observed that boys and girls showed interest in different learning activities. Given that the teachers observed different levels of motivation and engagement, humane education students should have the opportunity to select the animal topic or the activity of their choosing when possible. Children should be supplied with classroom materials and activity choices, as well as topics addressing both familiar and less familiar species, so as to motivate and engage both boys and girls.

When teachers implemented humane education, they received different reactions depending on where their students lived. Culture was a powerful force, according to teachers in this study. The findings of this study showed that educators could expect to navigate children's conflicted emotions pertaining to animals if the teacher's message was antithetical to a message they received at home. One of the teachers viewed this reality as an even more important reason to implement humane education. In explaining its relevance for the students in the high-crime area where she taught, she stated, "Because children learn what they live. The children in my area where I teach—there are some, it's not all, but I would say 80 percent, who don't know what empathy is because they don't see it, and they don't feel it. I think the humane education is more important than ever."

Participants believed school administration could be helpful or harmful upon learning of the teacher's interest in delivering humane education. The degree of implementation of humane education in each school depended on the principals' encouragement or discouragement. Public school instructors in the study were subject to more invasive oversight by school administration, particularly as it related to their accountability for teaching skills that were to be tested on state exams. Participants in private schools were able to work with more academic freedom, in part because their students do not take those state exams, and their administrators did not convey the same type of pressure as did public school administrators. The private school educators reported more elaborate humane education projects than public school teachers. For example, one educator worked with her students in their private school on a bill to ban so-called wild animal circuses, and she took them to meet their local councilman. When asked about her principal's permission, she said, "I had to get permission from the parents, but no, I didn't actually tell the administration—they just knew I was going on a trip. They never, ever questioned me." This extra action of visiting with a local politician was easier to plan with private school students, and it made the lesson more impactful. Private school educators observed some of their students following up on the humane education lessons after school hours, on their own time. Public school educators in the study did not report their students following up to the same degree. The implication is that schools providing more academic freedom and less restrictive administrative oversight allow for better implementation of humane education and will provide students with greater motivation and engagement.

Case-study participants recognized that humane education's delivery depended on navigating the realities of the school community, which included administrators, colleagues, and parents. This case study indicated that these diverse influences were factors in the successful delivery of humane education. Some of the obstacles to incorporating humane education fully throughout the curriculum were related to parents' concerns about certain topics. To receive the full benefits of humane education's impact on student motivation and engagement, teachers should engage parents in humane education and explain its purpose through communication at home. Prior studies have suggested further examination of the reciprocal influence of the students, their teachers, and their parents on student motivation and engagement.[28] Parents' influence could enhance the lessons taught in school, particularly if they are supportive of humane education's goals.

Educators also found that relationships with colleagues were impacted by humane education. One of the teachers recalled that a child study team member—a person who is in charge of the individualized educational placement of special education students, but who is not a teacher supervisor—tried to get her fired because of her forays into humane education. Knowing that the humane educator discussed her own veganism, the colleague admonished her, saying, "You shouldn't be telling your personal life anyway." The study participant disagreed that personal anecdotes were inappropriate, but her colleague did succeed in making her unsure of her teaching. "No students should leave my room feeling bad about themselves. I don't think I make them feel bad," she equivocated. Another teacher in the study was similarly insecure about colleagues' reactions to his teaching of humane education. "What I've learned throughout the years is that bringing your own personal opinion to the school, or to any worksite, you know, is very risky." He felt it was necessary to be seen as approachable. If he was seen as too "militant," he felt that a colleague might complain to a school administrator. In spite of some concerns, generally, the teachers wanted to invite their peers to understand why they viewed humane education as important.

[28] Maaike C. Engels et al., "Behavioral Engagement, Peer Status, and Teacher–Student Relationships in Adolescence: A Longitudinal Study on Reciprocal Influences," *Journal of Youth and Adolescence* 45, no. 6 (December 2016): 1192–1207, https://doi.org/10.1007/s10964-016-0414-5; Kerstin Helker and Marold Wosnitza,"The Interplay of Students' and Parents' Responsibility Judgements in the School Context and Their Associations with Student Motivation and Achievement," *International Journal of Educational Research* 76 (2016): 34–49, https://doi.org/10.1016/j.ijer.2016.01.001.

Researchers recognize that study in the field of humane education is not keeping pace with its use.[29] The findings of this study showed that teachers should utilize components of humane education to increase students' enjoyment and, therefore, their intrinsic motivation. While the interaction of various factors impacted the degree to which students were motivated and engaged by humane education, all of the case study participants perceived humane education to be positive.

21.7 Conclusions

In summary, findings from the current study largely supported humane education's motivational effects on children. The purpose of this phenomenological case study was to explore teacher perceptions regarding how humane education curriculum influenced student motivation and engagement. In all of the cases, teachers perceived that when topics were tailored to include animal welfare, the educators noticed enhanced student interest.

The findings from this study revealed that teachers believed humane education could effectively increase student motivation and engagement. Culture, gender, and school leadership should be considered when teachers design and deliver the curriculum. Educators agreed that a variety of methods to deliver humane education should be used.

It should be noted that there are several limitations to the study reported here. A possible limitation was that participants' responses might have been influenced by a desire to advocate for the benefits of humane education. The lived experiences might not have been accurately conveyed by the participants. Additionally, the analysis of themes was subjective, given possible biases in the interpretation. There was an assumed gender bias in the interview question about boys' and girls' reactions to humane education. The number of teachers who teach humane education is limited compared to the general education teaching population. Small sample size was a limitation; however, the in-depth qualitative information provided can be a springboard for future research questions in humane education that can include larger sample sizes for increased generalization. The study did not include interviews with teachers located outside the United States, and therefore, it is not possible to determine

[29] Signal, Arbour, and Taylor, "Teaching Kindness," 136.

whether perceptions about student motivation and humane education are impacted by international differences. Further limitations of the study have been outlined elsewhere.[30]

This research served to broaden the knowledge base surrounding humane education, allowing educators to structure a better program meant to motivate and engage students. Future researchers might improve upon this study by selecting students to examine their perspectives on humane education and elicit their perceptions of their own motivation and engagement. Additionally, given the numerous teachers and animal protection organizations who develop their own humane education lessons, researchers should examine a particularly motivating curriculum to identify its specific components and define which aspects of its variables are the most effective. New investigations should examine larger samples of teachers of both binary genders, additional grade levels, and diverse socioeconomic schools. If the study is conducted over a longer period of time, teachers will be able to produce more artifacts to triangulate future case study results. Further research exploring humane education could potentially provide some useful insights into the ways in which the curriculum can develop student curiosity and interest, making the program a more desirable choice for educators.

Appendix: Interview Protocol Instrument[31]

The interview protocol consisted of ten questions. The themes that arose per participant per question number are noted in this table.

[30] O'Connor, "A Qualitative Case Study of Teacher Perceptions."

[31] O'Connor, "A Qualitative Case Study of Teacher Perceptions."

Participants	*1*	*2*	*3*	*4*	*5*	*6*	*7*	*8*
Interview questions								
1. What is your experience teaching?	T6			T1, T4	T1		T1	T1
2. What does humane education mean to you?	T1			T1		T5	T1	
3. Please describe your experience with humane education	T1, T6	T5		T2, T3		T1, T2, T3	T2, T3, T5	T6
4. How do your students respond to your lessons including humane education?	T5	T4		T1	T2, T5		T3	T2
5. What differences have you noticed when you teach lessons including humane education as opposed to lessons not including humane education elements?	T2		T6		T3		T1	T2, T3
6. How do your students respond to different animal issues arising in your curriculum?	T3, T5	T3	T3	T6	T2	T2		T2
7. How do your boys respond differently than your girls when humane education is used?	T5	T5		T5	T4, T5	T4, T5	T5	T5
8. In general, how can you determine if your students are motivated to learn by humane education?	T4	T4			T1	T4		
9. What are some of the teaching techniques teachers should use to deliver humane education to effectively motivate and engage students?	T3	T1	T1, T3	T1	T1, T2, T6	T1, T2, T6	T1, T6	T6

(continued)

Participants	*1*	*2*	*3*	*4*	*5*	*6*	*7*	*8*
10. Thinking of the options available within humane education, what specific aspects of humane education motivate and engage students?	T5	T3	T3	T2, T3	T1			T4, T6

Key: Theme 1 (T1) = Teaching Techniques; Theme 2 (T2) = Safe Topics and Species; Theme 3 (T3) = Food and Animals on Farms; Theme 4 (T4) = Student and Teacher Connection
Theme 5 (T5) = Age, Gender, and Culture of Students; Theme 6 (T6) = Administrator, Colleague, and Parent Reaction
Participant 1 = Elementary/middle school health and physical education instructor (public school)
Participant 2 = Elementary all-subjects instructor (public school)
Participant 3 = Elementary, high school, and prison instructor (public school)
Participant 4 = Elementary/middle school psychologist and humane education elective class instructor (private school)
Participant 5 = Elementary all-subjects instructor (public school)
Participant 6 = Elementary primary-grade all-subjects instructor, former science district coordinator (public school)
Participant 7 = High school English, history, and business instructor (public, vocational school)
Participant 8 = Elementary primary-grade all-subjects instructor (private school)

Bibliography

Academy of Prosocial Learning. "Humane Education Training: Academy of Prosocial Learning." APSL. Accessed August 5, 2019. https://www.prosocialacademy.org/.

Akhtar, Aysha, and Carl Safina. *Our Symphony with Animals: On Health, Empathy, and Our Shared Destinies*. New York: Pegasus Books, 2019.

Barraza, Laura. "Attitudes to Animal Dilemmas: An Exploratory Comparison between Mexican and English Children." *International Electronic Journal of Environmental Education* 5, no. 2 (2015): 40–61. https://doi.org/10.18497/iejee-green.64245.

Beirne, Piers, and Meena Alagappan. "A Note on Pedagogy: Humane Education Making a Difference." *Journal for Critical Animal Studies* 5, no. 2 (2007).

Bloomberg, Linda Dale, and Marie Volpe. *Completing Your Qualitative Dissertation: A Road Map from Beginning to End*. 3rd ed. Thousand Oaks: SAGE Publications, 2016.

Bone, Jane. "The Animal as Fourth Educator: A Literature Review of Animals and Young Children in Pedagogical Relationships." *Australasian Journal of Early Childhood* 38, no. 2 (2013): 57–64. https://doi.org/10.1177/183693911303800208.

Cheon, Sung Hyeon, and John Marshall Reeve. "A Classroom-Based Intervention to Help Teachers Decrease Students' Amotivation." *Contemporary Educational Psychology* 40 (2015): 99–111. https://doi.org/10.1016/j.cedpsych.2014.06.004.

Coker, Jeffrey, and Desiree Porter. "Student Motivations and Perceptions across and within Five Forms of Experiential Learning." *The Journal of General Education* 65, no. 2 (2017): 138–56. https://doi.org/10.5325/jgeneeduc.65.2.0138.

Daly, Beth, and Suzanne Suggs. "Teachers' Experiences with Humane Education and Animals in the Elementary Classroom: Implications for Empathy Development." *Journal of Moral Education* 39, no. 1 (2010): 101–12. https://doi.org/10.1080/03057240903528733.

Drissner, Jürgen, Hans-Martin Haase, and Katrin Hille. "Short-Term Environmental Education—Does It Work? An Evaluation of the 'Green Classroom.'" *Journal of Biological Education* 44, no. 4 (2010): 149–55. https://doi.org/10.1080/00219266.2010.9656215.

Durlak, Joseph A., Roger P. Weissberg, Allison B. Dymnicki, Rebecca D. Taylor, and Kriston B. Schellinger. "The Impact of Enhancing Students' Social and Emotional Learning: A Meta-Analysis of School-Based Universal Interventions." *Child Development* 82, no. 1 (2011): 405–32. https://doi.org/10.1111/j.1467-8624.2010.01564.x.

Engels, Maaike C., Hilde Colpin, Karla Van Leeuwen, Patricia Bijttebier, Wim Van Den Noortgate, Stephan Claes, Luc Goossens, and Karine Verschueren.

"Behavioral Engagement, Peer Status, and Teacher–Student Relationships in Adolescence: A Longitudinal Study on Reciprocal Influences." *Journal of Youth and Adolescence* 45, no. 6 (December 2016): 1192–207. https://doi.org/10.1007/s10964-016-0414-5.

Faver, Catherine A. "School-Based Humane Education as a Strategy to Prevent Violence: Review and Recommendations." *Children and Youth Services Review* 32, no. 3 (2010): 365–70. https://doi.org/10.1016/j.childyouth.2009.10.006.

Fenstermacher, Gary D. "A Philosophical Consideration of Recent Research on Teacher Effectiveness." *Review of Research in Education* 6 (January 1, 1978): 157–85. https://doi.org/10.2307/1167245.

Francis, Marj, Kathy Paige, and David Lloyd. "Middle Years Students' Experiences in Nature: A Case Study on Nature-Play." *Teaching Science* 59, no. 2 (June/July 2013): 20–30. https://search.informit.com.au/documentSummary;dn=687559195442456;res=IELHSS.

Goddard, Anna, and Mary Gilmer. "The Role and Impact of Animals with Pediatric Patients." *Pediatric Nursing* 41, no. 2 (March/April 2015): 65–71. https://pdfs.semanticscholar.org/fd07/5d7ea6121852bebeeb2ffa8b0a56cbb72fa0.pdf?_ga=2.146038003.679191520.1565094097-521204368.1565094097.

Gunduz, Nuket, and Cigdem Hursen. "Constructivism in Teaching and Learning; Content Analysis Evaluation." *Procedia—Social and Behavioral Sciences* 191 (2015): 526–33. https://doi.org/10.1016/j.sbspro.2015.04.640.

Helker, Kerstin, and Marold Wosnitza. "The Interplay of Students' and Parents' Responsibility Judgements in the School Context and Their Associations with Student Motivation and Achievement." *International Journal of Educational Research* 76 (2016): 34–49. https://doi.org/10.1016/j.ijer.2016.01.001.

Hidi, Suzanne, and Judith M. Harackiewicz. "Motivating the Academically Unmotivated: A Critical Issue for the 21st Century." *Review of Educational Research* 70, no. 2 (2000): 151–79. https://doi.org/10.2307/1170660.

Institute for Humane Education. Home page. March 11, 2019. Accessed August 6, 2019. https://humaneeducation.org/.

Legault, Lisa, Isabelle Green-Demers, and Luc Pelletier. "Why Do High School Students Lack Motivation in the Classroom? Toward an Understanding of Academic Amotivation and the Role of Social Support." *Journal of Educational Psychology* 98, no. 3 (2006): 567–82. https://doi.org/10.1037/0022-0663.98.3.567.

Mega, Carolina, Lucia Ronconi, and Rossana De Beni. "What Makes a Good Student? How Emotions, Self-Regulated Learning, and Motivation Contribute to Academic Achievement." *Journal of Educational Psychology* 106, no. 1 (2014): 121–31. https://doi.org/10.1037/a0033546.

Migliarese, Nicole. "Researching the Child–Nature Connection." California State Parks' Children in Nature Campaign white paper, March 2008. https://www.parks.ca.gov/pages/24914/files/microsoft%20word%20-%20copy%20of%20white%20paper%20for%20kids%20and%20nature%20campaign%20ready%20to%20pub2%20single%20space.pdf.
O'Connor, Julie. "A Qualitative Case Study of Teacher Perceptions of the Motivation of Students in Humane Education." PhD diss., Northcentral University, 2018. ProQuest.
Pajares, Frank. "Teachers' Beliefs and Educational Research: Cleaning Up a Messy Construct." *Review of Educational Research* 62, no. 3 (1992): 307–32. https://doi.org/10.2307/1170741.
Ross, Mitchell, Helen Perkins, and Kelli Bodey. "Academic Motivation and Information Literacy Self-Efficacy: The Importance of a Simple Desire to Know." *Library & Information Science Research* 38, no. 1 (2016): 2–9. https://doi.org/10.1016/j.lisr.2016.01.002.
Sharma, Rajesh Kumar. "Constructivism—An Approach to Enhance Participatory Teaching Learning." *GYANODAYA—The Journal of Progressive Education* 7, no. 2 (2014): 13–17. https://doi.org/10.5958/2229-4422.2014.00003.6.
Signal, Tania, R. Arbour, and Nik Taylor. "Teaching Kindness: The Promise of Humane Education." *Society & Animals* 17, no. 2 (2009): 136–48. https://doi.org/10.1163/156853009x418073.
Usher, Alexandra. "What Nontraditional Approaches Can Motivate Unenthusiastic Students?" Washington, DC: Center on Education Policy, 2012.
Weil, Zoe. *Power and Promise of Humane Education*. Gabriola Island, BC: New Society Publishers, 2013.
Yin, Robert K. *Case Study Research: Design and Methods*. London: Sage Publications, 2014.

CHAPTER 22

The Five Freedoms for Animals as a Basic Learning Requirement for All Young Learners

Louise van der Merwe

Topic of Research Project

The Five Freedoms for Animals, as endorsed by the World Organisation for Animal Health (WOAH), should be incorporated in the South African National Schools Curriculum as a fundamental learning requirement for all young learners along with their ABCs and 123s.

L. van der Merwe (✉)
The Humane Education Trust, Cape Town, South Africa
e-mail: avoice@yebo.co.za

C. Linzey and A. Linzey (eds.), *The Palgrave Companion to Humane Education*, The Palgrave Macmillan Animal Ethics Series,
https://doi.org/10.1007/978-3-032-19795-5_22

22.1 Relevant Background Information

Environmental sustainability is entrenched in the South African Constitution, which in turn underpins and is reflected in the national schools curriculum.

In 2001 and 2002, with South African struggle icon Kader Asmal as Minister of Education, a decision was taken to revise the national schools curriculum with a view to the inclusion of environmental education in the curriculum. This process was known as the National Environmental Education Project, or NEEP. The Humane Education Trust (HET) was invited by the Department of Education to participate in NEEP and assist with the incorporation of animal welfare as part of environmental education. The revised curriculum subsequently came into being in 2012 and included multiple opportunities for the teaching of animal welfare. Environmental education[1] (which includes animal welfare/humane education) is integrated across the South African curriculum for Foundation and Intermediate Phase learners, as part of the compulsory learning areas of Life Skills and Life Orientation.[2] According to the most recent government census statistics, around 14.5 million learners attended the 24,900 schools in 2022.[3] The number of teachers is recorded at 450,993.[4]

However, since 2012, it has become apparent that teachers often are not equipped with enough background knowledge about animal welfare issues to teach them effectively. This is exacerbated by the following factors:

[1] Humane Education News, "Humane Education to be Included in New National Schools Curriculum," *Humane Education News* (Spring 2002), 2, https://view.publitas.com/p222-12894/humane-edu-news-spring-2002/page/2-3.

[2] Matseliso Lineo Mokhele, "Integrated Environmental Teaching in South Africa," *Perspectives in Education*, 29, no. 4 (December 2011): 78–86.

[3] South African Government, "Education," *South African Government*, accessed September 20, 2024, https://www.gov.za/about-sa/education; Statista, "Number of Schools in South Africa in 2022, by Sector," *Statista*, accessed September 20, 2024, https://www.statista.com/statistics/1262871/number-of-schools-in-south-africa-by-sector/.

[4] Statista, "Number of Teachers in Education in South Africa in 2022, by Province," *Statista*, accessed September 20, 2024, https://www.statista.com/statistics/1262709/number-of-teachers-in-education-in-south-africa-by-province/.

- A Westernized education system that, for generations, has stifled the development of emotional intelligence (EI) because EI has been associated with illogicality and weakness, in contrast to subjects such as science, technology, mathematics, and, more recently, robotics and coding.
- The generally prevailing societal sense that animals exist for human use and are undeserving of respect in their own right.
- The prevailing generalized belief that compassion toward animals is an inappropriate "white" or "Western" indulgence in a country where there has been and continues to be so much human suffering.

The United Nations' development goals[5] state that by 2030 all learners should acquire the knowledge and skills needed to promote sustainable development, including Sustainable Development Goal (SDG) 4.7 (a culture of peace and nonviolence) and SDG 12.8 (a lifestyle in harmony with nature). Furthermore, on September 18, 2023, the United Nations officially adopted an addition to the Charter on the Rights of the Child and decreed that children are to be protected from all forms of violence—be it domestic, societal, or inflicted on animals. In Section GC26:35, the UN decrees that "children are to be protected from all forms of physical and psychological violence, whether in their home or in society, and from exposure to violence, such as domestic violence or violence inflicted on animals."[6]

The Humane Education Trust is of the view that sustainable development will simply never be achieved unless the skills inherent in our understanding of modern EI education—namely, empathy and compassion in a post-anthropocentric world—are nurtured and developed to become vital components in all decision-making.

[5] United Nations, "The 2030 Agenda for Sustainable Development," *United Nations*, 2015, https://sustainabledevelopment.un.org/content/documents/21252030%20Agenda%20for%20Sustainable%20Development%20web.pdf; See also UNESCO, "Education 2030: Global Education Cooperation Mechanism," *UNESCO*, https://sdg4education2030.org/the-goal.

[6] United Nations, "General Comment No. 26 (2023) on Children's Rights and the Environment, with a Special Focus on Climate Change," Convention on the Rights of the Child, *United Nations*, August 22, 2023, GC26:35, accessed September 20, 2024, https://www.naturebased.education/_files/ugd/22c4ff_fd316d78c25548c3ad03de570964192e.pdf.

Post-democracy South Africa is experiencing unprecedented violence. The scourge of brutality, including domestic violence, rape, child rape, and murder, as well as animal abuse, is pervasive across the country.[7] The seminal resource book on intersections of violence, *Child Abuse, Domestic Violence, and Animal Abuse: Linking the Circles of Compassion for Prevention and Intervention*, features strongly in all our resources for teachers.[8] Coauthor of the book Phil Arkow, who chaired the Animal Abuse and Family Violence Prevention Project at the Latham Foundation in the United States, explains: "The abuse of animals is often the first step on the slippery slope of desensitization, the first step down that slope of a lack of empathy and violence. All too often animals are the first victims and what should be seen as a red flag or warning marker is readily dismissed by parents and teachers as oh well, boys will be boys, or it's only a rabbit, what's the big deal?"[9]

Against this background, and in light of the intersection of domestic violence, child abuse, and animal abuse, HET embarked on a pilot research project, as set out in the following pages, to assess the impact on learners of teaching the most basic of animal welfare principles—namely, "The Five Freedoms for Animals." The Five Freedoms for Animals, as endorsed by the World Organisation for Animal Health (WOAH) and by many governments internationally, including the South African government, are as follows:

- Freedom from hunger and thirst.
- Freedom from discomfort.
- Freedom from pain, injury, and disease.

[7] Marianne Merten, "Official Statistics Prove War on Women Is Real and Pretty Words Are Mere Lip Service," *Daily Maverick*, September 4, 2019, https://www.dailymaverick.co.za/article/2019-09-04-official-statistics-prove-war-on-women-is-real-and-pretty-words-are-mere-lip-service/.

[8] Frank Ascione and Phil Arkow, *Child Abuse, Domestic Violence, and Animal Abuse: Linking the Circles of Compassion for Prevention and Intervention* (West Lafayette, IN: Purdue University Press, 1999).

[9] Phil Arkow quoted in Humane Education News, "The All-Africa Humane Education Summit 2003. Spotlight on What the Speakers Said," *Humane Education News* (Summer 2003–2004), 5, https://view.publitas.com/p222-12894/humane-edu-news-summer-2003-2004/page/4-5.

- Freedom to express normal behavior.
- Freedom from fear and distress.[10]

The HET's pilot project was underpinned by a landmark judgment in South Africa's Constitutional Court,[11] which ruled as follows:

- Animals have intrinsic value as individuals.
- Animals are sentient beings capable of suffering and experiencing pain.
- Guardianship of the interests of animals reflects constitutional values and the interests of society at large.
- Animal protection safeguards the moral status of humans and the degeneration of human values.

The Department of Education has given support to values-driven learning. As head of the Western Cape Education Department's Safe School Programme, Eugene Daniels stated: "Our Safe Schools Programme is not only about the elimination of crime and violence in the school environment, it's about values. You can't actually address crime and violence unless you look into the hearts and minds of people."[12]

The HET contends that learning about the Five Freedoms for Animals as part of the learning area known as "Life Skills," in the foundation phase of schooling (ages 5–9), entrenches basic values such as ethics, integrity, and empathy into learner perceptions. Although the focus in the teaching of the Five Freedoms in the foundation phase is on companion animals, this focus forms fertile groundwork for the tackling of wider animal welfare issues in the intermediate phase (ages 10–12) and the senior

[10] Humane Education Trust, "The Five Freedoms for Animals," *Nature Based Education*, accessed September 20, 2024, https://www.naturebased.education/new-5-freedoms.

[11] National Society for the Prevention of Cruelty to Animals v. Minister of Justice and Constitutional Development and Another (CCT1/16), [2016] ZACC 46, http://www.saflii.org/za/cases/ZACC/2016/46.html.

[12] Eugene Daniels quoted in Animal Voice, "South Africa Takes the World Lead in Teaching," *Animal Voice* (May-August, 2000), 2, https://d91fe348-4a25-4dc0-a30d-96976402b0e1.filesusr.com/ugd/186d0e_79e7c160581545a9b503a135edf71448.pdf; See also Humane Education Supplement, "Psychological Evaluation of Humane Education in SA Schools Is Believed to Be a World First," *Humane Education Supplement* (September 2000), 1, https://view.publitas.com/p222-12894/humane-education-sep-2000/page/1.

phase (ages 13–17). In connection with this, the HET has developed curriculum-aligned multilingual resources, including teachers' guides to humane education, for the Life Skills and Life Orientation learning areas in all three phases.

Inclusion of the Five Freedoms for Animals in education is all the more urgent and imperative as the world enters an age where anthropocentric learning[13] is at odds with our new understanding of our fragile world. In his book *Earth Emotions*, geoscientist and eco-philosopher Glenn Albrecht writes that the tragedy is that as humans become more and more comfortable with artificial intelligence and robotic companions, each subsequent generation separates further from nature, widening the gulf, so that an impoverished nature becomes accepted as the norm. Little wonder then, he says, that children grow up with no understanding or empathy for the natural world, nor an understanding of their place in it. He says this is why education urgently needs to equip schoolchildren with knowledge of the science of symbiosis, making them the first generation to walk out of the Anthropocene (age of man) epoch and into the Symbiocene (the interconnectedness of life) epoch.[14]

Daniel Goleman, former senior editor at *Psychology Today*, is one of many contemporary thinkers exploring the immense social cost of being "emotionally tone deaf."[15] In his book *Emotional Intelligence: Why It Can Matter More than IQ*, page 19, Goleman says, "There are times when the fabric of society seems to unravel at ever-greater speed, when selfishness, violence, and a meanness of spirit seem to be rotting the goodness of our communal lives."[16] He says there is a worldwide trend for the present generation of children to be more troubled emotionally than the last: more lonely and depressed, more angry and unruly, more nervous and prone to worry, more impulsive and aggressive. "If there is a remedy," he says, "I feel it must lie in how we prepare our young for life. At present

[13] Karin Murris, *The Posthuman Child* (London: Routledge, 2016).

[14] Glenn Albrecht, *Earth Emotions* (Ithaca: Cornell University Press, 2018).

[15] Daniel Goleman, *Emotional Intelligence: Why It Can Matter More than IQ* (London: Bloomsbury, 1995).

[16] Goleman, *Emotional Intelligence*, 19.

we leave the emotional education of our children to chance. One solution is a new vision of what schools can do to educate the whole student, bringing together mind and heart in the classroom."[17]

22.2 The HET'S First Pilot Research Project Based on the Five Freedoms for Animals (2017–2018)

In 2017, the HET, in cooperation with the principal of Forest Heights Primary School, Mrs. Anthea Bosman, provided resources to enable a class of grade 3 learners at this school to receive a comprehensive understanding of the Five Freedoms for Animals.

The grade 3 class comprised forty-two learners, both boys and girls, and the average age was nine years. Situated about forty kilometers outside Cape Town, the school serves a community with socioeconomic challenges that include poverty, unemployment, domestic violence, depression, drug addiction, suicide among young adults, environmental degradation, and animal neglect and cruelty—including organized dogfighting.

Synopsis of the Method of Teaching the Five Freedoms for Animals

The school year in South Africa is divided into four terms. At the beginning of the first term of 2018, the following resources were supplied to the class teacher, Mrs. Vivienne Rutgers:

- A board game using felt cutouts and a book titled *Caring Kids*, comprising six stories on various aspects of animal welfare, to stimulate discussion on responsible guardianship of companion animals and animal welfare issues in general.
- A variety of posters depicting the Five Freedoms for Animals to stimulate discussion.
- A puppet play script titled *The Five Freedoms for Animals*, dealing with fairness and justice toward all living beings (duration of puppet play: seven minutes).
- Hand puppets representing characters in the puppet play.

[17] Goleman, *Emotional Intelligence*, 21.

Over a period of three terms (ten weeks per term), at a rate of two hours and fifty-five minutes per week (five 35-minute Life Skills lessons per five-day week), the pupils were taught the Five Freedoms program using the aforementioned resources, in order to accomplish the following:

- Familiarize the learners with various aspects of animal welfare issues.
- Familiarize the learners with the concepts of the Five Freedoms.
- Ensure learners' verbal and written understanding of these concepts.
- Ensure that the learners can read the script of the puppet play with fluency and understanding.
- Ensure that each learner attained proficiency in each of the various roles in the play (ten character roles in all).
- Ensure that learners understand the meaning of *sentience* as something shared by humans and animals, with specific attention to the power of the words that we unconsciously use to minimize, commodify, and denigrate not only animals but also each other and that underpin bullying behavior.

By the end of the third term, the learners showed significant proficiency in their understanding of the Five Freedoms and in their performance of the puppet play.[18]

Follow-Up on the HET'S First Pilot Research Project (2017–2018) After One Year

Of significance, one year later, although these learners had progressed to Grade 4 in that year, and had not received any further instruction on the Five Freedoms, they all retained significant memory of the Five Freedoms as demonstrated by the results of the following questionnaire:

[18] Humane Education, "Five Freedoms Dog Puppet Play," *YouTube*, March 1, 2023, https://www.youtube.com/watch?v=h_0HW79OQDc. See also Animal Voice, "Calling all Life Skills and Life Orientation Teachers," *Animal Voice* (April 2018), 18–19, https://d91fe348-4a25-4dc0-a30d-96976402b0e1.filesusr.com/ugd/186d0e_af536943197f4fe8a5f3ecd4fa6c2eed.pdf.

Question 1

"When you think back on the Five Freedoms for Animals, which freedom is the first that comes to your mind?"

Of the forty-two learners, thirty-four replied, "Freedom from hunger and thirst"; six replied, "Freedoms from pain, injury and disease"; and two replied, "Freedom from discomfort."

Question 2

"Do you think it is important for learners to know about the Five Freedoms for Animals? Please give a reason for your answer."

Of the forty-two learners, 100% replied in the affirmative. Reasons included the following:

- "Because the Five Freedoms are something we need to know."
- "Because if you help animals you will become very kind and help people also."
- "Because the Five Freedoms make me feel like a hero."
- "Because the Five Freedoms make us better people."
- "Because I know now how to treat my dog."
- "Because the Five Freedoms make you into a champion."
- "Because people will become kind."
- "Because some people will become good from this learning."
- "Because now I can teach my uncle about the Five Freedoms when he beats his donkey."
- "Because I love the puppet show and everyone says I should be an actress when I grow up."

22.3 The HET'S Second Pilot Research Project Based on the Five Freedoms for Animals (2019)

On the strength of the success of the HET's 2017–2018 pilot project, the HET was fortunate enough to be given a grant by the Latham Foundation to widen the focus of this project. The US-based Latham Foundation is a leading proponent internationally of the imperative to make the skills of empathy, kindness, and compassion central to education. Latham states that kindness to animals is the first and truly foundational step that supports all the other steps above it in the building of compassionate and empathic communities, countries, and nations and a new world ethos.[19]

Initially, all 457 learners in the foundation phase of Forest Heights Primary School participated in the entry-level questionnaire. The foundation phase includes grades R–3, ages 5–9. However, despite reassurances from the teachers concerned, only three classes of forty-six learners each consistently incorporated the program for a full year, significantly reducing our database for this research.

Results of Entry-Level Questionnaires for the Learners in the HET's 2019 Pilot Project

The first entry-level question for all 457 learners was "Do animals have feelings?" Fifty-five percent said yes; 45 percent said no. The second entry-level question for all 457 learners was "Do you think your companion animal is happy?" Fifty-eight percent said yes; 42 percent said no.

Follow-Up on the HET'S 2019 Pilot Project After Three Terms

At the time of this chapter's writing, only forty-six learners in one grade 2 class, with an average age of eight years, were ready for a follow-up questionnaire. Some were absent. They were asked whether learning about the freedoms had made a difference in the way they relate to animals. Their replies, set out in the following list, were taken individually and not in the presence of peers.

- "I look to see if my dogs have clean water and fresh food before I go to school."
- "I know now that animals have feelings. I try to treat them with respect."
- "I know my puppy cares for my whole family. If I fall he licks me. I love and respect all animals."
- "I know my dog gets hungry too so I make sure there is enough food for my dog."

[19] Latham Foundation, "General Information," *Latham Foundation for the Promotion of Humane Education*, accessed September 20, 2024, https://www.latham.org/grants/general-information/.

- "When my dog whimpers I see he looks sad and his tail is between his legs. He is showing me he is unhappy or hungry."
- "I will not hurt my dog anymore."
- "My dog doesn't like to be inside all the time; he wants to play with a tennis ball. He shows normal behavior. I am his guardian."
- "I allow my dog to jump up on me when I come from school. I know he expresses normal behavior."
- "I am a better guardian."
- "I don't tease my dog to make him angry anymore."
- "If I get a new cat I will give him a good life this time. I am more responsible."
- "I take my dog to the mobile vet because he also needs care."
- "I won't leave the gate open because my dog can run away or get knocked down. I will be responsible."
- "I know my cat also has feelings and so I see that her basket and blanket are clean and her bedding washed. Now she is more comfortable."
- "If my dog looks scared when I get home I pick him up and cuddle him because he is still a puppy."
- "I won't tease my cat anymore."
- "I whistle and then the cows come back. If one is lost I go with my uncle to look for him."
- "I saw an injured dog and I told his guardian to take him to the vet. I know my parents are my guardians."
- "I tell my friends not to tease my dog because then he will bite them because he is scared."
- "I look for ticks and fleas and take my dog to the veterinarian."
- "When I saw a stray kitten, I took it home and fed it and I will find a nice home for it."
- "I am kinder to my dog."
- "My dog feels scared when children throw stones at her. I'm kinder now."
- "Someone hit my dog with a stone. I went to tell his mom. My dog was bleeding and we took him to the mobile vet."
- "I see that animals are cute and can be happy. I understand that they need to be free."
- "My dog jumps on me and licks my face. I allow him to jump for joy."

- "When someone teases my cat I don't laugh at them. I tell them that my cat is scared of them."
- "My dog used to be on a short chain so my dad bought a long chain. I run with my dog for exercise."
- "I used not to give my dog food in the morning. Now I do. I also look to see if he has clean water."
- "I tell my friends about the Freedom from pain, injury and disease and that they must not throw stones at animals."
- "I allow my dog to jump on me and be joyful."
- "If I get a new dog I will let him come inside this time."
- "I used to hit my dogs when they fought for food in the bowl. Now I put their food into different bowls."
- "I don't tease my dog anymore because he gets cross."
- "Before I go to bed I check if my dog is in his kennel."
- "I know that when my dog keeps on barking he is being teased and sometimes he is feeling pain."
- "I tell my cousins about the Freedoms and how they must respect animals."
- "I know my dog has feelings. When his tail is between his legs, it shows he is scared and when his tail wags he is happy."
- "I won't hurt animals. I won't tie any dogs on a short chain. Now I know that when my puppy chases me to play, it is normal behavior."
- "I will not tease animals or throw stones at them."
- "Now I don't throw stones at dogs anymore."

Empirical Evidence of Development of Empathy

Vivienne Rutgers, supervising and co-teaching the program in the Grade 2 class of forty-six learners, said the pilot project had achieved a breakthrough in education in the sense that the learners began to show the early manifestations of empathy for nonhuman beings. She reported to the Humane Education Trust:

> An example is the discussion I overheard in the playground recently. A dove was sun-bathing nearby, extending its wing to catch the rays of warmth. Two nine-year-old boys saw this and the one said to the other: "See! That's Freedom from Pain, Injury and Disease." "No," argued the other. "That's Freedom to Express Normal Behaviour." They amicably

argued the point. Our Five Freedoms pilot project had opened a new window of curiosity for them that would otherwise have remained closed. It was really heart-warming. Without the programme, they probably would not have even noticed the bird or, based on my experience, they may well have decided to chase it away by throwing stones at it.

On another occasion recently, we unlocked the classroom for the day's schoolwork to begin, only to find a little bird had somehow been shut in all night. It perched on the rail above one of the windows and was clearly traumatised by the sudden surge of 46 children into the room. Before the introduction of the Caring Classrooms programme, I believe the children would have caused pandemonium trying to catch the bird. Instead, they heard me... I said "Sit quietly, I'll open the window, and we will be patient while the bird finds its way out." The little bird took a significant amount of time to do this but as it hopped nearer to the open window, one child whispered "nearly there," as encouragement to the bird. To hear that whisper, all 46 children had to have been as quiet as mice. Then several of the children whispered "nearly there" with every hop the bird took in the right direction. And we all whooped when it finally flew out. That "nearly there" whisper of encouragement was empathy in action across species, and I hadn't seen it before.[20]

22.4 Conclusion: Reforming Education to Produce First-Class Humans

Education reform to include the development of EI and its core component of empathy must become a priority and be placed at the core of Life Skills and Life Orientation curricula as a matter of urgency. German data scientist Andreas Schleicher, head of the education division at the Organisation for Economic Cooperation and Development, believes our education system is a relic of the industrial age. He explains: "The kinds of things that are easy to teach, and maybe easy to test, are precisely the kinds of things that are easy to digitize and to automate. The advent of AI should push us to think harder about what makes us human—our

[20] Vivienne Rutgers quoted in Animal Voice, "Can Empathy be Taught?" *Animal Voice's Education Rewired* (July 2019), 12–13, https://33742caf-bd46-4a0b-b7fa-4aa628 32f1ef.usrfiles.com/ugd/22c4ff_a7827e6481bc42a3aa7fc6faee958551.pdf.

capacity for empathy and creativity."[21] If we do not, he says, the world will be educating "second-class robots and not first-class humans."[22]

Simon Baron-Cohen, Professor of developmental psychopathology at the University of Cambridge, agrees. He says there is growing evidence for the argument "that instead of using the term evil, we should talk about reduced (or even absent) empathy."[23] He continues: "The critical role of empathy in our society has been overlooked. Empathy is the most valuable social resource in our world.... It is puzzling that in school or parenting curricula empathy figures hardly at all, and in politics, business, the courts, or policing, it is rarely, if ever, on the agenda. The erosion of empathy is a critical global issue of our time."[24]

The HET contends that ours is the first era that cannot escape the consequences of our foolhardy neglect of emotional intelligence, seen everywhere in our callous and reckless abuse of our environment and the animals who are part of the living world. Our own road to healing starts with those who are at our mercy—the animals.

The learners' responses in the HET's pilot projects demonstrate that learning the Five Freedoms for Animals is a positive experience and reinforces the learners' sense of self-esteem, self-reliance, and assertiveness and instills in them a sense of responsibility for the world in which we all live, instilling goodwill and hope for a kinder future. *However, the Five Freedoms program requires support and political will from the Department of Education in the form of teacher workshops to assist teachers to prepare for and embrace a post-Anthropocene era of education.*

The foundation phase, those years in a child's life where neuroscience shows humans' brains to be the most malleable, is a crucial window of opportunity to teach the basic principles of ethics and integrity so that they become second nature. The Five Freedoms for Animals, as endorsed

[21] Andreas Schleicher, "How Changing the PISA Test Could Change How Kids Learn," *Quartz*, February 22, 2019, https://qz.com/1540222/how-changing-the-pisa-test-could-change-how-kids-learn/.

[22] Schleicher, "How Changing the PISA Test Could Change How Kids Learn."

[23] Simon Baron-Cohen, *The Science of Evil* (New York: Basic Books, 2011), quoted in *Animal Voice*, June 2012: 8–9, https://d91fe348-4a25-4dc0-a30d-96976402b0e1.filesusr.com/ugd/186d0e_5c53d7b34f8a4e529e9e04981b6632b7.pdf.

[24] Baron-Cohen, *The Science of Evil*, quoted in *Animal Voice*, December 2013: 9, https://d91fe348-4a25-4dc0-a30d-96976402b0e1.filesusr.com/ugd/186d0e_482d31be185843b8b7e0ab97f004f787.pdf.

by the World Organisation for Animal Health, is an essential component of early childhood development in schools and educational institutions the world over, a basic learning requirement for all young learners, along with their 123s and ABCs.

Bibliography

Albrecht, Glenn. 2018. *Earth Emotions*. Ithaca: Cornell University Press.

Animal Voice. 2018. "Calling all Life Skills and Life Orientation Teachers." *Animal Voice*, April. https://d91fe348-4a25-4dc0-a30d-96976402b0e1.filesusr.com/ugd/186d0e_af536943197f4fe8a5f3ecd4fa6c2eed.pdf.

Animal Voice. 2019. "Can Empathy Be Taught?" *Animal Voice's Education Rewired*, July. https://33742caf-bd46-4a0b-b7fa-4aa62832f1ef.usrfiles.com/ugd/22c4ff_a7827e6481bc42a3aa7fc6faee958551.pdf.

Animal Voice. 2000. "South Africa Takes the World Lead in Teaching." *Animal Voice*, May–August. https://d91fe348-4a25-4dc0-a30d-96976402b0e1.filesusr.com/ugd/186d0e_79e7c160581545a9b503a135edf71448.pdf

Ascione, Frank, and Phil Arkow. 1999. *Child Abuse, Domestic Violence, and Animal Abuse: Linking the Circles of Compassion for Prevention and Intervention*. West Lafayette, IN: Purdue University Press.

Baron-Cohen, Simon. 2011. *The Science of Evil*. New York: Basic Books.

Goleman, Daniel. 1995. *Emotional Intelligence: Why It Can Matter More Than IQ*. London: Bloomsbury.

Humane Education. 2023. "Five Freedoms Dog Puppet Play." *YouTube*, March 1. https://www.youtube.com/watch?v=h_0HW79OQDc.

Humane Education News. 2002. "Humane Education to Be Included in New National Schools Curriculum." *Humane Education News*, Spring. https://view.publitas.com/p222-12894/humane-edu-news-spring-2002/page/2-3.

Humane Education News. 2003. "The All-Africa Humane Education Summit 2003. Spotlight on What the Speakers Said." *Humane Education News*, Summer 2003–2004. https://view.publitas.com/p222-12894/humane-edu-news-summer-2003-2004/page/4-5.

Humane Education Supplement. 2000. "Psychological Evaluation of Humane Education in SA Schools Is Believed to Be a World First." *Humane Education Supplement*, September. https://view.publitas.com/p222-12894/humane-education-sep-2000/page/1.

Humane Education Trust. "The Five Freedoms for Animals." *Nature Based Education*. Accessed September 20, 2024. https://www.naturebased.education/new-5-freedoms.

Latham Foundation. "General Information." *Latham Foundation for the Promotion of Humane Education*. Accessed Spetmeber 20, 2024. https://www.latham.org/grants/general-information/.

Merten, Marianne. "Official Statistics Prove War on Women Is Real and Pretty Words Are Mere Lip Service." *Daily Maverick*, September 4, 2019. https://www.dailymaverick.co.za/article/2019-09-04-official-statistics-prove-war-on-women-is-real-and-pretty-words-are-mere-lip-service/.

Mokhele, Matseliso Lineo. 2011. "Integrated Environmental Teaching in South Africa." *Perspectives in Education* 29 (4, December): 78–86.

Murris, Karin. 2016. *The Posthuman Child*. London: Routledge.

Schleicher, Andreas. 2019. "How Changing the PISA Test Could Change How Kids Learn." *Quartz*, February 22. https://qz.com/1540222/how-changing-the-pisa-test-could-change-how-kids-learn/.

Statista. "Number of Schools in South Africa in 2022, by Sector." *Statista*. Accessed September 20, 2024. https://www.statista.com/statistics/1262871/number-of-schools-in-south-africa-by-sector/.

Statista. "Number of Teachers in Education in South Africa in 2022, by Province." *Statista*. Accessed September 20, 2024. https://www.statista.com/statistics/1262709/number-of-teachers-in-education-in-south-africa-by-province/.

South African Government. "Education." *South African Government*. Accessed September 20, 2024. https://www.gov.za/about-sa/education.

UNESCO. "Education 2030: Global Education Cooperation Mechanism." UNESCO. https://sdg4education2030.org/the-goal.

United Nations. 2023. "General Comment No. 26 (2023) on Children's Rights and the Environment, with a Special Focus on Climate Change." Convention on the Rights of the Child. *United Nations*, August 22. GC26:35. Accessed September 20, 2024. https://www.naturebased.education/_files/ugd/22c4ff_fd316d78c25548c3ad03de570964192e.pdf.

United Nations. 2015. "The 2030 Agenda for Sustainable Development." *United Nations*. https://sdgs.un.org/publications/transforming-our-world-2030-agenda-sustainable-development-17981.

CHAPTER 23

An Unconscious Bias Toward Non-human Animal Exploitation in the British Schooling System

Laura Diamond

23.1 Introduction

This chapter explores how an unconscious bias for non-human animal exploitation has been, and continues to be, projected onto our school children daily, so much so that it has inevitably led to an adult population who, en masse, struggle to even contemplate veganism as a genuine concept.

My viewpoints and conclusion are drawn from my experiences as a qualified primary school teacher with SENCO (special educational needs coordinator) experience, as a vegan-inclusion consultant working with parents and/or carers and schools for the past two years, and as a parent of a vegan child currently moving through the British schooling system.

L. Diamond (✉)
Primary Veducation, Manchester, UK
e-mail: laurajanediamond26@gmail.com

C. Linzey and A. Linzey (eds.), *The Palgrave Companion to Humane Education*, The Palgrave Macmillan Animal Ethics Series,
https://doi.org/10.1007/978-3-032-19795-5_23

I aim to demonstrate and explain how prevalent the subtle indoctrination toward non-human animal exploitation occurs in schools nationally, including how this excludes the vegan child, and to suggest ways that the Department for Education could improve their curriculum while simultaneously supporting the vegan learners within their "inclusive" institutions.

23.2 Early Years Education

Before I delve into the lessons of non-human animal exploitation that our children unknowingly absorb, it is important to explain how other animals fit into the curriculum. As it stands currently, the only place where non-human animals are featured within England's statutory Early Years and Foundation Stage (EYFS) curriculum is under the heading "Understanding the World," where learners aged anywhere from birth to five years should be taught how to "make observations of animals and plants and explain why some things occur, and talk about changes."[1]

In Key Stage One, where the learners are between four and seven years old, and in Key Stage Two, where the learners are between seven and eleven, non-human animals feature in the curriculum only under the subject heading of "Science" in one of two subheadings: either "Animals, including humans" or "Living things and their habitats."[2] It is here that the key information children are to be taught by teachers is outlined. We shall return to this curriculum later in the chapter after first examining the EYFS.

Upon a learner's first entrance into the mainstream schooling system, they are welcomed into the EYFS setting, where "children learn and develop well in enabling environments, in which their experiences respond

[1] UK Department for Education, "Statutory Framework for the Early Years Foundation Stage: Setting the Standards for Learning, Development and Care for Children from Birth to Five" (Department of Education, 2014), accessed September 25, 2019, https://assets.publishing.service.gov.uk/government/uploads/system/uploads/attachment_data/file/299391/DFE-00337-2014.pdf.

[2] UK Department for Education, "National Curriculum in England: Science Programmes of Study" (Department for Education, September 11, 2013, last updated May 6, 2015), https://www.gov.uk/government/publications/national-curriculum-in-england-science-programmes-of-study.

to their individual needs and there is a strong partnership between practitioners and parents and/or carers."[3] It is also the first time that they get to learn about "The Farm." This is not a statutory requirement, although it is recommended in a non-statutory guidebook[4] that complements the statutory curriculum. Teaching and learning about animal farms are in fact a choice made by each educator who delivers the EYFS curriculum because it is universally thought to be the best way that learners can make observations of other animals.

Having been an EYFS educator myself, I can vouch for the fact that "The Farm" may as well be statutory. This is because it is so prevalent as a topic or theme within every one of our EYFS settings—so much so that if it were suddenly made into a statutory topic, there would be no notable difference. And with so many resources available online, the reason educators are drawn to it is obvious. They are busy people, bogged down with excess administration and targets to achieve, who rely massively on each other for models of great practice. Resource sharing is one great way to support one another so that educators do not feel as though they are constantly reinventing the wheel. The issue, I believe, with all learners discovering "The Farm" the instant that they enter formal education is that it presents the animal farming industry as a natural way of life for other animals when in fact it is an unnatural, human-constructed enterprise. These lessons misrepresent what happens there and create a falsified belief that this is where the farmed animals belong. The fact is that these manufactured habitats known as animal farms are filled with unimaginable suffering and cruelty. But with schools' messaging on animal farms delivered repeatedly and annually, generation after generation of learners begin to see them as part of everyday life. Normal, natural, and necessary. This happens before learners are even at an age where they can question if this treatment of non-human animals aligns with their belief system, and so the cycle continues.

New learners at the point of entering school have very little knowledge and understanding of the world, which is why one of the "seven areas for development" is literally called "understanding of the world,"[5] yet

[3] UK Department for Education, "Statutory Framework."

[4] UK Department for Education, "Development Matters" (Department of Education, 2023), accessed September 18, 2024, https://www.gov.uk/government/publications/development-matters--2.

[5] UK Department for Education, "Statutory Framework."

without a moment's hesitation, we are teaching them that animal farms are happy and friendly places, which is utterly contradictory to the truth. This not only begins to strip young learners of their innocence and the innate love for other animals with which they were born, but also sows the seed for non-human animal exploitation being seen as an essential, but hidden, way of life. Young learners' views of animal farms become warped, and their cognitive dissonance is firmly planted as they see no wrongdoing in the curated versions of animal farms that they learn about or visit on trips.

In many cases, around Easter time, chicken hatcheries are brought into the EYFS, much to the detriment of the vegan learner, and at Christmas-time perhaps a reindeer will visit too; add a school pet here, a goldfish there, and before you know it, the children do not blink an eye when penguins are invited to perform in the town centre where they live. The lessons supposedly taught on how to care for and look after other animals by putting them into anxiety-inducing and unnatural situations get lost. When the overall focus is on the humans' enjoyment as opposed to the non-human animals' welfare or rights, this is where the education stops. Yes, these young learners have observed and been able to talk about other animals, so the criteria set out for them to achieve their EYFS goal have been ticked off, but at what cost? They certainly have not become proficient in compassion or empathy, which are two qualities lacking in our modern-day society.

23.3 Primary School: Key Stage One and Key Stage Two

As learners progress further with their formal education and enter into primary school between the ages of four and eleven, they will learn about non-human animals through the science topic "Animals, including humans," but this topic does not inform these developing minds as well as you might hope. Learners are simply taught to separate, categorize, and find differences in the animal kingdom. Nowhere does the curriculum state that non-human animals are sentient, loving, intuitive, and intelligent or that they have similar maternal and paternal instincts as we humans do. Furthermore, the curriculum does not state what veganism is and what beliefs ethical vegans hold, and so without these qualitative conversations, the work stays within the boundaries of a quantitative approach.

Within each academic year, there are six half-terms, each with an average of seven weeks, and only one hour in each of those weeks is dedicated to teaching science. With only two out of the six science topics per half-term containing other animals, with one concentrating more on habitats than on their inhabitants, this does not amount to very much attention to other animals, and even this estimate is generous. As well-intentioned as the curriculum may seem, with some inclusion of non-human animals in its biology syllabus, realistically, the educators who deliver this curriculum will be concentrating more on the pressure to show that learning has taken place, rather than on how meaningful the content is. This is how Ofsted, a non-ministerial department of the UK government responsible for inspecting schools, will judge them. Consequently, a categorization worksheet with a clear learning intention and desired outcome, such as the message that humans are at the top of the food chain, is the obvious choice. Even if educators were so inclined, there is no real wiggle room for conversations about non-human animal sentience or for the questioning of societal norms. Additionally, when planning their lessons, individual educators bring with them their own set of beliefs and life experiences, including a detachment from the non-human animal world in the form of cognitive dissonance and a conscious or unconscious desire to use other animals for personal benefit; most probably taught to them while they too were in school. With all of these points in mind, it is certainly no surprise to me that our primary schools are not currently in a position to deliver either a humane or vegan-inclusive education.

Umbrella topics are favored in British primary schools so that the teaching of all subjects, whether physical education, music, numeracy, or religious education, can be interconnected with a recurrent theme. Six such topics are chosen throughout each academic year, with the most popular ones being Space, the Great Fire of London, the Romans, and Growing Plants.

Let us say, perhaps, that during the first autumn half-term in Year 1 (four- and five-year-olds), a history element is chosen, and pirates are the focus; then all subjects taught in that half-term will have a pirate theme. This could look something like painting and making ships in art and design class, writing a pirate story in literacy, singing pirate songs in music class, and drawing pirate maps in geography class. As you walk into a classroom, everything would be pirate-related and decorated in that manner to enhance the learning experience.

I am an advocate for umbrella themes and wholeheartedly believe that there is a place for them in our schools. The only issue I see is when an educator decides to pick a science-based element of the curriculum and then the whole half-term is based on learning about one particular non-human animal, or group of animals. This is primarily because the cohort will spend that full half-term focusing on that one particular non-human animal, or group of animals, with the ultimate goal of then organizing a trip to a zoo, aquarium, or safari park.

Lessons will almost certainly have catchy names to hook learners in, including, more often than not, "Zoo-Do," "Under the Sea," or "Let's Go on Safari." Young minds will enjoy a plethora of books, activities, and videos normalizing these non-human animal entertainment industries and will not think twice about the cages, tanks, or boundaries that they are held in against their will, and which cause irreparable psychological damage.

For a vegan learner in the classroom, seven weeks can feel like a long time to study something they oppose. Being part of this biased learning environment may be seen as a form of indirect exclusion, as their only options are to refuse the trip and receive an unauthorized absence or to participate in something that contradicts their values.

For the learner who is not vegan, these topics will seem exciting, and the fact that they are working toward an awesome trip will, theoretically, lead them to become more invested in and enthusiastic about their learning. Incidentally, I am an advocate for school trips and do believe that they add value to the deeper understanding of a topic at hand. However, a trip to an unnatural habitat where non-human animals are caged and unable to follow their instinctual patterns of behavior seems much less of an educational experience to me. Impressionable minds are once again being shown that other animals do not mind being used for entertainment, and this is simply not the case.

If young learners are not prompted to question the morality of these entertainment houses, and for the past six or seven weeks they have been learning intently all about different aspects of the zoo, aquarium, or safari park that they are due to visit, then it will not even occur to them that their existence might very well be wrong. A half-term full of subtle indoctrination comes to a head on the day of the trip. In addition, their teacher is someone whose word they trust, and it would be rude, perhaps unheard of, for them as mere children to disagree or see conflict in the message that has been presented to them. Some learners might—those who will probably end up vegetarian or vegan later on in life—but they are currently in the vast minority and may very well have their suspicions squashed by an educator who simply wants everyone to enjoy themselves.

23.4 The Language of Subtle Indoctrination

At first glance, these songs may seem harmless, but they present certain ideas that warrant closer examination. For instance, "Old MacDonald" gives the impression that he loves his animals and that they, in turn, love him. However, this is misleading, as one wouldn't typically send a loved one to slaughter. The songs also carry messages of dominance, implying that humans have the right to take what isn't ours, such as wool from a sheep's back. Similarly, in "Once I Caught a Fish Alive," the message suggests it's perfectly acceptable to drag a fish from the sea with a hook, disregarding the fish's natural instinct to return to water. "Going on a Lion Hunt" is about hunting—a practice that many now view as unnecessary and barbaric in modern times.

While these songs are woven into our cultural fabric, when we reflect on their deeper meanings, we uncover a more troubling narrative that reinforces the notion that other animals exist for our use, rather than with us as fellow beings. Nursery rhymes like "Three Blind Mice," where the farmer's wife cuts off the animals' tails with a knife, and "Ding Dong Bell," about a cat being drowned in a well, also contain violent imagery toward animals. Though not everyone may interpret these songs in the same way, it is important to consider the ethical messages they convey to children, whose values and views on animals are still forming. Reflecting on this can play a crucial role in fostering a more compassionate outlook toward non-human animals.

There is also a case for encouraging educators to leave idioms out of the classroom, such as "kill two birds with one stone," "flog a dead horse," and "there's more than one way to skin a cat." There is simply no need for this type of violence to be brought into a classroom setting. Again, using phrases such as these sends a subtle subliminal message that using non-human animals is acceptable, even though these phrases might be dropped into a conversation without a thought. Times have changed. These idioms are as outdated as the aforementioned nursery rhymes, and neither should feature in modern, inclusive settings designed for young learners, especially in settings that include a vegan learner.

23.5 Duty of the Curriculum

At this point, it is worth noting the value of involving non-human animals as a muse to teach key skills through. Children love animals; they are often a child's first and best friend. The sound of a non-human animal may have been the first noise a child made on demand, and their cots and beds are likely filled with soft, stuffed piggies and bunnies to cuddle. It is understandable, then, that educators want to use non-human animals as a theme for half-term topics, as they help keep learners engaged and interested.

However, for the curriculum to become vegan-inclusive and to reduce its unconscious bias toward the use and exploitation of non-human animals, learners should be encouraged to appreciate their local environments and support the wildlife that lives there. Natural habitats such as forests, ponds, swamps, and rivers, along with opportunities for learners to explore them, should be incorporated into lessons, with school trips to these valuable places integrated into the curriculum. The national curriculum needs a transformative overhaul, so young learners can fully grasp the impact of their daily actions on the world around them. At its core, it must champion compassion and empathy, shaping new generations that lead with these values in everything they do. A humane education.

Our youth should be taught to care for non-human animals in their natural habitats by taking responsibility for the environment and learning about animal sentience, social behaviors, and their roles in the interdependence of all species, habitats, and ecosystems.

This is not to say that children should not learn about other animals typically featured in safari-themed topics, such as rhinoceroses, elephants, or hippopotamuses. They absolutely should, but much can be learned through internet clips and videos, which can still inspire awe if used and taught effectively. Consider how much children love learning about dinosaurs, even though they have never seen one in person.

Educators may believe they are giving their learners unforgettable experiences by taking them to zoos, aquaria, or safaris, but in reality, they are depriving them of a true connection with nature—one that comes from fostering a deep respect for the natural world. The fact that these magnificent animals are endangered, and that some individuals, after going through our education system, would pay to hunt and kill them, is evidence that this theory holds weight. With the currently used lessons and school trips, learners experience no real sense of discovery regarding the sentience, thoughts, feelings, family, or social structures of these non-human animals; at the end, they have simply learned how to categorize them as herbivorous or omnivorous mammals—job done! Had more people been exposed in their childhood to a humane education that values all animals, including free-roaming animals, and respects their individual needs, then perhaps those in captivity would have been left to their own devices, living in their natural habitats without fear of extinction. Perhaps there would be no animal prisons in the form of zoos, aquaria, or safaris, leaving room for the development of a much more holistic curriculum. As a side note, I would mention the circus here too, but thankfully, it is likely that a worldwide ban on animal circuses is becoming a reality at great speed.

It is the duty of the Department for Education to implement statutory guidance requiring educators to teach future generations about the deep connection between themselves and other living beings, from ants and birds to crabs, seahorses, sharks, mice, and badgers. Each of these animals, like humans, has the capacity to experience life, and all play a vital role in their ecosystems. By fostering respect and compassion for all non-human animals and their habitats, children can learn that all creatures deserve consideration and protection. The Department for Education will be doing our children a great disservice if it does not move toward a more humane curriculum—one that is not only inclusive of vegan children but also fair, balanced, and protective of our environment, health, well-being, and all non-human animals with whom we share this planet.

23.6 School Dinners

There is still one further dimension of the unconscious bias toward non-human animal exploitation imposed on children in British primary schools, and that is the food aspect. Many people who are conscious about not eating other animals are eco-conscious, or follow a humane education, often adopt a vegan diet as an ethical default choice to minimize harm to other animals. This is why I use the vegan example when explaining how difficult it can be to eat ethically in schools. Part of the science subheading, "Animals, including humans," explores lessons about what foods are considered healthy or unhealthy. Without delving too deeply into this area, it is safe to say that meat, fish, eggs, and dairy feature quite prominently in these lessons. This message is then replicated in the school canteen, as shown in the Department for Education's mandatory guidance on school foods:

Meat, fish, eggs, beans, and other non-dairy sources of protein:

- A portion of food from this group every day.
- A portion of meat or poultry on three or more days each week.
- Oily fish once or more every three weeks.
- For vegetarians, a portion of non-dairy protein on three or more days each week.
- A meat or poultry product (manufactured or homemade, and meeting the legal requirements) no more than once each week in primary schools and twice each week in secondary schools (applies across the whole school day).[6]

Noticeably, the word "vegan" is completely missing in this extract and in fact is not mentioned once within the entire document. In addition, in order for a school to fully comply with this mandatory guidance, a lot of animal products must be served up, with at least one portion of food from this group every day. It is no wonder that vegan parents and/or carers are meeting resistance when they request vegan menus at school. Schools are well within their rights to reject this request based on mandatory dietary guidance that does not acknowledge the existence of veganism. It should be harder for schools to ignore the legal side of the request, being that veganism is a protected belief, and service providers have a responsibility not to discriminate under the Equality Act of 2010.[7] Nonetheless, many schools, including the one my own daughter attends, do initially refuse. They look at the mandatory guide on school foods,[8] cannot find the word "vegan" there, and then do not see it as their personal duty to provide

[6] UK Department for Education, "School Food Standards: A Practical Guide for Schools, Their Cooks and Caterers" (Department for Education, March 26, 2019), accessed September 24, 2019, https://assets.publishing.service.gov.uk/government/uploads/system/uploads/attachment_data/file/786339/School-Food-Standards-Guidance.pdf.

[7] UK Government Equalities Office and Equality and Human Rights Commission (EHRC), "Equality Act 2010: Guidance" (Government Equalities Office and EHRC, February 27, 2013), accessed September 25, 2019, https://www.gov.uk/guidance/equality-act-2010-guidance.

[8] UK Department for Education, "School Food Standards."

vegan food. There is currently no national consistency whatsoever where vegan food provisions are concerned, and in fact, the availability can differ vastly from one school to the next, let alone from one borough to the next.

I argue that there should be an overhaul of the school food standards, not only to include appropriate meals for vegan learners but also to offer alternatives for the other learners in school who may fancy something different. With veganism growing exponentially, an obesity crisis rising in our schools, and many food allergies and religious mandates to follow, perhaps the safest and best option for all schools would be to remove all animal products in favor of adopting a whole-foods and plant-based diet that is nondiscriminatory and fully inclusive. This would run alongside a humane education beautifully, complementing the idea that it is possible to "create a better world through education."[9]

23.7 Conclusion

Schools can achieve a humane education in several different ways, by omitting topics that promote non-human animal exploitation and replacing them with a vegan-inclusive education; by creating new songs and rhymes that reflect our modern-day society and using language that shows respect for each other and for non-human animals; by holding deep discussions on the interconnectivity of all lives on earth and inducing a love for nature; and, finally, by serving inclusive and healthy daily meals that will marry with the lessons taught about healthy food in class.

As holistic and achievable as this peaceful ethos may seem, such changes are unfortunately in the hands of the government and would take many years to implement. We can be optimistic, though, as there is talk that personal, social, health, and economic (PSHE) education may soon become a statutory part of the curriculum, and religious education is now referred to as "religion and worldviews," adding the potential for veganism to be introduced as a stand-alone philosophical topic. Naturally, all of this would be more achievable and probable if the Department for Education were to implement the statutory recommendations made in this chapter. Until then, it is up to ethical, vegan parents and/or carers and humane education advocates to ask for inclusion, school by school,

[9] Institute for Humane Education, home page, accessed September 25, 2019, https://humaneeducation.org/.

and for vegan learners to stay strong despite the deluge of misinformation and bias toward non-human animal exploitation that they are exposed to daily.

Bibliography

Institute for Humane Education. Home page. Accessed September 25, 2019. https://humaneeducation.org/.

UK Department for Education. "Development Matters." Department of Education, 2023. Accessed September 18, 2024. https://www.gov.uk/government/publications/development-matters--2.

UK Department for Education. "National Curriculum in England: Science Programmes of Study." Department for Education, September 11, 2013. Last updated May 6, 2015. https://www.gov.uk/government/publications/national-curriculum-in-england-science-programmes-of-study.

UK Department for Education. "School Food Standards: A Practical Guide for Schools, Their Cooks and Caterers." Department for Education, March 26, 2019. Accessed September 24, 2019. https://assets.publishing.service.gov.uk/government/uploads/system/uploads/attachment_data/file/786339/School-Food-Standards-Guidance.pdf.

UK Department for Education. "Statutory Framework for the Early Years Foundation Stage: Setting the Standards for Learning, Development and Care for Children from Birth to Five." Department of Education, 2014. Accessed September 25, 2019. https://assets.publishing.service.gov.uk/government/uploads/system/uploads/attachment_data/file/299391/DFE-00337-2014.pdf.

UK Government Equalities Office and Equality and Human Rights Commission (EHRC). "Equality Act 2010: Guidance." Government Equalities Office and EHRC, February 27, 2013. Accessed September 25, 2019. https://www.gov.uk/guidance/equality-act-2010-guidance.

CHAPTER 24

Effect of Humane Education on Middle School Reading Comprehension

Julie O'Connor and William Ellery Samuels

24.1 Introduction

A primary goal of public school education is to develop and maintain a literate population.[1] Many schools administer state-sponsored annual exams to measure reading achievement. Whereas new and diverse teaching methods and reading comprehension materials are implemented annually in schools, the bar by which success is measured does not vacillate. Standardized, uniform measurement of reading success is the

[1] Tondra L. Loder, "Why We Can't Leave Public Schools Behind: The Inseparable Legacy of Public Education and American Democracy," *Educational Researcher* 35, no. 5 (2006): 30–35, https://journals.sagepub.com/home/edr.

J. O'Connor (✉)
School of Education, Touro College, New York, NY, USA
e-mail: julie.oconnor@aplnj.org

W. E. Samuels
School of Nursing, Hunter College, The City University of New York, New York, NY, USA
e-mail: s@muels.email

C. Linzey and A. Linzey (eds.), *The Palgrave Companion to Humane Education*, The Palgrave Macmillan Animal Ethics Series,
https://doi.org/10.1007/978-3-032-19795-5_24

framework for much of the public educational system in the United States.[2]

A Humane Society of the United States study demonstrated that 99.3 percent of children aged three to thirteen wanted to have pets.[3] This overwhelming interest children have in animals can lead to many positive results, including academic ones.

To determine whether humane education, or learning about kindness to animals, could improve reading comprehension scores, this study utilized a randomized control trial design to compare students who completed literacy assessments that either did or did not contain humane-education-related content. Students in the experimental group were administered a series of five humane education reading measures, while those in the control group received high-interest reading passages that did not have animal protection themes. The scores are analyzed in this chapter.

Standards-Based Education and Reading Comprehension

The Common Core State Standards (CCSS) Initiative of 2010 set out to address the issue of nationwide standards consistency in the United States. The CCSS are a set of goals and expectations developed by the National Governors Association.[4] Although not every state in the United States uses the CCSS, their goals and objectives are universally recognized as necessary for students to meet. Therefore, almost every state has developed something very similar to the CCSS. For example, when New Jersey revised its standards to no longer follow the CCSS in 2016, after initially adopting them in 2010, it was noted that the changes were minimal, with the state maintaining 84 percent of the standards.[5] This research will

[2] 114th Congress, *Every Student Succeeds Act: Conference Report (to Accompany S. 1177)* (2015), https://www.govtrack.us/congress/bills/114/s1177.

[3] Humane Society of the United States, *The Empathy Connection* (HSUS, 2004), http://www.humanesociety.org/assets/pdfs/abuse/empathy-connection.pdf.

[4] National Governors Association (NGA) Center for Best Practices and Council of Chief State School Officers (CCSSO), *Common Core State Standards* (Washington, DC: NGA Center for Best Practices and CCSSO, 2010).

[5] Adam Clark, "N.J. Revises, Renames Common Core Academic Standards," *NJ.com*, May 4, 2016.

explore the comprehension component of reading against the goals set forth within the CCSS as determined by established educational research.

In tandem with the CCSS are higher-order thinking skills, which are the goal of standards-based literacy instruction and its aligned assessments.[6] According to the CCSS developers, the standards are "based on rigorous content and application of knowledge through higher-order thinking skills."[7]

For accurate assessment of reading skills, it is important that the content be unpredictable.[8] Therefore, standards-aligned assessments need to be text-dependent by nature. We followed this strategy in our experiment in that we assessed students' reading skills via standards-aligned questions that were related directly to the passage the student was reading at the time. In doing so, we ensured that the animal and non-animal passages were equal and that they assessed the same reading standards. The questions were designed to meet educational objectives that included analysis and evaluation. Those skills are considered among the more advanced educational objectives, more complex than memorization.[9]

Humane Education Curricula

Humane education can be defined as a "form of character education that uses animal-related stories, lessons, and activities to foster respect, kindness, and responsibility in children's relationships with both animals and people."[10] The Academy of Prosocial Learning further defines humane education as encouraging growth through developing empathy as well as problem-solving skills related to people, animals, the planet, and the

[6] Karen Tankersley, "Higher-Order Thinking," in *Literacy Strategies for Grades 4–12: Reinforcing the Threads of Reading* (Alexandria: ASCD, 2005).

[7] NGA Center for Best Practices and CCSSO, *Common Core State Standards.*

[8] E. D. Hirsch Jr., "What Do Reading Comprehension Tests Mainly Measure? Knowledge," *American Educator*, Spring 2006, https://www.aft.org/periodical/american-educator/spring-2006/what-do-reading-comprehension-tests-mainly.

[9] Lorin Anderson, David Krathwohl, and Benjamin Bloom, *A Taxonomy for Learning, Teaching, and Assessing: A Revision of Bloom's Taxonomy of Educational Objectives* (Boston: Allyn & Bacon, 2001).

[10] Catherine A. Faver, "School-Based Humane Education as a Strategy to Prevent Violence: Review and Recommendations," *Children and Youth Services Review* 32, no. 3 (2010): 365–70, 10.1016/j.childyouth.2009.10.006.

intersectionality among them.[11] A humane-education-based curriculum has been found to motivate children because of its inclusion of animal protection in classroom lessons.[12]

When animal protection issues are included in academic content, students can be motivated to pay closer attention to lessons.[13] That motivation may lead to the students answering questions more accurately, in that they attend to the questions more meaningfully and consider their answers more carefully. Given that academic content can boost student motivation,[14] teachers should identify what areas students enjoy exploring. Common-Core-aligned reading assessments using humane education stories may yield higher scores than their non-humane education counterparts. Educators could benefit from understanding whether humane education can be utilized to improve standards-aligned reading assessments.

Hypothesis: Including animal protection content in the passages will result in better performance on these items.

24.2 Methods

Participating students were asked to read five Common-Core-standards-aligned reading passages and answer text-based questions requiring higher-order thinking skills.

This randomized control trial with constraints used five passages with humane education themes on the experimental group and five non-humane education passages on the control group. To study whether humane education themes have an effect, we developed passages that differed only in the passage content: experimental-group passages addressed animal protection issues, whereas the control-group passages did not. The passages were otherwise equated in terms of reading level,

[11] Academy of Prosocial Learning, home page, accessed July 1, 2021, https://www.prosocialacademy.org.

[12] Julie O'Connor, "A Qualitative Case Study of Teacher Perceptions of the Motivation of Students in Humane Education" (PhD diss., Northcentral University, 2018), ProQuest (10,744,993).

[13] O'Connor, "Qualitative Case Study."

[14] Kaylene C. Williams and Caroline C. Williams, "Five Key Ingredients for Improving Student Motivation," *Research in Higher Education Journal*, no. 12 (2011): 1–23, http://aabri.com/manuscripts/11834.pdf.

length, and so on, as subsequently described. The randomized control design allowed us to reduce bias when testing the effectiveness of humane education on improving standards-based reading comprehension scores.

Non-humane education passages were obtained from previous fifth-grade CCSS-aligned state exams as well as test preparation materials from New Jersey, New York, and Texas. High-interest and student-relevant topics were purposefully selected among the test preparation materials that different states provided. Non-humane education passage topics included basketball coaching, toy inventions, bubbles, and the science of listening to music.

One of the humane education reading passages originated from a New Jersey CCSS assessment preparation guide, and the remaining four passages were adapted from the TeachKind humane education division of the animal protection group People for the Ethical Treatment of Animals (PETA).[15] The humane education passages focused on the following themes: domesticated animals, wildlife, animals on farms, and animals in entertainment.

The experiment was conducted in the first half of a pre-COVID-19 academic year. The assessments were given on five separate days over the course of two months, and students were given unlimited time to complete their tests, although most children were done with each one in less than twenty minutes.

Lexile level assesses how difficult a text is to comprehend. The two main criteria assessed are word frequency and sentence length in determining Lexile level.[16] The Flesch-Kincaid grade level is determined by a readability test that also assesses how difficult a passage is to understand. The number that results corresponds with grade levels in the United States.[17] The word count of the text also becomes part of the calculation for the Flesch-Kincaid level.[18] For this research, experimental and control

[15] People for the Ethical Treatment of Animals, "K–12 Lesson Plans and Classroom Tips for Compassionate Teachers" (PETA, last modified 2021), https://www.peta.org/teachkind/.

[16] Colleen Lennon and Hal Burdick, "The Lexile Framework as an Approach for Reading Measurement and Success" (MetaMetrics, 2004).

[17] J. Peter Kincaid et al., "Derivation of New Readability Formulas (Automated Readability Index, Fog Count and Flesch Reading Ease Formula) for Navy Enlisted Personnel" (Institute for Simulation and Training, 1975), 10.21236/ada006655.

[18] Kincaid et al., "Derivation of New Readability Formulas."

passages were matched using their Flesch-Kincaid grade level and their Lexile level to ensure commensurate readability. Each partnered passage was matched using similar word count. All content was nonfiction. The text-based multiple-choice questions were matched both in the design of their wording and in the specific CCSS they addressed.

Participating sixth-grade students were given fifth-grade Lexile-level reading passages. Lessening the effect of some students' struggle in reading grade-level material allowed the humane education content to be further analyzed in the side-by-side comparison of the two texts. The CCSS Lexile levels for grade 5 range from 875 to 1010.[19] All questions were aligned to grade 5 CCSS English Language Arts and Literacy "reading for information" (RI) anchor standards. The assessed standards were as follows: RI 5.1, "Quote accurately from a text when explaining what the text says explicitly and when drawing inferences from the text"; RI 5.2, "Determine two or more main ideas of a text and explain how they are supported by key details; summarize the text"; RI 5.3, "Explain the relationships or interactions between two or more individuals, events, ideas, or concepts in a historical, scientific, or technical text based on specific information in the text"; RI 5.4, "Determine the meaning of general academic and domain-specific words and phrases in a text relevant to a grade 5 topic or subject area"; RI 5.5, "Compare and contrast the overall structure (e.g., chronology, comparison, cause/effect, problem/solution) of events, ideas, concepts, or information in two or more texts"; RI 5.8, "Explain how an author uses reasons and evidence to support particular points in a text, identifying which reasons and evidence support which point(s)."[20]

The first set of matched texts were titled "Bubblology" and "The Amazing Penguin Rescue," with both texts asking questions addressing Common-Core anchor standards RI 5.1, RI 5.2, RI 5.4, and RI 5.8. The next set of matched texts were "Coach Motivates Her Girls" and "Miss Willie's Bucket List," containing questions that aligned with standards RI 5.1, RI 5.3, and RI 5.5. The third testing wave's readings were "High Volume" and "Rescued Camels Meet Their Soulmates," with questions asking about RI 5.1 and RI 5.3. The fourth pairing was "The Stories

[19] Mary Doman, "Lexile Levels: What Parents Need to Know," Scholastic, accessed July 11, 2021, https://www.scholastic.com/parents/books-and-reading/reading-resources/book-selection-tips/lexile-levels-made-easy.html.

[20] NGA Center for Best Practices and CCSSO, *Common Core State Standards*.

Behind 'Toy Story's' Beloved Characters" and "Freddy on the Run," both addressing RI 5.1, RI 5.2, and RI 5.3. Last, the fifth pair of texts was "An Accidental Toy" and "Florence the Cat," containing questions that addressed standards RI 5.1, RI 5.2, RI 5.3, and RI 5.4.

Participants

Participants included 192 sixth-grade elementary students in New Jersey, aged twelve to thirteen years; ninety-three students were randomly assigned to the control group, and ninety-nine to the experimental. Covariates included English language learner status, special education status, gender, and grade retention status.

Analyses

We tested the effect of including animal-based themes in the reading passages with multilevel models—also known as hierarchical linear models—that included not only the group membership (experimental versus control) but also both demographic variables (gender, English language learner status, etc.) and relevant characteristics of the passage (namely, Flesch-Kincaid grade level and word count; Lexile level was not included given the multicollinearity with Flesch-Kincaid grade level). These models can account for the nested nature of data like those analyzed here better than, for example, analyses of covariance can.[21]

We attempted both to nest passage content pairs within students and to nest students within their class at the school, but the models did not converge on stable sets of parameters when both levels of nesting were used. We therefore decided to only nest passage content pairs within students and not account for class membership in the results presented herein. We conducted all of the analyses presented here when nesting students in class (and not considering the nested nature of the passages); these alternate analyses generated very similar patterns of results, with no differences in which terms were significant. Continuous variables were transformed into *z*-scores before being included in the models. All other variables were entered as dummy variables, with values of either 0 or 1.

[21] Duncan Anderson et al., *A Practitioner's Guide to Generalized Linear Models*, 3rd ed. (Arlington, VA: Casualty Actuarial Society Discussion Paper Program, 2007.)

Among the advantages of doing this is that we can compare the magnitude of the effects between terms in the model. The β-weight for each term in the model is in fact the partial correlation between that term and the given outcome; we can therefore compare magnitudes of the effects and not simply rely on the *p*-values for insights.

Gender was coded so that 0 denoted male, and 1 denoted female. Individualized education program (IEP) status was coded so that 0 indicated that the student did not have an IEP, and 1 indicated that the student had an IEP; no distinction was made between students with different diagnoses that resulted in their being granted an IEP. English language learner (ELL) status was similarly coded so that 0 indicated that the student was not an English language learner, and 1 indicated that they were. Students who were not eligible for free or reduced-price school lunches were coded as 0 for lunch status; those who were eligible for either free or reduced-price lunches were coded as 1. Finally, ethnicity was transformed into a set of dummy variables (with 0 = not a member of a given ethnic group and 1 = is a member of that group); ethnicities were coded by the school so that students could be a member of only one of the ethnic groups; we decided to set European American as the referent group and so included dummy variables for the other groups in the models.

24.3 Results

Summary of Reading Passages

The reading difficulties of the various passages (i.e., Flesch-Kincaid grade level, Lexile level, and word count) are presented in Table 24.1. As can be seen in that table, the reading difficulties for the passages varied, but the difficulties of the experimental and control passages in each pair were quite similar.

Descriptive Statistics

Table 24.2 presents the number of students in the experimental and control groups categorized by the demographic variables collected about them. This table also provides χ^2-tests of independence, testing different proportions of each demographic variable between the groups. For example, the χ^2-score for the test of gender found that there were significantly more boys in the control group than in the experimental group

Table 24.1 Passage reading difficulty

	Flesch-Kincaid Grade level		*Word count*		*Lexile level*	
Content Pair	*Control*	*Experimental*	*Control*	*Experimental*	*Control*	*Experimental*
Accidental Toy (Cntl) Florence Cat (Exp)	7.7	7.6	601	615	1000L–1100L	1000L–1100L
Basketball (Cntl) Miss Willie (Exp)	6.7	6.8	755	756	920L	900L–1000L
Bubblology (Cntl) Penguin (Exp)	5.7	5.7	632	648	880L	800L–900L
High Volume (Cntl) Camel (Exp)	7.3	7.3	729	740	980L	900L–1000L
Toy Story (Cntl) Freddy (Exp)	5.9	6	771	760	800L–900L	800L–900L

"Cntl" denotes the control-group passage; "Exp" denotes the experimental-group passage

($\chi^2 = 6.06$, $df = 1$, $p = 0.014$). Although not all of these demographic variables differed significantly between the groups, we included them all in the final model to partial out the effects they might have on the tests of group differences.

The correlations between the predictors used in the models are presented in Table 24.3. Flesch-Kincaid scores and word counts were included as continuous variables. All other variables were dummy variables, so the correlations that included them were either point-biserial correlations (with Flesch-Kincaid or word count scores) or φ coefficients (with other dichotomous variables). Most of the correlations were small; instances when they were not were in some correlations between reading scores (Flesch-Kincaid & Lexile $r = 0.943$) but not with word count (Lexil & word count $r = -0.384$; Flesch-Kincaid & word count =

Table 24.2 Demographic characteristics of students by experimental group

		Counts			*Tests of Independence*		
Demographic	*Condition*	*Control*	*Experimental*	*Total*	χ^2	df	p
Gender	Male	49	42	91	6.06	1	0.014
	Female	44	57	101			
Has IEP?	No	78	83	161	0.75	1	0.39
	Yes	15	16	31			
ELL?	No	91	94	185	6.87	1	0.0088
	Yes	2	5	7			
Eligible for free/reduced school lunch?	No	50	51	101	0.06	1	0.80
	Yes	43	48	91			
Ethnicity*	African American	4	3	7	14.84	3	0.0020
	Asian American	8	3	11			
	European American	39	35	74			
	Latin American	42	58	100			
Total		93	99	192			

– 0.183) and between lunch status and some race/ethnicities (lunch status & white φ = – 0.195; lunch status with Latin φ = 0.222) and between some race/ethnicities (largest φ between white & Latin = – 0.831).

The correlations with group are of special interest. None of these were large (largest φ between group & Latin = 0.115), but none were significant. This general lack of correlation between predictors—especially with group—indicates that most contribute unique information to the model.

We also determined the mean outcome scores for the experimental and control groups and calculated the number and standard deviations for those scores as well as the 95 percent confidence intervals. The confidence intervals provided an assessment of the differences between the groups on each outcome—without factoring out any of the demographic variables. As such, we were provided with an initial set of tests of whether there were group differences that might be worth further investigation. We saw that the confidence intervals for the total scores did not overlap: For example, the mean total score for the control group

Table 24.3 Correlations between variables. Bold-faced correlations are significant at $\alpha = 0.05$

	Group[a]	*FKRL*	*Lexile*	*Word count*	*Gender*[b]	*IEP*	*ELL*	*Lunch Status*	*Asian*	*Black*	*White*
Flesch-Kincaid Reading Level	0.026										
Lexile	– 0.003	**0.943**									
Word Count	0.056	– 0.183	– **0.384**								
Gender[b]	0.084	– 0.013	– 0.017	0.013							
IEP	– 0.030	– 0.020	– 0.015	0.007	0.066						
ESL	0.097	0.008	0.006	< 0.001	– 0.098	– 0.006					
Lunch Status	– 0.005	< 0.001	0.004	– 0.002	0.139	0.135	0.101				
Asian	– 0.087	< 0.001	0.005	– 0.002	– 0.118	– 0.095	0.089	0.016			
Black	– 0.003	0.007	0.003	0.023	– 0.088	0.074	– 0.038	– 0.103	– 0.046		
White	– 0.077	– 0.011	– 0.011	– 0.001	– 0.016	– 0.037	– 0.045	– **0.195**	– **0.154**	– **0.193**	
Latin	0.115	0.008	0.007	– 0.007	0.102	0.052	0.018	**0.222**	– **0.200**	– **0.250**	– **0.831**

[a]Control = 0, Experimental = 1; [b] Male = 0, female = 1

(2.63) was not within the 95 percent confidence interval for the experimental group (3.083.26), suggesting that those scores were significantly different between the groups. Among the reading standard-level scores, there appeared to be significant group differences for passages addressing standards RI 5.1, 5.2, 5.5, and 5.8; the means for standards RI 5.3 and 5.4 were not clearly different.

24.4 Total Reading Competence

Table 24.4 summarizes the results of two multilevel models investigating the relationships between the relevant predictors and students' total reading competency scores. The first set of columns in this table summarize the "base" model that included all of the demographic variables but not group membership (i.e., whether the student was randomly assigned to the experimental or control group). The second set of columns indicate the changes to the model when group membership was added.

Most importantly, Table 24.4 shows that group membership significantly predicted total reading competence ($t_{577.6} = 9.76$, $p < 0.001$). The β-weight ($\beta = 0.55$) further suggests that this effect was not only significant but even quite large after the factoring out of the demographic variables. The zero-order correlation between group and total reading competence was smaller but also significant ($r_{pb} = 0.26$, $df = 840$, $p < 0.001$).

In addition to the term for group membership being itself significant (and to the β-weight being quite large), adding the group term to the model significantly improved the fit of the entire model to the data (difference in Bayesian Information Criteria [BICs] = $\chi^2 = 80.57$, $df = 1$, $p < 0.001$). Altogether, these lines of evidence support our conclusion that use of animal-themed passages was strongly associated with improved performance on reading competence for these middle school students—especially when the effects of demographic covariates were removed.

Considering those demographic variables, Table 24.4 indicates that the Flesch-Kincaid grade level and word count were both significantly related to total reading competence; this surely comes as no surprise to anyone. Whether the student had an IEP and whether the student was an English language learner were also significantly related to reading competence, as was whether the student was eligible for free or reduced school lunches;

Table 24.4 Summary of multilevel models predicting total reading competence

	Base model				*Final model*			
Model term	*B*	df	t	p	*β*	df	t	p
Flesch-Kincaid level	– 0.113	841.1	– 3.45	< 0.001	– 0.122	840.9	– 3.92	< 0.001
Word count	– 0.268	841.0	– 8.20	< 0.001	– 0.284	841.5	– 9.13	< 0.001
Gender	0.129	841.6	2.23	0.026	– 0.037	841.4	– 0.64	0.522
IEP status	– 0.542	841.6	– 5.74	< 0.001	– 0.544	841.0	– 6.07	< 0.001
ELL Status	– 0.469	841.6	– 2.77	0.006	– 0.691	841.6	– 4.25	< 0.001
Lunch Status	– 0.111	841.6	– 1.70	0.089	– 0.149	841.1	– 2.40	0.017
Asian	0.365	841.6	2.58	0.010	0.275	841.2	2.04	0.042
Black	– 0.019	841.6	– 0.11	0.911	– 0.227	841.6	– 1.40	0.162
Latin	0.041	841.6	0.68	0.497	– 0.142	841.0	– 2.34	0.019
Group					0.552	577.6	9.76	< 0.001

Flesch-Kincaid readability level and word count were transformed into *z*-scores. For gender, 0 = male, 1 = female. For IEP status, 0 = no IEP, 1 = has IEP. For ELL, 0 = not ELL, 1 = is ELL student. For lunch status, 0 = not eligible, 1 = eligible for free/reduced-priced lunch. For ethnicities, 0 = is not a member of the given ethnic group, 1 = is a member of that ethnic group all of these relationships were in the—often unfortunately—expected directions.

The race/ethnicity of the students also were sometimes significantly related to reading competence. These tests were relative to the comparison category: white students. We found that the reading competence of Black students did not significantly differ from that of white students; Asian students' competences tended to be significantly better; and Latin students' were significantly worse.

We must, however, interpret at least some of these demographic effects with care since some of them changed in significance between the base and final model; these included gender, lunch status, and Latin ethnicity. (Even some of the demographic variables that did not change significances were still found to have rather different β-weights between the two models.) As Table 24.2 shows, the counts of all three of these demographic variables differed significantly between the experimental and control groups. Therefore, we caution against reading too much into the changes in these significances.

24.5 Discussion

The purpose of this study was to determine whether humane education reading passages would positively affect performance on standards-aligned reading comprehension assessments. We predicted that animal protection themes in reading passages would engage students in the content, and their answers to related questions would be more accurate than the answers of their counterparts who received non-animal protection topics about which to read. This would contribute to the eventual development of better standards-based English language arts programs and assessments.

Variables and Covariates in Reading Comprehension

Flesch-Kincaid grade level, word count of the passage, and English language learner (ELL) status had significant influences on the students' reading scores. However, by including those factors in the model, we were able to isolate the effects of those variables from the other terms in the model—including group membership. It is evident that humane education passages were related to an improvement in reading comprehension.

The effect of the humane education theme was independent of reading level, word count, or ELL status.

The word count significantly predicted students' scores, playing an even larger role than the passage's reading level—namely, its Flesch-Kincaid grade level. This was interesting given that the Flesch-Kincaid grade level (FKGL) demonstrates the education level needed in order to understand the text. It is worth noting that word count is related to other reading-level measures. The FKGL uses a formula that includes calculating the word count per sentence in addition to the syllable count per word.[22] Whereas the FKGL was significant to scores, it was not as significant as the word count. Additionally, in the equation to determine the Lexile level, the measurements of word frequency and sentence length are incorporated.[23] Yet the word count alone became a more significant variable than either the Lexile or the Flesch-Kincaid. Supporting the importance of text length, Baştuğ and Keskin[24] found that students who reread short texts performed better in terms of reading rate and reading accuracy than those rereading long texts.

English language learner (ELL) status had a significant influence on the student reading score. This was not surprising given that ELLs, in most cases, have a deficit in English academic vocabulary[25]as compared to their native English-speaking counterparts. There is a linear relationship between the percentage of vocabulary a reader has and the degree to which the reader comprehends what he or she reads.[26] The size of an ELL student's vocabulary has predicted reading achievement for older learners as well. The performance of college-level ELL students on the reading section of the Test of English for International Communication

[22] Kincaid et al., "Derivation of New Readability Formulas."

[23] Lennon and Burdick, "The Lexile Framework."

[24] Muhammet Baştuğ and H. Keskin, "The Role of Text Length in Repeated Reading," *European Journal of Education Studies* 6, no. 3 (September 2014), https://oapub.org/edu/index.php/ejes/index.

[25] Deanna L. Nisbet and Evie R. Tindall, "A Framework for Explicit Vocabulary Instruction with English Language Learners," *Kappa Delta Pi Record* 51, no. 2 (2015): 75, 10.1080/00228958.2015.1023141.

[26] Norbert Schmitt, Xiangying Jiang, and William Grabe, "The Percentage of Words Known in a Text and Reading Comprehension," *Modern Language Journal* 95, no. 1 (2011): 26–43, 10.1111/j.1540-4781.2011.01146.x.

was related to the size of their English vocabulary.[27] In addition, the more one reads, the more vocabulary one tends to acquire, thus enabling one to read more difficult, higher-grade-level text. Reading is important to the process of vocabulary acquisition in older children and adolescents.[28] Perhaps reading more passages with humane education themes will increase vocabulary, which in turn will raise student reading levels.

Gender, lunch status, and race/ethnicity were infrequently significant. The few differences that do exist likely reflect societal and economic effects outside of academic per se.

The order or timing of the passages also was not significant. The students did not decrease their performance over time. In fact, the performance of students in the experimental group improved throughout the exam. A qualification to this could be the content itself. Students appear more interested when learning about dogs.[29] The passage about dogs happened to be found later in the experiment.

Reading scores also improved significantly over the course of the exam among the student in the control group—but to a lesser degree than among students in the experimental group. This can be explained because students' reading should improve over time as they are exposed to more standards-based instruction. That the control-group students did not improve as much as the experimental-group students may further suggest that the intervention of humane education was effective.

Humane Education Improving Academic Results

Although this research was conducted on intermediate-level students, there is supportive evidence that children across grade levels demonstrate more academic engagement when the topic of animals is introduced. In one experiment, Swedish preschool children's desire to learn was found to be positively affected when pets were used for educational purposes, even though the animals were not physically in the classroom, and they were

[27] Hui-Hua Chiang, "English Vocabulary Size as a Predictor of TOEIC Listening and Reading Achievement among EFL Students in Taiwan," *Theory and Practice in Language Studies* 8, no. 2 (2018): 203, 10.17507/tpls.0802.04.

[28] Dawna Duff, J. B. Tomblin, and Hugh Catts, "The Influence of Reading on Vocabulary Growth: A Case for a Matthew Effect," *Journal of Speech, Language, and Hearing Research* 58, no. 3 (2015): 854, 10.1044/2015_jslhr-l-13-0310.

[29] O'Connor, "Qualitative Case Study."

just used as the basis of educational conversations.[30] An explanation for this is that children have a natural affinity toward animals that is observed as early as eleven months.[31]

Another reason that the experimental group achieved higher academic outcomes may be that the topics served a deeper purpose to which some students may have responded. Humane education serves dual education functions for children, in that academic content is delivered while compassion toward animals is engendered. Students in the experimental group learned about the issues of undomesticated animals in captivity, proper companion animal treatment, and the impacts of oil spills on wildlife. Such topics may resonate with students, who may see these issues as more important than those typically covered in reading comprehension passages. Generally, when students see their learning as part of a larger, important context, they achieve better academic results. A previous study showed that when students were motivated by broader educational purposes, their cumulative grade point average increased.[32]

There is evidence that the value of learning about animals affects students deeply. Ascione and Weber[33] found that students who were exposed to animal-related curricula behaved more empathically to fellow humans, even a year after the curricula's end. Additionally, humane education has been shown to result in positive outcomes pertaining to the quality of students' relationships with animals, including bonding and friendship.[34]It stands to reason that since this experiment contained

[30] Agneta Svensson, "The Impact of the Animals on Children's Learning and Their Development—A Study of What Children Learn from and with Pets: The Example of Dog and Cat," *Problems of Education in the 21st Century*, no. 59 (2014): 77–85.

[31] Vanessa LoBue et al., "Young Children's Interest in Live Animals," *British Journal of Developmental Psychology* 31, no. 1 (2012): 57–69, 10.1111/j.2044-835x.2012.02078.x.

[32] Hsun-yu Chan and Xueli Wang, "Interact for What? The Relationship between Interpersonal Interaction Based on Motivation and Educational Outcomes among Students in Manufacturing Programs at Two-Year Technical Colleges," *Community College Review* 44, no. 1 (2015): 26–48, 10.1177/0091552115612560.

[33] Frank R. Ascione and Claudia V. Weber, "Children's Attitudes about the Humane Treatment of Animals and Empathy: One-Year Follow Up of a School-Based Intervention," *Anthrozoös* 9, no. 4 (1996): 188–95, 10.2752/089279396787001455.

[34] Christine Y. Tardif-Williams and Sandra L. Bosacki, "Evaluating the Impact of a Humane Education Summer-Camp Program on School-Aged Children's Relationships with Companion Animals," *Anthrozoös* 28, no. 4 (2015): 587–600, 10.1080/08927936.2015.1070001.

reading topics about compassion toward animals, students would have been more personally impacted and paid those readings increased attention.

It is interesting to note that there was no difference between the scores of the boys and the girls in the experimental group. Research shows that girls have more compassion toward animals. Eckardt Erlanger and Tsytsarev[35] explored the role of empathy and personality characteristics in attitudes toward the treatment of animals in university students. Female students scored higher on empathy than their male peers and expressed more positive attitudes toward animals.[36] However, in a study of the attitudes of teachers who used humane education with their students, the educators believed that the boys and girls were both equally affected by lessons about animal welfare, with some noticing a difference when the students got older.[37] Most of the students in the experiment were eleven and twelve years old at the time of the study. The results demonstrate that both binary genders can benefit from the intervention of humane education in language arts.

Common Core and Higher-Order Thinking Skills

The questions in this study aligned with the "reading for information" Common-Core anchor standards, and they required higher-order thinking skills. Some of the questions that students answered involved inferring, summarizing, and determining the motivation of characters in the texts. Children who received humane education passages scored higher on those questions than did their non-humane education counterparts. Those higher-order thinking tasks required the students to think on a level that was more advanced than memorization or parroting back information. Therefore, the experimental intervention of humane education was better at eliciting the skills necessary for analyzing and evaluating, which are the foundations to problem solving. This is in line with stated

[35] Ann C. Eckardt Erlanger and Sergei V. Tsytsarev, "The Relationship between Empathy and Personality in Undergraduate Students' Attitudes toward Nonhuman Animals," *Society & Animals* 20, no. 1 (2012): 21–38, 10.1163/156853012 × 614,341.

[36] Eckardt Erlanger and Tsytsarev, "Relationship between Empathy and Personality."

[37] O'Connor, "Qualitative Case Study."

goals of humane education. Humane education should develop critical thinking, problem solving, and perspective-taking.[38] Therefore, it is unsurprising that the experiment results confirm that passages containing humane education themes elicit those abilities in students.

Limitations

An important limitation of this study was that it occurred in one school district. More varied samples of participants may increase the evidence of humane education's impact on diverse student populations. Further, the reading passages did not match the waves insofar as increasing difficulty. Wave 1 was not an easier reading level or a lower word count than wave 5. Therefore, it is not possible to determine whether the experimental-group students would have improved their scores from one wave to the next on increasingly challenging assessments. Moreover, given the limitations of designing a test that requires text-based questions, there is an inherent inability to make it a completely sound measure. However, the validity of the study is supported by the fact that the passages and questions were matched as closely as possible: assessing the same standards, matching quantifiable reading-difficulty measures, using the same font, having the same number of graphics and pictures, and so on.

24.6 Conclusion

This study suggests that humane education can have a positive effect on standards-based reading comprehension. This can be measured through use of standards-aligned reading passages with text-based questions requiring higher-order thinking skills. Although the study was not designed to determine *why* humane education affected the students, we believe it was through changes in student motivation. This study builds on prior research showing that animals are important to young people and that humane education has the potential to improve student performance. This research provides support for the idea that student interest in animals and animal topics may be used to improve academic performance in language arts assessments.

[38] Stephanie Itle-Clark and Erin Comaskey, "A Proposal for a Humane Pedagogy," *International Journal of Humane Education* 1, no. 1 (2020): 1–31, https://www.prosocialacademy.org/ijhe.

Future research in this area should compare the effects of humane education themes in reading assessments for both younger and older students. Reading assessments containing varied animal topics and species should be studied to determine their influence, as to which areas of humane education intrigue students more and increase their performance. Humane education-themed fiction passages should be assessed in addition to the nonfiction ones in this research. Future studies should take place over a longer period of time in an attempt to determine cumulative effects of humane education on reading performance.

Bibliography

Academy of Prosocial Learning. "Home Page." Accessed July 1, 2021. https://www.prosocialacademy.org.

Anderson, Duncan, Sholom Feldblum, Claudine Modlin, Doris Schirmacher, Ernesto Schirmacher, and Neeza Thandi. *A Practitioner's Guide to Generalized Linear Models*. 3rd ed. Arlington, VA: Casualty Actuarial Society Discussion Paper Program. 2007.

Anderson, Lorin, David Krathwohl, and Benjamin Bloom. *A Taxonomy for Learning, Teaching, and Assessing: A Revision of Bloom's Taxonomy of Educational Objectives*. Boston: Allyn & Bacon, 2001.

Ascione, Frank R., and Claudia V. Weber. "Children's Attitudes about the Humane Treatment of Animals and Empathy: One-Year Follow Up of a School-Based Intervention." *Anthrozoös* 9, no. 4 (1996): 188–95. https://doi.org/10.2752/089279396787001455.

Baştuğ, Muhammet, and H. Keskin. "The Role of Text Length in Repeated Reading." *European Journal of Education Studies* 6, no. 3 (September 2014). https://oapub.org/edu/index.php/ejes/index.

Chan, Hsun-yu, and Xueli Wang. "Interact for What? The Relationship between Interpersonal Interaction Based on Motivation and Educational Outcomes among Students in Manufacturing Programs at Two-Year Technical Colleges." *Community College Review* 44, no. 1 (2015): 26–48. https://doi.org/10.1177/0091552115612560.

Chiang, Hui-Hua. "English Vocabulary Size as a Predictor of TOEIC Listening and Reading Achievement among EFL Students in Taiwan." *Theory and Practice in Language Studies* 8, no. 2 (2018): 203. https://doi.org/10.17507/tpls.0802.04.

Clark, Adam. "N.J. Revises, Renames Common Core Academic Standards." *NJ.com*, May 4, 2016.

Doman, Mary. "Lexile Levels: What Parents Need to Know." Scholastic. Accessed July 11, 2021. https://www.scholastic.com/parents/books-and-reading/reading-resources/book-selection-tips/lexile-levels-made-easy.html.

Duff, Dawna, J. B. Tomblin, and Hugh Catts. "The Influence of Reading on Vocabulary Growth: A Case for a Matthew Effect." *Journal of Speech, Language, and Hearing Research* 58, no. 3 (2015): 853–64. https://doi.org/10.1044/2015_jslhr-l-13-0310.

Eckardt Erlanger, Ann C., and Sergei V. Tsytsarev. "The Relationship between Empathy and Personality in Undergraduate Students' Attitudes toward Nonhuman Animals." *Society & Animals* 20, no. 1 (2012): 21–38. https://doi.org/10.1163/156853012x614341.

Faver, Catherine A. "School-Based Humane Education as a Strategy to Prevent Violence: Review and Recommendations." *Children and Youth Services Review* 32, no. 3 (2010): 365–70. https://doi.org/10.1016/j.childyouth.2009.10.006.

Hirsch, E. D., Jr. "What Do Reading Comprehension Tests Mainly Measure? Knowledge." *American Educator*, Spring 2006. https://www.aft.org/periodical/american-educator/spring-2006/what-do-reading-comprehension-tests-mainly.

Humane Society of the United States. *The Empathy Connection*. HSUS, 2004. http://www.humanesociety.org/assets/pdfs/abuse/empathy-connection.pdf.

Itle-Clark, Stephanie, and Erin Comaskey. "A Proposal for a Humane Pedagogy." *International Journal of Humane Education* 1, no. 1 (2020): 1–31. https://www.prosocialacademy.org/ijhe.

Kincaid, J. Peter, Robert P. Fishburne, Richard L. Rogers, Brad S. Chissom. "Derivation of New Readability Formulas (Automated Readability Index, Fog Count and Flesch Reading Ease Formula) for Navy Enlisted Personnel." Institute for Simulation and Training, 1975. https://doi.org/10.21236/ada006655.

Lennon, Colleen, and Hal Burdick. *The Lexile Framework as an Approach for Reading Measurement and Success*. MetaMetrics, 2004.

LoBue, Vanessa, Megan Bloom Pickard, Kathleen Sherman, Chrystal Axford, and Judy S. DeLoache. "Young Children's Interest in Live Animals." *British Journal of Developmental Psychology* 31, no. 1 (2012): 57–69. https://doi.org/10.1111/j.2044-835x.2012.02078.x.

Loder, Tondra L. "Why We Can't Leave Public Schools Behind: The Inseparable Legacy of Public Education and American Democracy." *Educational Researcher* 35, no. 5 (2006): 30–35. https://journals.sagepub.com/home/edr.

National Governors Association (NGA) Center for Best Practices and Council of Chief State School Officers (CCSSO). *Common Core State Standards*. Washington, DC: NGA Center for Best Practices and CCSSO, 2010.

Nisbet, Deanna L., and Evie R. Tindall. "A Framework for Explicit Vocabulary Instruction with English Language Learners." *Kappa Delta Pi Record* 51, no. 2 (2015): 75–80. https://doi.org/10.1080/00228958.2015.1023141.

O'Connor, Julie. "A Qualitative Case Study of Teacher Perceptions of the Motivation of Students in Humane Education." PhD diss., Northcentral University, 2018. ProQuest (10744993).

114th Congress. *Every Student Succeeds Act: Conference Report (to Accompany S. 1177)*. 2015. https://www.govtrack.us/congress/bills/114/s1177.

People for the Ethical Treatment of Animals. "K–12 Lesson Plans and Classroom Tips for Compassionate Teachers." PETA, last modified 2021. https://www.peta.org/teachkind/.

Schmitt, Norbert, Xiangying Jiang, and William Grabe. "The Percentage of Words Known in a Text and Reading Comprehension." *Modern Language Journal* 95, no. 1 (2011): 26–43. https://doi.org/10.1111/j.1540-4781.2011.01146.x.

Svensson, Agneta. "The Impact of the Animals on Children's Learning and Their Development—A Study of What Children Learn from and with Pets: The Example of Dog and Cat." *Problems of Education in the 21st Century*, no. 59 (2014): 77–85.

Tankersley, Karen. "Higher-Order Thinking." In *Literacy Strategies for Grades 4–12: Reinforcing the Threads of Reading*. Alexandria: ASCD, 2005.

Tardif-Williams, Christine Y., and Sandra L. Bosacki. "Evaluating the Impact of a Humane Education Summer-Camp Program on School-Aged Children's Relationships with Companion Animals." *Anthrozoös* 28, no. 4 (2015): 587–600. https://doi.org/10.1080/08927936.2015.1070001.

Williams, Kaylene C., and Caroline C. Williams. "Five Key Ingredients for Improving Student Motivation." *Research in Higher Education Journal*, no. 12 (2011): 1–23. http://aabri.com/manuscripts/11834.pdf.

CHAPTER 25

Raising Consciousness About Animals Across the Curriculum: A Personal Journey

Maureen O'Sullivan

Contents

M. O'Sullivan (✉)
School of Law, National University of Ireland, Galway, Ireland

C. Linzey and A. Linzey (eds.), *The Palgrave Companion to Humane Education*, The Palgrave Macmillan Animal Ethics Series,
https://doi.org/10.1007/978-3-032-19795-5_25

25.1 Introduction

I lecture in law at the University of Galway in Ireland, and this chapter describes how I have introduced issues related to animals and the law in courses that traditionally would not have accommodated this subject. In the absence of a dedicated course on animal law at my university and given that this area is at the cutting edge, I made the decision to include a topic on personhood in one of my modules, Sociology of Law. I also was able to introduce in another module, Industrial and Intellectual Property Law, aspects of the rights of animals in topics related to both copyright and patents. The examination of animal-related issues is more fully fledged in Sociology of Law than in Industrial and Intellectual Property, but the aim is to raise consciousness among students about novel legal topics, such as the rights of animals, and allow students to make their own decisions about the material. My aim is not to evangelize: I take my duties as a lawyer seriously, and students are there first and foremost to learn to be good advocates. Lecturers do have some freedom, however, in selecting material, and animal law is quite interesting from a legal viewpoint, given that new rights are being identified and recognized, and so in a sense, the sands are shifting.[1] I believe strongly in training budding lawyers to see all sides of a case so that if they pursue further studies in this area, they will know the counterarguments to their advocacy. Furthermore, if they agree with animal rights, they will augment their knowledge in these courses, and if not, at least they will become acquainted with another way of seeing the matter and perhaps will open their minds. I have received positive feedback from students: they tell me that they appreciate the information and also the manner in which it is delivered. I have never encountered resistance to this enlightenment. Vegan students apparently speculate that I may too be vegan, and some of my students have told me that because of my classes, they decided to give up meat. Some students are aware of my fellowship with the Oxford Centre for Animal Ethics and my involvement with the Vegetarian Society of Ireland, but these affiliations stay in the background in my classes so that students do not feel pressured to think in a particular way.

In the following section, I relate how material on the rights of animals can be included in intellectual property topics and follow thereafter with

[1] I will introduce a module entitled Animal Law for undergraduate students in January 2025. This will be the first Animal Law course in a law school in Ireland.

an account of how I integrate a case study on personhood in my Sociology of Law class. This overview might be useful for other teachers and lecturers who are grappling with how to teach a topic of interest where there does not appear to be a framework. Necessity is the mother of invention, as they say, and where there is a will, a way can be found.

25.2 Personhood in Industrial and Intellectual Property

Industrial and Intellectual Property is a one-semester course of five European Credit Transfers for undergraduate law students and is taught over twelve weeks in two-hour slots. Unfortunately, we do not have seminars or tutorials, but students often venture by to talk to me about the material during my office hours. Some recent cases concerning certain rights of animals have arisen in the area of copyright and patents, and I integrate these into my teaching in the area. Ireland is a very small jurisdiction, and awareness of the international scene is useful for students. In any case, in a globalized world, intellectual property, while legislated at the national level, also is regulated regionally and internationally. Cases and legislation in other common-law jurisdictions cast a shadow over Irish courts—in the terrain of intellectual property, Ireland certainly is not an island.

The issue that has arisen in copyright on the international stage is whether animals can own their own artworks.[2] This matter arose from the experience of a freelance photographer, one David Slater, who was taking photographs in the Indonesian jungle some years ago. While he was having a rest, a macaque grabbed his camera and took a selfie, which Slater subsequently published in a book of wildlife photography.[3] The photograph appeared on Wikimedia and other online news sites shortly thereafter. Slater asked Wikimedia to take down the photograph, but the site refused, claiming that the shot was in the public domain. Slater contested this and asserted ownership of the photo. In copyright law the owner of a photograph is not necessarily the proprietor of the camera

[2] I deal with this in some detail in a previous article: Maureen O'Sullivan, "Irish Artistic Copyright Law: A Menagerie of Holy Cows and Turtle Doves?," *Intellectual Property Quarterly* 1 (2015): 31–61, 50–51.

[3] Reuters, "Monkey Who Took Selfie Cannot Sue for Copyright, Court Rules." *Irish Times*, April 24, 2018, https://www.irishtimes.com/news/world/us/monkey-who-took-selfie-cannot-sue-for-copyright-court-rules-1.3472802.

but rather the person who takes the photo. Because the monkey had taken the photo, Wikimedia responded that the photographer had no infringeable rights. Soon afterward the US Copyright Office made a statement clarifying that animals cannot own any artworks that they create, requiring that the author be human.[4] Subsequently, the nonprofit organization People for the Ethical Treatment of Animals (PETA) took a case on behalf of the macaque, seeking to assert his ownership over the photograph and claiming damages from Slater.[5] This ultimately was not successful; the district judge in a federal court in San Francisco stated that while Congress could extend rights to animals, there was no evidence of an intention to do so in the Copyright Act.[6] PETA then appealed this judgment, and the parties subsequently agreed to a settlement. However, the Court of Appeals for the Ninth Circuit affirmed the judgment of the lower court that monkeys cannot own copyright in their own works.[7]

What students learn from this aspect of copyright is twofold. First, most students are not aware that animals such as gorillas and elephants can paint, and I show them slides of paintings done by such animals. We discuss an art prize for apes that is sponsored by the Humane Society

[4] U.S. Copyright Office, *The Compendium of US Copyright Office Practices*, Third Edition, chap. 300, accessed August 16, 2024, https://www.copyright.gov/comp3/chap300/ch300-copyrightable-authorship.pdf. At paragraph 306, it states:

> The US Copyright Office will register an original work of authorship, provided that the work was created by a human being. The copyright law only protects "the fruits of intellectual labor" that "are founded in the creative powers of the mind." Trade-Mark Cases, 100 U.S. 82, 94 (1879). Because copyright law is limited to "original intellectual conceptions of the author," the Office will refuse to register a claim if it determines that a human being did not create the work. Burrow-Giles Lithographic Co. v. Sarony, 111 U.S. 53, 58 (1884).

The text continues: "For representative examples of works that do not satisfy this requirement," readers are directed to see Sect. 313.2 of the compendium.

[5] *Naruto v. David John Slater* et al., No. 3:2015cv04324—Document 45 (N.D. Cal. 2016), https://cases.justia.com/federal/district-courts/california/candce/3:2015cv04324/291324/45/0.pdf?ts=1454149106.

[6] Andrew Chung, "Monkey Cannot Own Copyright to 'Selfie,' U.S. Judge Says," Reuters, January 7, 2016, https://www.reuters.com/article/us-usa-california-monkey-idUSKBN0UL28620160107.

[7] *Naruto v. Slater*, 888 F.3d 418, 426 (2018), https://www.courthousenews.com/wp-content/uploads/2018/06/SlaterNaruto.pdf.

of the United States to show that this is a genuine area of concern.[8] Second, it is educational for my class to realize that there is a movement to recognize rights to one's own works, be the owner human or nonhuman. It legitimizes the area and gives students food for thought. I believe it to be an exercise not only in learning but also in consciousness-raising.

The second area in which the rights of animals are broached with students is patents and biotechnology. We look at this topic from an international perspective, and I include material on an ongoing discussion regarding the rights of hybrid beings created in the laboratory, especially in the United States. Long before there was any regulation in place, biotechnologists were mixing genetic material across species and seeking patents on their "inventions," ever since the discovery of the DNA helix in the 1950s. This practice raised a number of constitutional issues discussed in academic literature[9] and culminated in an intervention by activists who wished to arrest this practice, known as the "Newman-Rifkin" patent[10] in the late 1990s. Jeremy Rifkin, a biotech campaigner, and Stuart Newman, a scientist, applied to the US Patent and Trademark Office (USPTO) for a number of patents on "humanzees," "humice," and other combinations of human and animal genes. For Rifkin and Newman, the expected outcome of the application was either that they would gain a patent, in which case they would refuse to license the technology and instead would promulgate a public debate during the subsequent twenty-year duration of the patent, or that conversely, the USPTO would refuse the patent and, they hoped, give reasons as to why. It did not turn out quite that way because though the patents were refused seven years after the application, the justifications given were somewhat obscure. One provided reason was

[8] Harriet Gibsone, "Chimpanzee Wins $10,000 Prize for Abstract Painting," *Guardian*, August 30, 2013, https://www.theguardian.com/artanddesign/2013/aug/30/chimpanzee-wins-10000-dollars-abstract-painting.

[9] There are a plethora of articles, among which the following are most informative: Margo A. Bagley, "Patent First, Ask Questions Later: Morality and Biotechnology in Patent Law," *William and Mary Law Review 45* (2003): 469; Ryan Hagglund, "Patentability of Human–Animal Chimeras," *Santa Clara High Technology Law Journal* 25, no. 1 (2012): 51, https://digitalcommons.law.scu.edu/cgi/viewcontent.cgi?article=1477&context=chtlj; and Thomas A. Magnani, "The Patentability of Human–Animal Chimeras," *Berkeley Technology Law Journal* 14 (1999).

[10] Rick Weiss, "U.S. Denies Patent for a Too-Human Hybrid," *Washington Post*, February 13, 2005, A03, http://www.washingtonpost.com/wp-dyn/articles/A19781-2005Feb12.html.

privacy infringement—because patent law could potentially restrict the patent object's right to procreate, given that the patent would allow the owner to exclude others from making the invention. Moreover, there was a possible violation of the Thirteenth Amendment to the Constitution prohibiting slavery. Patent law permits owners to prevent others from using the invention, and the term "use" can also mean "employ." This would resemble involuntary servitude and could also restrict the right to travel. The deputy commissioner for patents, John Doll, said, "I don't think anyone knows in terms of crude percentages how to differentiate between humans and nonhumans," and he acknowledged that guidance from Congress or the courts would be welcome.[11] Students are enthralled at the magnitude of the ethical dilemmas raised and come to appreciate keenly how similar animals and humans actually are.

Pronouncements about such patents being unconstitutional are not within the jurisdiction of the USPTO because only the superior courts can interpret the constitution. Legislation introduced in 2011 now makes it impossible to get patents on "human organisms,"[12] although defining what this means also will fall within the purview of the courts. This legislation is not retrospective and therefore does not affect those patents on human-animal hybrids and chimeras that have already been granted. This raises profound questions about what it means to be human and, indeed, animal. Many students now join my master's course titled Advanced Intellectual Property Law and Development because this has piqued their interest in biotechnology, morality, and patents,[13] and I plan to develop this topic further in the future.

This second study in Industrial and Intellectual Property appears to bring students into the area of science fiction—except that this matter is very real. The discovery of the DNA helix in the 1950s allowed scientists to unlock genetic codes, and species barriers, which ironically some animal rights activists abhor, were broken down, allowing beings to be built either through hybridization, where two parents from different species give rise to offspring, or else through the creation of chimeras. Chimeras have a multitude of parents and distinct organs from different sources: a

[11] Weiss, "U.S. Denies Patent."

[12] The America Invents Act, 2011, s.33(a), https://www.uspto.gov/sites/default/files/aia_implementation/human-organism-memo.pdf.

[13] Maureen O'Sullivan, *Patents, Morality and Biotechnology: A Deliberative and Participatory Paradigm for Reform* (London: Routledge, 2019).

sheep might be engineered to contain a human liver that will be used in transplantation, with the animal put into the food chain after the nearly human organ has been excised. The science is now so complex that a group of European scientists coined a new term to reflect this ontological conundrum: chimbrids.[14] These beings who do not exist in nature may in time delineate a new rights frontier in addition to the area of animal rights. This creates a challenge with an opportunity to explore the boundary between animals and human beings.

25.3 Where Do We Set the Boundary Between Humans and Animals?

Most undergraduate law students learn about compulsory subjects such as contract, company, criminal, and constitutional law, along with others. When I was pursuing my degree, human rights was an optional subject that I did not take. There was no course on animal rights, welfare, or law, and this has not changed to date.[15] Regardless of whether students study human rights, even the name itself suggests that rights are the preserve of human beings alone, so in the classroom I start my inquiry at that point. I ask students to reflect on whether human beings are different from animals or whether we are all primates. It may be that in Ireland, many students have inherited an idea about animals not having souls through religious instruction, so the frontier is established early on and not often questioned. Many students also come from farming backgrounds, and their livelihoods may depend on the use and killing of animals.[16]

We first consider the ideas of René Descartes, the French philosopher whose phrase "I think, therefore I am"[17] gave rise to the notion that the ability to reason defined his humanity and set him apart. Descartes opined

[14] Jochen Taupitz and Marion Weschka, *CHIMBRIDS—Chimeras and Hybrids in Comparative European and International Research: Scientific, Ethical, Philosophical and Legal Aspects* (Berlin: Springer Science and Business Media, 2009).

[15] I will introduce Ireland's first module on Animal Law at University of Galway in January 2025.

[16] A telling tale of modern farming is provided in John Connell's *The Cow Book* (London: Granta, 2018).

[17] René Descartes, *Discourse on the Method of Rightly Conducting the Reason and Seeking Truth in the Sciences* (1637), trans. John Veitch (Project Gutenberg, 1995, last updated 2022), https://www.gutenberg.org/files/59/59-h/59-h.htm.

that animals lacked this faculty, were automatons, and could rightfully be controlled by human beings. This meant that humans could do unto animals as we wished—animal reactions were only reflexes, and Descartes conducted on them very cruel experiments, denying their pain. In class we also discuss how the book of Genesis casts an influence on our treatment of animals, in that humans appear to have been given dominion over the earth. However, it is worth noting that Catholicism does not focus as deeply on the Old Testament as on the New, so perhaps notions of divinely sanctioned human dominance over animals are not deeply entrenched. I introduce students to a sample of the works of Professors Andrew Linzey[18] and Clair Linzey[19] on interpreting Christianity as not sanctioning animal use and abuse as traditionally understood. This gives us an opportunity to mention the ills of industrialized farming, which was not practised when Genesis was written, meaning biblical support for modern farming practices is absent. We see some slides of collateral damage to animals during natural disasters such as floods where farming is industrialized and where animals cannot escape. Often thousands of animals are abandoned to their deaths, locked in and deprived of liberty. The material I use is not gory but rather informative only, and many students are moved by this.

During this discussion, we also touch on the ethics of other religious traditions such as Hinduism, Buddhism, and Jainism, which follow, to greater or lesser degrees, the principle of ahimsa or "do no harm." We talk about how this principle is tied up with belief in reincarnation, karma, and responsibility for our behavior. This may involve the avoidance of any harm to animals or strict dietary rules.

Students are often surprised to discover that the iconic lawyer Jeremy Bentham also wrote about animals[20] and held that the significant line that should be drawn between humans and animals was at the ability to suffer or sentience. If an animal could feel pain, she should not be tortured, Bentham argued, although his theories fall short of advocating

[18] Andrew Linzey, *Christianity and the Rights of Animals* (London: SPCK, 1987; New York: Crossroad, 1989).

[19] Andrew Linzey and Clair Linzey, eds., *The Routledge Handbook of Religion and Animal Ethics* (London: Routledge, 2018).

[20] Cass R. Sunstein, "Introduction: What are Animal Rights?" in *Animal Rights: Current Debates and New Directions*, Cass R. Sunstein and Martha C. Nussbaum, eds. (New York: Oxford University Press, 2004).

for animal liberation. We also note the recent laws in Switzerland that set out detailed animal welfare, and the work of the animal rights defense lawyer Antoine Goetschel is noted. I make students aware of the international database of animal rights laws that Goetschel has established, Global Animal Law.[21] We discuss the late Steven Wise's work later.

From a legal point of view, we identify that at present, most human beings are classified as persons, which gives them liberty, whereas most animals are classified as property, which puts them under the control of human beings. Even though Article 13 of the Treaty on the Functioning of the European Union recognizes that animals are sentient beings, they are still currently classified as property and not persons. This appears strange to students when we talk about the fact that corporations have legal personhood, which means they have legal rights and duties: the right to sue and be sued and to enter into contracts. The US Supreme Court, for instance, has held that a corporation is "capable of being treated as a citizen of [the State which created it], as much as a natural person."[22] Companies have privacy rights and can sue for defamation, as did McDonalds in the 1990s in the so-called McLibel case in the United Kingdom that it filed against two activists who criticized their products.[23]

We then interrogate the legal recognition of the circle of humanity and show that it has not always been all-embracing toward humanity itself. Students come to appreciate that race and gender have been used as barriers to human emancipation. For instance, in the US case of *Standing Bear v. Crook*,[24] Native American rights to habeas corpus were recognized by Judge Dundy, who stated that "An Indian is a person," showing that this matter was not regarded as self-evident. Indeed, in an article in *Punch's Magazine* in 1862, the Irish were identified as the "missing" link between apes and men.[25] This is a very powerful point in Ireland, as it taps into our collective consciousness as a former colony. Racism against

[21] Global Animal Law, accessed April 9, 2020, https://www.globalanimallaw.org/who-we-are.html.

[22] *Louisville, C. & C.R. Co. v. Letson*, 2 How. 497, 558, 11 L.Ed. 353 (1844).

[23] *McDonald's Corporation v. Steel & Morris*, [1997] EWHC QB 366.

[24] *U.S. ex Rel. Standing Bear v. Crook*, 25 F. Cas. 695 (D. Neb. 1879), https://casetext.com/case/us-ex-rel-standing-bear-v-crook.

[25] Ronnie O'Gorman, "A Time When the Irish Were Not Welcome," *Galway Advertiser*, October 22, 2015, https://www.Advertiser.Ie/Galway/Article/80454/A-Time-When-The-Irish-Were-Not-Welcome.

Indigenous peoples still abides, and they are often dehumanized by being compared to monkeys or apes. They can be denied standing, and Aborigines in Australia were often described as "flora and fauna" until 1967, when they were included in the census for the first time.[26]

Women have not fared much better, historically. For instance, the Married Women's Property Act 1882 in the United Kingdom allowed married women to own their own property after marriage for the first time. Before this, for centuries marriage signified a civil death for women: they, their possessions, and their children became the property of their husbands within the marriage contract. Feminist literature abounds with explorations of how women are still commodified and denied their humanity both in society and in the legal system. A particularly vivid portrayal can be seen in the magazine *Hustler*, which once depicted women as mincemeat. Students grasp that human rights are not the preserve of all humanity, and in my experience, this makes them receptive to hearing that we may need to revise our laws relating to animals.

Having seen an image of women ground up as meat, we then turn our attention to the topic of meat consumption and acknowledge that in most societies the consumption of human flesh is taboo. Depending on the country, we also have cultural taboos about eating certain animals, even though the reasons may not be so clear,[27] and such taboos vary from country to country. For our laws to be legitimate, their premise should be reasonable, so that premise needs to be questioned. The cozy divide may erode where genetically engineered beings containing large quantities of human DNA enter the food chain.

The next part of this class considers how laws on separate legal personality or personhood have been extended to nonhuman subjects. This shows the glaring lacuna of sentient beings not having rights of their own while corporations, which are inanimate entities, are entitled to rights enjoyed by humans. Moreover, such rights are now also being extended to

[26] Paul Daley, "It's 50 Years Since Indigenous Australians First 'Counted.' Why Has So Little Changed?," *Guardian*, May 18, 2017, https://www.theguardian.com/inequality/2017/may/18/50-years-since-indigenous-australians-first-counted-why-has-so-little-changed-1967-referendum.

[27] People in Ireland and the United Kingdom were horrified some years ago when it was discovered that horse meat was present in many processed meat products. Felicity Lawrence, "Horsemeat Scandal: The Essential Guide," *Guardian*, February 15, 2013, https://www.theguardian.com/uk/2013/feb/15/horsemeat-scandal-the-essential-guide.

natural objects, such as rivers, mountains, and areas of ecological importance. Advocacy of this novel approach toward the natural world arose in the area of environmental law with Christopher Stone's seminal article on the matter of standing or legal rights for trees, which was published in the 1970s.[28] His theory influenced Justice William O. Douglas, who gave a dissenting judgment in *Sierra Club v. Morton*[29] in which he accepted the idea of granting personhood beyond the human circle to natural objects. This raises a question as to whether we protect natural objects for human-centered interests or whether natural objects should have interests that are independent—for instance, a right to be left alone. Recognizing standing for natural objects would require the establishment of a guardianship paradigm for the environment. This is comparable to the role of the trustee in bankruptcy for the endangered corporation. Students tend to be quite knowledgeable about our environmental crisis and are very receptive to learning about different legal structures that can help to improve the status of the natural world and, indeed, animals.

Stone's thesis has been actualized in New Zealand, which granted the River Whanganui personhood status in 2017. This status also has been extended to some other natural objects such as woods and mountains and involves an agreement between the government and the Iwi tribe.[30] State-appointed trustees work together with representatives from the Iwi, and both are obliged to act in the interests of the river. This has led to the development of a sort of bill of rights for the river, which has separate legal standing and cannot be owned outright.

The Andean nations of Ecuador, Bolivia,[31] and Colombia[32] each have introduced some rights for nature in constitutional or legislative form within the past two decades. When tested in court, these rights have

[28] Christopher Stone, "Should Trees Have Standing? Toward Legal Rights for Natural Objects," *Southern California Law Review* 45 (1972): 450–501.

[29] *Sierra Club v. Morton*, 405 U.S. 727 (1972).

[30] Te Awa Tupua (Whanganui River Claims Settlement) Act 2017.

[31] John Vidal, "Bolivia Enshrines Natural World's Rights with Equal Status for Mother Earth," *Guardian*, April 10, 2011, https://www.theguardian.com/environment/2011/apr/10/bolivia-enshrines-natural-worlds-rights.

[32] Mari Margil, "Our Laws Make Slaves of Nature: It's Not Just Humans Who Need Rights," *Guardian*, May 23, 2018, https://www.theguardian.com/commentisfree/2018/may/23/laws-slaves-nature-humans-rights-environment-amazon.

been recognized and upheld,[33] although implementation can prove difficult.[34] The trend toward extending personhood beyond the human remit is growing. These developments are very enlightening for the students, many of whom feel a great affinity with nature and are concerned about environmental damage. They gain an appreciation of the interests of animals within this context, if they have not already considered the matter.

Most vegans and vegetarians at some point encounter questions about whether plants also should be considered as suffering through our consumption of them, and I just touch on this issue by identifying the works of Michael Marder, a professor of philosophy at the University of the Basque Country who writes extensively on this topic. We discuss this in the context of the tension between compassion and survival. Students also learn that there have been and are political parties that advocate for the rights of animals, such as in the Netherlands, with its Party for the Plants and Party for the Animals, and these are becoming more commonplace across Europe. Such parties have now been registered in the United Kingdom[35] as well as in Ireland.[36] We note how cultural festivals such as bullfighting have become controversial in the last few years and now attract many protesters.

25.4 Some Evidence for a Rethink on Our Legal Treatment of Animals

The foregoing material is designed to encourage students to question legal paradigms that are usually unquestioned in an undergraduate curriculum in terms of both their scope and their legitimacy. Singer and

[33] Hugo Echeverría, "Rights of Nature: The Ecuadorian Case," *Revista Esmat* 9, no. (2017), https://www.researchgate.net/publication/333708922_RIGHTS_OF_NATURE_THE_ECUADORIAN_CASE.

[34] Ashley Westerman, "Should Rivers Have Same Legal Rights as Humans? A Growing Number of Voices Say Yes," NPR, August 3, 2019, https://www.kcur.org/post/should-rivers-have-same-legal-rights-humans-growing-number-voices-say-yes#stream/0.

[35] Animal Welfare Party, accessed April 9, 2020, https://www.animalwelfareparty.org/.

[36] Bernie English, "New Political Party to Launch in Limerick," *Limerick Post*, March 1, 2019, https://www.limerickpost.ie/2019/03/01/new-political-party-to-launch-in-limerick/.

Cavalieri's *Great Ape Project*[37] is a very useful text because it contains a plethora of evidence-based material showing that some animals are not distinguishable from humans in characteristics that we may erroneously believe to be the case, such as speech. The great apes include chimpanzees, bonobos, orangutans, and gorillas. Cavalieri has since advocated that it be acknowledged that cetaceans can also recognize speech and language.[38]

Pressure is now being brought to bear on amusement parks that use cetaceans in their shows, with people calling on them to desist from the practice. This has grown out of publicity in the wake of the film *Blackfish*, about an orca, Tilikum, and one of the women he killed. Public opinion became mobilized, and support for the parks has fallen off, causing bans and changes of practice.

The culmination of this topic, personhood, in my Sociology of Law module is the work of the late Steven Wise and the Nonhuman Rights Project (NhRP). Cases taken on by Wise and NhRP have begun to accumulate since I started teaching in this area, and it is convincing for students that judges have agreed with Wise's arguments, even if they feel constrained in the legal paradigm within which they operate. Students have already been introduced to the concept of habeas corpus in the abstract, so it is enlightening to see it being relied on in practice. Wise, who passed away in 2024, and the NhRP have sought to have habeas corpus applied to all chimpanzees living in New York State and elsewhere. Publicity around the case of Tommy the chimpanzee earned that case a full front-page exposé in the magazine of the *New York Times* in 2014.[39]

[37] Paola Cavalieri and Peter Singer, eds., *The Great Ape Project* (New York: St. Martin's Press, 1994).

[38] Paola Cavalieri, "Given All We Know of Whales, Why Resume the Slaughter?," *Guardian*, April 6, 2010, https://www.theguardian.com/commentisfree/2010/apr/06/whale-hunting-moratorium-japan-iceland.

[39] Charles Siebert, "Should a Chimp Be Able to Sue Its Owner?," *New York Times*, April 23, 2014, https://www.nytimes.com/2014/04/27/magazine/the-rights-of-man-and-beast.html.

25.5 Some Change Coming for Animals' Status: The Elephant in the Room?

Consideration of the status of animals arose in the Canadian case of *Reece* et al. *v. Edmonton*,[40] which concerned an elephant held in captivity in 2011. In this case, the chief justice of Alberta acknowledged in a dissenting judgment that animals' interests were not sufficiently represented in animal welfare laws because humans tend to prioritize their own interests. Laws designed to protect animals are therefore interpreted narrowly.[41] The judge also gave some consideration to whether animals should be afforded some form of legal standing in their own right.[42] Even though the Canadian Supreme Court denied leave to appeal, the zoo officials decided that Lucy, the elephant in the case, would be the last elephant at the zoo.[43]

In the wake of several cases taken on behalf of chimpanzees, discussed subsequently, the Nonhuman Rights Project instituted proceedings on behalf of three elephants in Connecticut and New York in 2017. The elephants had been used in a circus. Habeas corpus claims made on their behalf were dismissed on the grounds that they relied on human rights precepts that animals could not enjoy.[44] The NhRP appealed unsuccessfully. However, the organization subsequently took another case involving an elephant by the name of Happy who is housed alone in the Bronx Zoo in New York. While denying Happy the relief of habeas corpus on the grounds of a lack of precedent, the judge, Justice Alison Y. Tuitt, acknowledged that Happy "is more than just a legal thing," that her treatment should include respect and dignity, and that she may merit liberty.[45] The ruling denying relief was appealed by the NhRP to the

[40] *Reece v. Edmonton*, 2011 ABCA 238.

[41] *Reece v. Edmonton*, 2011 ABCA 238.

[42] Tyler Totten, "Should Elephants Have Standing?," *Western Journal of Legal Studies* 6, no. 1 (2015), https://ojs.lib.uwo.ca/index.php/uwojls/article/view/5640.

[43] Ashifa Kassam, "The World's Coldest Elephant? Activists Demand Lucy's Removal from Canadian Zoo," *Guardian*, March 1, 2017, https://www.theguardian.com/world/2017/mar/01/lucy-the-elephant-canada-edmonton-valley-zoo.

[44] *The Nonhuman Rights Project, Inc. v. R. W Commerford & Sons Inc.*, 2017 WL 7,053,738 (Conn Sup Ct).

[45] *The Nonhuman Rights Project, Inc., on behalf of Happy, v. James J. Breheny and Wildlife Conservation Society* 2020 N.Y. Slip Op. 35,291 (N.Y. Sup. Ct. 2020), accessed

Court of Appeals in New York[46] which is the highest court in the state. Whilst the Court of Appeals upheld the ruling of the Supreme Court which declined to issue a writ of habeas corpus in Happy's favor, there were two powerful dissenting judgments from Wilson and Rivera JJ. The majority of the judges' principal grounds for refusing to extend habeas corpus to cover elephants was that this relief only applied to humans.[47] The judges believed that a decision in favor of the elephant would have a profound effect on our use of animals.[48] Justice Wilson, dissenting, did not assert that Happy is a person but he examined the origin and use of the writ of habeas corpus in some detail. He opined that it is a flexible device which can advance the law in its common law, rather than legislative function and therefore is not confined to benefiting human persons only. It has been used in the past to free slaves and women who had been confined by their husbands. Wilson J also examined the nature of rights and denied that rights always exist in conjunction with obligations as some humans cannot fulfil obligations. Therefore, animals could not be denied rights just because they were unable to honor obligations. Moreover, our knowledge of animals' cognitive abilities and sentience has increased, and the law needs to be updated to reflect this.

Rivera J who also dissented believed that habeas corpus should be available to Happy to challenge her confinement, regardless of the fact that she is not human. Animals should be able to make a habeas corpus petition in a bid for liberty. The judge did not equate animals with humans but the question for the court was that of unjust confinement. Happy had done nothing, apart from being born as an elephant, to warrant being deprived of her liberty. An expert in elephant behavior, Dr Joyce Poole was quoted extensively about elephant behavior and how similar in certain aspects it is to human behavior. Rivera J agreed with Wilson J that habeas corpus could be used to transfer Happy from one facility to another rather than setting her free. As she is a wild animal, not a domestic one she cannot

April 9, 2020, https://www.nonhumanrights.org/content/uploads/HappyFeb182020.pdf?fbclid=IwAR0EIp_Sf5yhAy8h2I_d0R4waBFRaaxFtkh_kKjRVB_ZH_6IDOMojgsuptk.

46 *Nonhuman Rights Project, Inc. v. Breheny*, 38 N.Y.3d 555 (2022) 197 N.E.3d 921, 176 N.Y.S.3d 533, 2022 N.Y. Slip Op. 03,859.

47 *Nonhuman Rights Project, Inc. v. Breheny*, 38 N.Y.3d 555 (2022) 197 N.E.3d 921, 176 N.Y.S.3d 533, 2022 N.Y. Slip Op. 03,859, II, [1].

48 *Nonhuman Rights Project, Inc. v. Breheny*, 38 N.Y.3d 555 (2022) 197 N.E.3d 921, 176 N.Y.S.3d 533, 2022 N.Y. Slip Op. 03,859, II, [13].

be set free because it could be dangerous for humans and potentially fatal for Happy.

These cases raise several novel issues for students and encourage them to "think outside the box."

25.6 Monkeying Around Personhood?

The NhRP also took a case on behalf of four chimpanzees: Tommy,[49] a chimpanzee who was being held in a shed; Kiko,[50] who was confined to a cement building on the property of his "owner"; and Hercules and Leo,[51] two chimpanzees who had been loaned to a university for use in research. The NhRP filed a habeas corpus petition on behalf of Tommy,[52] but the Appellate Division of the New York Supreme Court found against NhRP, holding that a chimpanzee is not a person entitled to the rights and protections afforded by the writ of habeas corpus.[53] The rationale for the decision was that in order to be eligible to avail oneself of habeas corpus relief, the subject must be capable of holding both rights and duties, and chimpanzees are considered incapable of both. This was despite the NhRP's argument that not everyone has the capacity to bear duties, yet those individuals do not forgo their rights as a result. Nonetheless, the court drew the line at habeas corpus extending beyond human beings.[54]

In another hearing, before Judge Jaffe of the New York Supreme Court, a grant of habeas corpus was denied to the aforementioned

[49] *People ex rel. Nonhuman Rights Project, Inc. v. Lavery*, 998 N.Y.S.2d 248, 124 A.D.3d 148, 2014 N.Y. Slip Op. 8531 (N.Y. App. Div. 2014).

[50] *Nonhuman Rights Project, Inc. ex rel. Kiko v. Presti*, 2015 N.Y. Slip Op. 85 (N.Y. App. Div. 2015).

[51] *The Nonhuman Rights Project, Inc. v. Stanley*, 2015 NY Slip Op 25,257 [49 Misc 3d 746], http://www.courts.state.ny.us/REPORTER/3dseries/2015/2015_25257.htm.

[52] *People ex rel. Nonhuman Rights Project, Inc. v. Lavery*, 998 N.Y.S.2d 248, 124 A.D.3d 148, 2014 N.Y. Slip Op. 8531 (N.Y. App. Div. 2014).

[53] *People ex rel. Nonhuman Rights Project, Inc. v. Lavery*, 998 N.Y.S.2d 248, 124 A.D.3d 148, 2014 N.Y. Slip Op. 8531 (N.Y. App. Div. 2014).

[54] *Nonhuman Rights Project, Inc. v. Lavery*, 152 AD3d 73, 78 [1st Dept 2017]; see *also People ex rel. Nonhuman Rights Project, Inc. v. Lavery*, 124 AD3d 148, 152 [3d Dept 2014], lv denied 26 NY3d 902 [2015].

Hercules and Leo because the judge held that she was bound by the decision in *Tommy* "for now," implying that this may change in the future. However, the outcome for these two chimpanzees was positive, in that, similar to the elephant in *Edmonton v. Reece*, they were both moved to a sanctuary. Kiko was not so fortunate at first instance, and he was denied transfer to a sanctuary in 2015 because his "immediate release" was not being sought by the NhRP.[55]

Subsequently, in the Appellate Division of the New York Supreme Court, which is New York State's highest court, the court declined to grant a writ of habeas corpus in a joint hearing on the cases of Tommy and Kiko,[56] upholding the lower court's judgment that the chimpanzees were incapable of enjoying rights and bearing duties. Justice Fahey wrote a concurrence, and while he did not depart from his fellow judges' views, he used the opportunity to offer his reflection on the matter. He observed that deep matters of ethics and policy would have to be resorted to rather than mere dictionary definitions. He also stated the following:

> In the interval since we first denied leave to the Nonhuman Rights Project, I have struggled with whether this was the right decision. Although I concur in the Court's decision to deny leave to appeal now, I continue to question whether the Court was right to deny leave in the first instance. The issue whether a nonhuman animal has a fundamental right to liberty protected by the writ of habeas corpus is profound and far-reaching. It speaks to our relationship with all the life around us. Ultimately, we will not be able to ignore it. While it may be arguable that a chimpanzee is not a "person," there is no doubt that it is not merely a thing.[57]

The judicial acknowledgment that a chimpanzee is not a thing is electrifying in the class I teach. Also of significant interest to the students is the realization that in some other jurisdictions, personhood for some apes and bears has already been recognized. The following are examples of such cases: Sandra, an orangutan in Argentina, was liberated under habeas

[55] *Nonhuman Rights Project ex rel. Kiko v. Presti*, 124 A.D.3d 1334, 999 N.Y.S.2d 652, 2015 N.Y. Slip Op. 85 (N.Y. App. Div. 2015).

[56] *Nonhuman Rights Project, Inc. v. Lavery*, (2017) NY Slip Op 04,574, http://www.nycourts.gov/reporter/3dseries /2017/2017_04574.htm.

[57] Justice Fahey, concurring opinion in *Nonhuman Rights Project, Inc. ex rel. Tommy v. Lavery*, 2018 N.Y. Slip Op. 3309 (N.Y. 2018), accessed April 9, 2020, http://www.nycourts.gov/ctapps/Decisions/2018/May18/M2018-268opn18-Decision.pdf.

corpus and moved to a sanctuary in Florida.[58] Chucho, a spectacled bear in Colombia, was recognized as a person on appeal,[59] although this decision was subsequently overturned by a higher court[60]; an appeal to Colombia's highest court, the Constitutional Court, was similarly unsuccessful.[61] Nonetheless, in Argentina a chimpanzee named Cecelia has been recognized as a person.[62] In Pakistan, the High Court has recognized that an elephant has a right to life and ordered his release into a sanctuary.[63] Cases of this nature are becoming more frequent and they raise many legal issues of interest to students.

After this class, I then pose some questions for the students to discuss in groups, as follows:

- Why do we assume that animals can be treated as property? (Should they have rights?)
- If apes remain property, should our use of them be restricted? (Does welfare work?)
- Should they be consumed or used in research into disease?
- What about other animals? Who should decide?
- What about the environment?
- What about the cost of change?
- Does our law need to be evidence-based in order to be legitimate? Where new evidence emerges, should we change our laws, and if

[58] Associated Press, "Orangutan Granted 'Personhood' Turns 34, Makes New Friend," AP News, February 16, 2020, https://apnews.com/f5bf0834c9363563815318ca37b3de8a.

[59] Saskia Stucki and Juan C. Herrera, "Habea(r)s Corpus: Some Thoughts on the Role of Habeas Corpus in the Evolution of Animal Rights," *International Journal of Constitutional Law*, June 2017.

[60] "El oso de la justicia: Chucho se queda en el zoológico de Barranquilla," *Semana*, August 18, 2017, https://www.semana.com/nacion/articulo/niega-corte-suprema-habeas-corpus-al-oso-de-anteojos-chucho/536929.

[61] "Colombia's Constitutional Court Denies Habeas Corpus for Andean Bear," *City Paper* (Bogota), January 23, 2020, https://thecitypaperbogota.com/news/colombias-constitutional-court-denies-habeas-corpus-for-andean-bear/23781.

[62] Lauren Choplin, "Chimpanzee Recognized as Legal Person," Nonhuman Rights Project Blog, December 5, 2016, https://www.nonhumanrights.org/blog/cecilia-chimpanzee-legal-person/.

[63] https://ecojurisprudence.org/wp-content/uploads/2022/02/PK_Pakistan-Rights-of-Animals_251.pdf.

so, how? Should it be left to judges, the legislature, or international bodies?

I examine this course through each student's reflective log and a written examination or an essay of 2,500 words, whereas Industrial and Intellectual Property is assessed by exam only. Students can do independent research projects and essays of 5,000 words if they choose, and every year I give the students essay titles on the issue of personhood and sometimes have a very good uptake in terms of numbers. This is an optional module for students.

25.7 Conclusion

I do not currently teach a course on animal law although this is set to change in January 2025 when I will introduce Ireland's first undergraduate Animal Law course in a school of law. Not having a designated course was a blessing in disguise for some years as I was able to develop material in other courses and this encouraged students' interest in themes relating to animals which I was able to fit into existing, mainstream courses. The Animal Law course will be optional, and probably, for the first few years at least, the numbers will be small, and I may be preaching to the converted. However, I still have the luxury of teaching several popular and well-established courses where I can introduce material with which students are unfamiliar, alongside this new course. At present, there is little academic work on animal law in Ireland, but things are changing, and I and a group of other lawyers are now busily trying to redress the deficit. The topic is novel and thought-provoking. I have supervised a Ph.D. in this area and a master's dissertation and am currently involved in writing and editing two books and some articles on topics pertaining to animal law and vegetarianism. I never planned to research, write on, or teach animal law and related issues, but sometimes providence redirects us somewhat, and this has certainly been true of my academic journey so far.

There has been a significant increase in interest in animal law among lawyers in the last decade. Several factors have contributed to this change. The use of social media has made the sharing of information about the abuse of animals much easier. In Ireland, most crimes against animals are not punished severely by the courts, the court cases are not recorded if heard in the lower courts and cases are only sometimes reported in local newspapers, so the gathering of information is laborious. However, there is now a website called Animal Law Ireland[64] which was set up by volunteers and this is addressing this deficit by consolidating and publishing in a centralized location a lot of material on the abuse of animals that up to now has been inaccessible.

We are a group of about 20 lawyers who have agreed to work together to document animal law, share information, seek support from our counterparts in the United Kingdom and other jurisdictions, educate the public and advocate for animals. Every year students will learn about the law and problems with enforcement.

Most importantly, there has been a profound societal shift, and many people are beginning to realize that animals are not mere things. Therefore, their status requires reevaluation. The national broadcaster, RTE has done several documentaries on animal abuse relating to the export of cattle, greyhound racing and the horse racing industry in recent times. Fur farming was banned in Ireland in 2022[65] and although hare coursing continues, most people want to see its abolition but at the moment, political will is weak. People are becoming more aware of the ills of industrialized farming and its negative effects on the environment, on the health of rivers and human health. Concern for the animals is also rising. Moreover, animals are gaining status in legislative initiatives and their sentience is now largely recognized. This has not translated into an improvement in their wellbeing to date, but times are changing, and one would hope that with greater information and awareness, new laws and their enforcement, the lot of animals will get better. The paradigm shift in which animal interests are considered will permeate all facets of society and this aspect of our educational curriculum will, hopefully, contribute positively toward constructing a more humane education.

Bibliography

Associated Press. "Orangutan Granted 'Personhood' Turns 34, Makes New Friend." AP News, February 16, 2020. https://apnews.com/f5bf0834c9363563815318ca37b3de8a.

Bagley, Margo A. "Patent First, Ask Questions Later: Morality and Biotechnology in Patent Law." *William and Mary Law Review* 45 (2003): 469.

[64] https://animallaw.ie/.

[65] https://www.rte.ie/news/ireland/2022/0404/1290402-fur-farming-ireland/.

Cavalieri, Paola. "Given All We Know of Whales, Why Resume the Slaughter?" *Guardian*, April 6, 2010. https://www.theguardian.com/commentisfree/2010/apr/06/whale-hunting-moratorium-japan-iceland.

Cavalieri, Paola, and Peter Singer, eds. *The Great Ape Project*. New York: St. Martin's Press, 1994.

Chung, Andrew. "Monkey Cannot Own Copyright to 'Selfie,' U.S. Judge Says." Reuters, January 7, 2016. https://www.reuters.com/article/us-usa-california-monkey-idUSKBN0UL28620160107.

Choplin, Lauren. "Chimpanzee Recognized as Legal Person." Nonhuman Rights Project Blog, December 5, 2016. https://www.nonhumanrights.org/blog/cecilia-chimpanzee-legal-person/.

"Colombia's Constitutional Court Denies Habeas Corpus for Andean Bear." *City Paper* (Bogota), January 23, 2020. https://thecitypaperbogota.com/news/colombias-constitutional-court-denies-habeas-corpus-for-andean-bear/23781.

Connell, John. *The Cow Book*. London: Granta, 2018.

Daley, Paul. "It's 50 Years Since Indigenous Australians First 'Counted.' Why Has So Little Changed?" *Guardian*, May 18, 2017. https://www.theguardian.com/inequality/2017/may/18/50-years-since-indigenous-australians-first-counted-why-has-so-little-changed-1967-referendum.

Descartes, René. *Discourse on the Method of Rightly Conducting the Reason and Seeking Truth in the Sciences*. 1637. Translated by John Veitch. Project Gutenberg, 1995, last updated 2022. https://www.gutenberg.org/files/59/59-h/59-h.htm.

Echeverría, Hugo. "Rights of Nature: The Ecuadorian Case." *Revista Esmat* 9, no. (2017). https://www.researchgate.net/publication/333708922_RIGHTS_OF_NATURE_THE_ECUADORIAN_CASE.

"El oso de la justicia: Chucho se queda en el zoológico de Barranquilla." *Semana*, August 18, 2017. https://www.semana.com/nacion/articulo/niega-corte-suprema-habeas-corpus-al-oso-de-anteojos-chucho/536929.

English, Bernie. "New Political Party to Launch in Limerick." *Limerick Post*, March 1, 2019. https://www.limerickpost.ie/2019/03/01/new-political-party-to-launch-in-limerick/.

Fahey, Justice. Concurring opinion in *Nonhuman Rights Project, Inc. ex rel. Tommy v. Lavery*, 2018 N.Y. Slip Op. 3309 (N.Y. 2018) Accessed April 9, 2020. http://www.nycourts.gov/ctapps/Decisions/2018/May18/M2018-268opn18-Decision.pdf.

Gibsone, Harriet. "Chimpanzee Wins $10,000 Prize for Abstract Painting." *Guardian*, August 30, 2013. https://www.theguardian.com/artanddesign/2013/aug/30/chimpanzee-wins-10000-dollars-abstract-painting.

O'Gorman, Ronnie. "A Time When the Irish Were Not Welcome." *Galway Advertiser*, October 22, 2015. https://www.Advertiser.Ie/Galway/Article/80454/A-Time-When-The-Irish-Were-Not-Welcome.

Hagglund, Ryan. "Patentability of Human–Animal Chimeras." *Santa Clara High Technology Law Journal* 25, no. 1 (2012): 51. https://digitalcommons.law.scu.edu/cgi/viewcontent.cgi?article=1477&context=chtlj.

Kassam, Ashifa. "The World's Coldest Elephant? Activists Demand Lucy's Removal from Canadian Zoo." *Guardian*, March 1, 2017. https://www.theguardian.com/world/2017/mar/01/lucy-the-elephant-canada-edmonton-valley-zoo.

Lawrence, Felicity. "Horsemeat Scandal: The Essential Guide." *Guardian*, February 15, 2013. https://www.theguardian.com/uk/2013/feb/15/horsemeat-scandal-the-essential-guide.

Linzey, Andrew. *Christianity and the Rights of Animals*. London: SPCK, 1987; New York: Crossroad, 1989.

Linzey, Andrew, and Clair Linzey, eds. *The Routledge Handbook of Religion and Animal Ethics*. London: Routledge, 2018.

Magnani, Thomas A. "The Patentability of Human–Animal Chimeras." *Berkeley Technology Law Journal* 14 (1999).

Margil, Mari. "Our Laws Make Slaves of Nature: It's Not Just Humans Who Need Rights." *Guardian*, May 23, 2018. https://www.theguardian.com/commentisfree/2018/may/23/laws-slaves-nature-humans-rights-environment-amazon.

O'Sullivan, Maureen. "Irish Artistic Copyright Law: A Menagerie of Holy Cows and Turtle Doves?" *Intellectual Property Quarterly* 1 (2015): 31–61, 50–51.

O'Sullivan, Maureen. *Patents, Morality and Biotechnology: A Deliberative and Participatory Paradigm for Reform*. London: Routledge, 2019.

Reuters. "Monkey Who Took Selfie Cannot Sue for Copyright, Court Rules." *Irish Times*, April 24, 2018. https://www.irishtimes.com/news/world/us/monkey-who-took-selfie-cannot-sue-for-copyright-court-rules-1.3472802.

Siebert, Charles. "Should a Chimp Be Able to Sue Its Owner?" *New York Times*, April 23, 2014. https://www.nytimes.com/2014/04/27/magazine/the-rights-of-man-and-beast.html.

Stone, Christopher. "Should Trees Have Standing? Toward Legal Rights for Natural Objects." *Southern California Law Review* 45 (1972): 450–501.

Stucki, Saskia, and Juan C. Herrera. "Habea(r)s Corpus: Some Thoughts on the Role of Habeas Corpus in the Evolution of Animal Rights." *International Journal of Constitutional Law*, June 2017.

Sunstein, Cass R. "Introduction: What are Animal Rights?" In *Animal Rights: Current Debates and New Directions*, edited by Cass R. Sunstein and Martha C. Nussbaum. New York: Oxford University Press, 2004.

Taupitz, Jochen, and Marion Weschka. *CHIMBRIDS—Chimeras and Hybrids in Comparative European and International Research: Scientific, Ethical, Philosophical and Legal Aspects*. Berlin: Springer Science and Business Media, 2009.

Totten, Tyler. "Should Elephants Have Standing?" *Western Journal of Legal Studies* 6, no. 1 (2015). https://ojs.lib.uwo.ca/index.php/uwojls/article/view/5640.

U.S. Copyright Office. *The Compendium of US Copyright Office Practices*, Third Edition, 01/28/2021. Accessed August 16, 2024. https://www.copyright.gov/comp3/chap300/ch300-copyrightable-authorship.pdf.

Vidal, John. "Bolivia Enshrines Natural World's Rights with Equal Status for Mother Earth." *Guardian*, April 10, 2011. https://www.theguardian.com/environment/2011/apr/10/bolivia-enshrines-natural-worlds-rights.

Weiss, Rick. "U.S. Denies Patent for a Too-Human Hybrid." *Washington Post*, February 13, 2005, A03. http://www.washingtonpost.com/wp-dyn/articles/A19781-2005Feb12.html.

Westerman, Ashley. "Should Rivers Have Same Legal Rights as Humans? A Growing Number of Voices Say Yes." NPR, August 3, 2019. https://www.kcur.org/post/should-rivers-have-same-legal-rights-humans-growing-number-voices-say-yes#stream/0.

Index

C. Linzey and A. Linzey (eds.), *The Palgrave Companion to Humane Education*, The Palgrave Macmillan Animal Ethics Series,
https://doi.org/10.1007/978-3-032-19795-5

N

www.ingramcontent.com/pod-product-compliance
Ingram Content Group UK Ltd.
Pitfield, Milton Keynes, MK11 3LW, UK
UKHW040819180926
475692UK00001B/1
9783032197948